T0296274

Inclusive Radio Communications for 5G and Beyond

Inclusive Radio Communications for 5G and Beyond

Edited by

Claude Oestges
ICTEAM - Electrical Engineering
School of Engineering - Ecole Polytechnique de Louvain
Université catholique de Louvain
Louvain, Belgium

François Quitin
Brussels School of Engineering
Université libre de Bruxelles
Brussels, Belgium

ACADEMIC PRESS
An imprint of Elsevier

This publication is based upon work from COST Action CA15104 (IRACON), supported by COST (European Cooperation in Science and Technology).

COST (European Cooperation in Science and Technology) is a funding agency for research and innovation networks. The goal of COST is to help connect research initiatives across Europe and enable scientists to grow their ideas by sharing them with their peers. See www.cost.eu for more information. The research project CA15104 (IRACON) on which this book is based was funded by the Horizon 2020 Framework Programme of the European Union.

www.cost.eu

 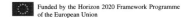

EUROPEAN COOPERATION
IN SCIENCE & TECHNOLOGY

Funded by the Horizon 2020 Framework Programme
of the European Union

Copyright © 2021 Elsevier Ltd. All rights reserved.

This book and the individual contributions contained in it are protected under copyright, and the following terms and conditions apply to their use in addition to the terms of any Creative Commons (CC) or other user license that has been applied by the publisher to an individual chapter:

Photocopying: Single photocopies of single chapters may be made for personal use as allowed by national copyright laws. Permission is not required for photocopying of chapters published under the CC BY license nor for photocopying for non-commercial purposes in accordance with any other user license applied by the publisher. Permission of the publisher and payment of a fee is required for all other photocopying, including multiple or systematic copying, copying for advertising or promotional purposes, resale, and all forms of document delivery. Special rates are available for educational institutions that wish to make photocopies for non-profit educational classroom use.

Derivative works: Users may reproduce tables of contents or prepare lists of chapters including abstracts for internal circulation within their institutions or companies. Other than for chapters published under the CC BY license, permission of the publisher is required for resale or distribution outside the subscribing institution or company. For any subscribed chapters or chapters published under a CC BY-NC-ND license, permission of the publisher is required for all other derivative works, including compilations and translations.

Storage or usage: Except as outlined above or as set out in the relevant user license, no part of this publication may be reproduced, stored in a retrieval system or transmitted in any form or by any means, electronic, mechanical, photocopying, recording or otherwise, without prior written permission of the publisher.

Permissions: For information on how to seek permission visit www.elsevier.com/permissions or call: (+1) 800-523-4069 x 3808.

Author rights: Authors may have additional rights in their chapters as set out in their agreement with the publisher (more information at http://www.elsevier.com/authorsrights).
Notice: Practitioners and researchers must always rely on their own experience and knowledge in evaluating and using any information, methods, compounds or experiments described herein. Because of rapid advances in the medical sciences, in particular, independent verification of diagnoses and drug dosages should be made. To the fullest extent of the law, no responsibility is assumed by the publisher for any injury and/or damage to persons or property as a matter of products liability, negligence or otherwise, or from any use or operation of any methods, products, instructions or ideas contained in the material herein. Although all advertising material is expected to conform to ethical (medical) standards, inclusion in this publication does not constitute a guarantee or endorsement of the quality or value of such product or of the claims made of it by its manufacturer.

Publisher: Mara Conner
Acquisitions Editor: Tim Pitts
Editorial Project Manager: Naomi Robertson
Production Project Manager: Paul Prasad Chandramohan
Designer: Matthew Limbert

Typeset by VTeX

Working together to grow libraries in developing countries

www.elsevier.com • www.bookaid.org

Contents

CHAPTER 6   5G and beyond networks 141

Silvia Ruiz, Hamed Ahmadi, Gordana Gardašević,
Yoram Haddad, Konstantinos Katzis, Paolo Grazioso,
Valeria Petrini, Arie Reichman, M. Kemal Ozdemir,
Fernando Velez, Rui Paulo, Sergio Fortes,
Luis M. Correia, Behnam Rouzbehani,
Mojgan Barahman, Margot Deruyck, Silvia Mignardi,
Karim Nasr, and Haibin Zhang

Kamran Sayrafian, Sławomir J. Ambroziak,
Dragana Bajic, Lazar Berbakov, Luis M. Correia,
Krzysztof K. Cwalina, Concepcion Garcia-Pardo,
Gordana Gardašević, Konstantinos Katzis,
Pawel Kulakowski, and Kenan Turbic

Klaus Witrisal, Carles Anton-Haro, Stefan Grebien,
Wout Joseph, Erik Leitinger, Xuhong Li,
José A. Del Peral-Rosado, David Plets, Jordi Vilà-Valls,
and Thomas Wilding

List of contributors

Hamed Ahmadi[1]
University College Dublin, Dublin, Ireland

Sławomir J. Ambroziak[1]
Gdańsk University of Technology, Gdańsk, Poland

Carles Anton-Haro[1]
Centre Tecnològic de Telecomunicacions de Catalunya, Barcelona, Catalonia, Spain

Dragana Bajic
University of Novi Sad, Novi Sad, Serbia

Titus Constantin Balan
Universitatea Transilvania Braşov, Braşov, Romania

Mojgan Barahman
Universidade de Lisboa, Lisbon, Portugal

Marina Barbiroli
University of Bologna, Bologna, Italy

Lazar Berbakov
Institute Mihailo Pupin, Belgrade, Serbia

Thomas Blazek
Silicon Austria Labs, Linz, Austria

Chiara Buratti[1]
University of Bologna, Bologna, Italy

Alister Burr
The University of York, York, United Kingdom

Laurent Clavier[1]
Université de Lille, Lille, France

Luis M. Correia
INESC-ID / IST, University of Lisbon, Lisbon, Portugal

---

[1] Chapter editors.

Krzysztof K. Cwalina
Gdańsk University of Technology, Gdańsk, Poland

Andreas Czylwik
University Duisburg-Essen, Duisburg and Essen, Germany

José A. Del Peral-Rosado
Universitat Autònoma de Barcelona, Barcelona, Spain

Margot Deruyck
Ghent University, Ghent, Belgium

Luca Feltrin
Ericsson, Stockholm, Sweden

Sergio Fortes
Universidad de Malaga, Malaga, Spain

Davy Gaillot
Université des Sciences et Technologies de Lille, Lille, France

Concepcion Garcia-Pardo
Universitat Politècnica de València, València, Spain

Gordana Gardašević
University of Banja Luka, Banja Luka, Bosnia and Herzegovina

Paolo Grazioso
Fondazione Ugo Bordoni, Rome, Italy

Stefan Grebien
Graz University of Technology, Styria, Austria

Ke Guan
Beijing Jiaotong University, Beijing, China

Yoram Haddad
Jerusalem College of Technology, Jerusalem, Israel

Katsuyuki Haneda[1]
Aalto University, Espoo, Finland

Danping He
Beijing Jiaotong University, Beijing, China

Ruisi He
Beijing Jiaotong University, Beijing, China

Wout Joseph
Ghent University, Ghent, Belgium

Konstantinos Katzis
European University Cyprus, Nicosia, Cyprus

Wim Kotterman[1]
Technische Universität Ilmenau, Ilmenau, Thuringia, Germany

Pawel Kulakowski
AGH University of Science and Technology, Kraków, Poland

Pekka Kyösti
University of Oulu, Oulu, Finland
Keysight Technologies, Oulu, Finland

Erik Leitinger
Graz University of Technology, Styria, Austria

Xuhong Li
Lund University, Lund, Sweden

Yang Miao
University of Twente, Enschede, Netherlands

Silvia Mignardi
University of Bologna, Bologna, Italy

Jose-Maria Molina-Garcia-Pardo
Universidad Politécnica de Cartagena, Cartagena (Murcia), Spain

Andreas F. Molisch[1]
University of Southern California, Los Angeles, CA, United States

Karim Nasr
University of Greenwich, London, United Kingdom

Claude Oestges[1]
Université catholique de Louvain, Louvain-la-Neuve, Belgium

Luis Orozco-Barbosa
Universidad de Castilla-La Mancha, Ciudad Real, Spain

M. Kemal Ozdemir
Istanbul Medipol University, Istanbul, Turkey

Juan Pascual-Garcia
Universidad Politécnica de Cartagena, Cartagena (Murcia), Spain

Rui Paulo
Universidade Beira Interior, Covilhã, Portugal

Valeria Petrini
Fondazione Ugo Bordoni, Rome, Italy

David Plets
Ghent University, Ghent, Belgium

François Quitin[1]
Université libre de Bruxelles, Brussels, Belgium

Piotr Rajchowski
Gdańsk University of Technology, Gdańsk, Poland

Arie Reichman
Ruppin Academic Center, Hadera, Israel

Behnam Rouzbehani
Universidade de Lisboa, Lisbon, Portugal

Richard Rudd
Plum Consulting, London, United Kingdom

Silvia Ruiz[1]
Universitat Politècnica de Catalunya, Barcelona, Catalonia, Spain

Moray Rumney
Rumney Telecom Limited, Edinburgh, United Kingdom

Kentaro Saito
Tokyo Institute of Technology, Tokyo, Japan

Sana Salous[1]
University of Durham, Durham, United Kingdom

Kamran Sayrafian[1]
National Institute of Standards & Technology, Gaithersburg, MD, United States

Erik Ström[1]
Chalmers University, Göteborg, Sweden

Jan Sykora[1]
Czech Technical University in Prague, Prague, Czech Republic

Emmeric Tanghe
Ghent University, Ghent, Belgium

Fredrik Tufvesson
Lund University, Lund, Sweden

Kenan Turbic
INESC-ID / IST, University of Lisbon, Lisbon, Portugal

Fernando Velez
Universidade Beira Interior, Covilhã, Portugal

Jordi Vilà-Valls
Institut Supérieur de l'Aéronautique et de l'Espace, Toulouse, France

Enrico Vitucci
University of Bologna, Bologna, Italy

Thomas Wilding
Graz University of Technology, Styria, Austria

Klaus Witrisal[1]
Graz University of Technology, Styria, Austria

Haibin Zhang
TNO, The Hague, Netherlands

Introduction

1

Claude Oestges[a,b] **and François Quitin**[c,b]

[a]*Université catholique de Louvain, Louvain-la-Neuve, Belgium*
[c]*Université libre de Bruxelles, Brussels, Belgium*

1.1 Technology trends and challenges

The demand for mobile connectivity is continuously increasing, and today, Mobile and Wireless Communications serve not only very dense populations of mobile phones and nomadic computers, but also the expected multiplicity of devices and sensors located in machines, vehicles, health systems, and city infrastructures. The concept of Inclusive Radio Communication Networks defines the technologies for supporting wireless connectivity for any rates, type of communicating units, and scenario, as promised by the 5th generation of radio communication networks. Spectral and spatial efficiency are key challenges, in addition to constraints such as energy consumption, latency, mobility, adaptability, heterogeneity, coverage, and reliability, amongst others. While many of these aspects are not especially new, the wireless Internet of Things beyond 2020 in particular will require revolutionary approaches in Radio Access Technologies, networks and systems in order to overcome the limitations of the current cellular deployments, the layered networking protocols, and the centralized management of spectrum, radio resources, services, and content. Theoretical foundations need to be fully revisited so that disruptive technologies may emerge.

In this context, this book aims to report on the scientific breakthroughs achieved within COST Action CA15104, also known as IRACON or COST-IRACON. The overreaching goal is to propose solutions for inclusive and multidimensional communication systems with a wide variety of devices, practical constraints, and real-world scenarios, addressing systems ranging from very simple transceivers and sensors, to smartphones and highly flexible cognitive radios. However, a number of issues should be addressed to meet this ambitious objective.

- The current knowledge of radio channels must be extended over a wider span of frequency bands not previously considered, to allow system designers to fully employ radio channel models to improve performance in coverage, coexistence,

---

[b] Chapter editors.

Inclusive Radio Communications for 5G and Beyond. https://doi.org/10.1016/B978-0-12-820581-5.00007-9
Copyright © 2021 Elsevier Ltd. All rights reserved.

data rate and capacity, delay and latency, reliability and dependability, setup and mobility, amongst many other metrics. Many aspects of the radio channel still require better understanding and modeling, such as depolarization, frequency dependence, vegetation influence, new materials, and interaction with human tissue. Antenna design and evaluation methods require new approaches for improved energy-efficiency, cost-effectiveness, and capabilities to analyze human-antenna and antenna-channel interactions. New modeling methods should also build upon progress in computational power, thereby extending the capabilities of ray-tracing tools.

- Looking beyond 5G technologies, the main challenges are to foresee the evolutions of radio communication in 5 to 10 years. This implies to discuss the implications of theoretical fundamentals (energy/capacity/latency trade-offs, short vs. long packets, non-Gaussian environment, time and space dependence, waveform design, etc.) and to propose novel and practical approaches for joint channel and network coding, interference mitigation, and overcoming technological limitations at higher frequencies.

- One of the major hurdles in wireless networks continues to be the rapid increase in demand for capacity. Network densification has been seen so far as the most efficient solution to meet this demand in addition to other improvements in spectrum allocation and its management. Some approaches in dense urban environments include mesh networks between base stations, deploying small-cells, necessarily combined with the introduction of Heterogeneous Network (HetNets) technologies. HetNets implies the coordinated planning and operation of multiple radio access technologies and of multiple cell layers. In addition to the increasing number of cells, this also means increased complexity. This complexity calls for new solutions, and research is needed both towards the further evolution of 3GPP standards, and mostly for the design of 5G-and-beyond, in order for future networks to deliver the necessary performance as well as cost-efficiency. The further development of cooperative techniques at the physical layer means more interaction between traditional physical and network level functions.

- Experimental platforms are instrumental in order to test communication or localization algorithms. While channel modeling is naturally confronted by measurements, PHY to NET layer solutions need specific test beds with real time implementation, taking into account all real-world parameters.

Specifically, in the area of channel measurements and modeling, accurate radio channel characterization and modeling efforts have included (i) new deployment scenarios, such as very high mobility (vehicles, trains, drones), different human body postures and tissues, harsh physical environments, ultra-dense device deployment, very-short links, highly directional front/back-haul links, among others, (ii) multiband and wideband channel modeling with carrier frequencies above UHF up to Terahertz, and (iii) 3D modeling with site-specificity. In addition, there is a need for antenna systems that are self-reconfigurable to optimize near/far-field interaction with channels in the deployment scenarios, under constraints of size and disturbances in the vicinity.

In view of a global design of new standards and more reliable interaction between layers,

- new PHY-layer algorithms and protocols for 5G-and-beyond should account for the inclusiveness of heterogeneous networks, links, and devices, considering practical scenarios to optimize capacity, energy-consumption and resource usage in future massive communications;
- the inclusiveness of future networks also calls for a different design of the communication system, in which the control information is better integrated with the transmission protocol to reach an optimal balance between the transmission overhead and PHY performance;
- there is a need to enable accurate position estimation, physical layer security, maximum capacity, energy efficiency, and other applications of these models for the upper layers.

For an efficient use of spectrum resources, it is obvious that future Radio Access Networks (RANs) must become real cognitive automated networks, which goes far beyond the self-organizing and self-optimizing network functionalities that have been introduced in LTE. Resource management in RANs must also evolve to broaden the meaning of resource sharing to multi-technology, multi-band, multi-service, and multi-link connectivity and content delivery. Future network architectures have to further consolidate the concepts of Cloud RAN and virtual nodes, leaving the base station centric approaches for a more flexible device-centric architecture. Eventually, the concept of moving networks has to be introduced to support connected nodes in highly dense traffic areas, where ad hoc, high rate and low latency connectivity among the moving nodes and to the fixed RAN rely on multiple clouds and multiple paths.

By addressing the heterogeneous nature of future massive connectivity of humans and machines with embedded intelligence, including the diversity of wireless channels used to create this connectivity, COST-IRACON has therefore provided a large variety of results to better tackle the unprecedented scale change and the superposition of networks and devices with drastically different capabilities that are required for the success 5G-and-beyond wireless networks.

1.2 Scope of the book

This books aims at covering all the aspects of Fifth Generation (5G) communications, from the radio communications to specific application scenarios, and can be roughly divided in three major parts. The first part investigates the radio channel in 5G-specific scenarios, the second part focuses on signal processing and algorithms for 5G networks, while the final part focuses on applications and services specific to 5G networks.

Chapter 2 presents definitions, methods and tool to model propagation in the radio channel for 5G communications. The radio channel is a fundamental aspect of any type of Radio Frequency (RF) communications, and providing accurate models of radio channel propagation allows to understand the most critical parameters of the radio channel and allows to exploit the radio channel to its fullest capacity. The new generation of 5G networks is challenging because of the large amounts of scenarios that need to be considered, from high-bandwidth low-mobility indoor cases to low-latency high-speed vehicular situations. This chapter starts by presenting the large amount of propagation environments that need to be considered in the context of 5G networks. It then proceeds to present existing wireless channel models, and proposes improvements and extensions to existing models to account for the new aspects of 5G networks. Finally, the latest research on algorithms for estimation of radio channel parameters is presented, providing state-of-the-art solutions for channel modeling.

Chapter 3 presents results from experimental channel measurements, and proposes channel models based on these results. The first part of the chapter summarizes a wide range of measurement scenarios and measurement campaigns for 5G networks. It then investigates millimeter-Wave (mmWave) and Terahertz channels, which are expected to be a major contributor of 5G networks for situations that require very high bandwidth with limited mobility. The next part of the chapter focuses on Multiple Input Multiple Output (MIMO) and massive MIMO channels. Using very large antenna arrays at the base station is one key enabler technology for providing high throughput to many users simultaneously, but such technology presents many modeling challenges, both for the radio channel as for the antenna array hardware. Fast time-varying channels, such as Vehicle-to-Everything (V2X) or railway communications are a particularly challenging aspect of 5G. Chapter 3 presents a wide range of channel measurements in high-speed situations, and proposes models both for sub-6 GHz and mmWave frequencies. Finally, the chapter concludes with measurements of Electro-Magnetic (EM) properties of materials to be used in channel simulators.

Chapter 4 presents Over-the-Air testing for 5G communications, an often-overlooked (albeit critical) item in wireless communications. The chapter starts by investigating field emulation for electrically large objects, which occurs in 5G because of the usage of large antenna arrays and the usage of mmWave systems. A sectored multi-probe approach is presented to overcome this problem. The chapter continues by examining emulation of mmWave channels, which present significant differences with sub-6 GHz channel emulation. Finally, the chapter presents some future challenges for emulation of wireless channel, such as emulation complexity or emulating human influence (which can be a deal-breaker for mmWave communications).

Chapter 5 presents the latest research on coding and signal processing for 5G wireless networks. It first presents the latest progress in signal processing algorithms and advanced waveforms design. It then proceeds to apply this three specific scenarios of 5G. The first is the case of distributed and cooperative Radio Access Network (RAN), where physically separated radio heads cooperate to improve network performances. The second scenario is the case of massive MIMO, where precoding presents a challenging problem. New algorithms for massive MIMO are presented, and the performances of massive MIMO systems are then evaluated. Finally, the third scenario is the one of full-duplex communications, where a single time-frequency slot is used for both uplink and downlink communications. New implementations are proposed to tackle the challenges of full-duplex communication systems.

Chapter 6 investigates the challenges at the network layer of 5G networks. It investigates a wide variety of topics and scenarios for 5G. The chapter starts by investigating ad-hoc and V2V networks, for which issues such as reliability will pose a major challenge. Spectrum management and sharing is investigated in the context of Internet of Things (IoT) and Machine To Machine (M2M) communications, as well as from a more general perspective. The problem of radio resource management and scheduling is then investigated, focusing on various scenarios (mesh networks, device-to-device communications, etc.). The next part of the chapter presents various solutions for heterogeneous and ultra-dense networks, such as small cells and mmWave cells. Cloud radio access networks and software-defined networks are also a major aspect of future networks, and the various techniques for these new generations of networks are presented. Finally, the chapter concludes on emerging type of networks and applications, such as UAV-based networks or smart grids.

Chapter 7 presents the latest research and solutions for IoT systems and networks. The chapter starts by presenting the different solutions for low-power wide area networks, before introducing Medium Access Control (MAC) and routing protocols for typical IoT networks. Vehicular communications will be one special application of IoT networks, and special hardware integration aspects of vehicular IoT networks are presented. Since power consumption is one of the major bottlenecks of IoT systems, energy harvesting and energy efficiency solutions are presented in detail. Specific network techniques (such as software-defined networks) are then investigated in the particular context of IoT networks.

Chapter 8 introduces a specific type of IoT system, namely IoT for healthcare applications. Smart and implantable sensors, as well as remote health monitoring will be a major driver of future IoT technology. First, research in wearable and implantable IoT technology is presented, including recent progress on human body phantoms and Specific Absorption Rate (SAR) measurements. Next, the networking implication of IoT for healthcare is investigated. Finally, the chapter presents a new type of in-body communication networks, i.e. nanocommunication networks, where research on cell-to-cell communications is presented.

Chapter 9 introduces an increasingly important service of wireless networks, i.e. localization and tracking. The improving performances of wireless networks allow for increasingly precise localization as a side-benefit. The chapter starts by introducing the fundamentals of wireless localization, before presenting the latest research on localization and tracking algorithms. Multipath channels are a well-known challenge for localization networks, but they can also be turned into an asset. Finally, localization for different type of systems is investigated, and different testbeds and prototyping activities are presented.

Radio propagation modeling methods and tools

2

Katsuyuki Haneda[a,b]**, Richard Rudd**[c]**, Enrico Vitucci**[d]**, Danping He**[e]**, Pekka Kyösti**[f]**,
Fredrik Tufvesson**[g]**, Sana Salous**[h]**, Yang Miao**[i]**, Wout Joseph**[j]**, and
Emmeric Tanghe**[j]

[a]*Aalto University, Espoo, Finland*
[c]*Plum Consulting, London, United Kingdom*
[d]*University of Bologna, Bologna, Italy*
[e]*Beijing Jiaotong University, Beijing, China*
[f]*University of Oulu, Oulu, Finland*
[g]*Lund University, Lund, Sweden*
[h]*University of Durham, Durham, United Kingdom*
[i]*University of Twente, Enschede, Netherlands*
[j]*Ghent University, Ghent, Belgium*

This chapter provides overview of fundamental definitions, tools, and new methods towards improved channel modeling reported in the Co-operation in Science and Technology (COST)-Inclusive Radio Communications (IRACON) Action for future wireless communications and networks. The overview first covers definitions of *propagation environments* as they determine most relevant propagation mechanisms to consider and model, and furthermore, guide our approach to channel modeling methods. This chapter then introduces new insights into popular approaches of channel modeling, i.e., *site-specific and geometry-based stochastic channel modeling*, where the latter particularly features canonical and standardized channel modeling approaches taken by the 3rd Generation Partnership Project (3GPP), COST, and International Telecommunication Union (ITU) communities. Finally, this chapter shed lights on new modeling approaches to *small-scale radio propagation behaviors*, covering plane wave propagation paths and distributed diffuse scattering.

---

[b] Chapter editor.

Inclusive Radio Communications for 5G and Beyond. https://doi.org/10.1016/B978-0-12-820581-5.00008-0
Copyright © 2021 Elsevier Ltd. All rights reserved.

2.1 Propagation environments
2.1.1 Introduction

Let us first put propagation environments in the context of the history of wireless communications so that the contributions of the COST-IRACON action become clear. The first experimental work on radio channels, by the likes of Hertz [CW95] and Bose [Eme97] was carried out over very short paths using ultra high frequency or microwave radiation in an indoor environment. The utility of lower frequencies was soon discovered by workers such as Marconi and for many years the focus shifted to much lower frequencies and paths of tens or hundreds of kilometers. The available bandwidth was very small and multipath effects only became noticeable in the context of ionospheric propagation. With the advent of television, wideband telephony, and data in the mid-twentieth century, higher frequencies were re-visited and it became necessary to pay attention to the dispersive nature of the radio channel. With the quest for bandwidth, carrier frequencies continued to rise and the millimeter wavelengths pioneered by Bose are now the subject of intense study in the context of 5G communications. The aim of the radio propagation activities within the COST-IRACON action has been to support the goal of 'inclusive' use of wireless connectivity in all environments by the development of improved channel models. Whereas traditional radio systems were required to address a fairly limited range of topologies (e.g., broadcast transmission from high sites to rooftop antennas, cellular base stations to hand-held terminals, a wireless LAN router to nomadic terminals within a building), future systems will be expected to accommodate a much more heterogeneous range of channels; this variety is reflected in the propagation environments that have been subject to scrutiny within the COST-IRACON action.

2.1.2 Outdoor environment

Historically, it has been the outdoor channel that has received most attention, with most early modeling effort concentrating on characterizing path loss using empirical expressions such as those due to Okumura [OOKF68] and Hata [Hat80]. Semideterministic models [Bul47,Dey66] attempted to account for diffraction around major terrain features, but local signal variability due to shadowing by trees, buildings, and other 'clutter' could only be treated statistically, with this 'location variability' generally characterized as a log-normal distribution. The 'physical-statistical' approach blends predicable physical effects, such as rooftop diffraction or reflection from walls, with statistical data such as building heights and orientation, to give simple but efficient models [WB88] as illustrated in Fig. 2.1.

With the advent of digital mobile communication, the wideband characteristics of the channel started to attract attention [Cox72]. Characterizing multipath with ever-greater resolution has been a focus for the last decades, first in two dimensions and now in three, as smaller operating wavelengths and greater computing power have enabled highly directional, dynamically-synthesized antenna beams to be realized. Much of the current work still centers on the statistical characterization, by measure-

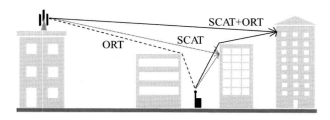

FIGURE 2.1

Contribution of scattered (SCAT) and over-roof-top (ORT) signals [VFB⁺18].

ment, of the spatial channel; such studies are, however, time-consuming and, with the recent explosion in the availability of high-resolution environmental data, the use of deterministic models to explore and characterize the channel is attractive. At lower frequencies, it is reasonable to expect that bulk diffraction effects due to buildings can be modeled with some accuracy as vector representations of the built environment are widely available.

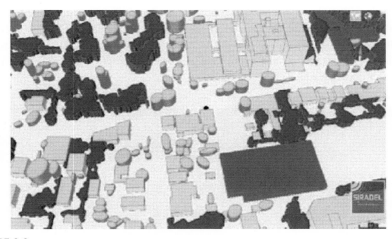

FIGURE 2.2

Vector representation of buildings and vegetation [CTS⁺16].

Major challenges remain, however, in characterizing building materials in terms of their electrical constants and surface roughness [FPK⁺18], both important for the modeling of reflection and scattering and hence of the multipath environment. Similarly, vegetation is inevitably less well described in environmental databases (although the situation is improving) and will often need to be treated statistically as exemplified in Fig. 2.2.

2.1.3 Indoor environment

The highest-bandwidth wireless services are disproportionately consumed in indoor environments. The indoor propagation channel is generally very dispersive with metal structures and fittings giving rise to a rich variety of clustered multipath components that can be problematic for increasingly broadband channels; a visual representation of multipath power arrival profile is shown in Fig. 2.3 [KIU+20]. While the outdoor environment is clearly varied, indoor spaces are even more diverse as seen in channels of a small domestic room, a factory, a tunnel, and an airport expected to show very different behavior. The sheer variety of indoor environments makes characterization by simple statistical models challenging. Deterministic approaches, such as ray tracing or even full-field methods [MVB+18] are attractive because of the limited dimensions, and hence modest computational effort involved. Set against this, however, is the same problem as mentioned in Section 2.1.2 of obtaining sufficiently detailed data on the constituent materials and their electrical properties. Moreover, items such as furniture may move (and are unlikely to be included in computer-aided design models), while human bodies may significantly change the RF environment. For the most part, it is unlikely that prediction models will be used for the operational planning of indoor radio networks; much work is therefore aimed at the development of statistical models for use in physical layer design and standardization activities. With the trend towards increasing system bandwidths, understanding the distribution and clustering of multipath components is a major area of research, while the drive to exploit higher frequencies has led to debate, and conflicting evidence, regarding the frequency dependence of Diffuse Scattering (DiS).

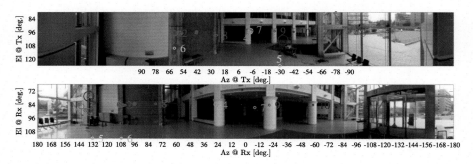

FIGURE 2.3

Visualization of spatial energy distribution in an indoor environment [KIU+20].

2.1.4 Outdoor-to-indoor environment

There is an increasing expectation that radio systems will provide seamless connectivity as the user moves in and out of buildings. The ability to make realistic predictions of the loss suffered by signals penetrating building fabric is therefore important not only for the estimation of the indoor service that may be provided by outdoor base stations, but also in the assessment of interference between systems located on different sides of the indoor-outdoor interface. It is therefore unfortunate that building loss is poorly-characterized, partly due to the wide variability within and between buildings. Concern has also been expressed that the increased use of energy-efficient construction practices may be causing building entry loss to increase [Han16].

Previous work [RML$^+$18] has shown that the trend of loss with frequency is non-monotonic, with the lowest and highest frequencies suffering the greatest loss, presumably due to the small size of building apertures in terms of wavelength at the low-frequency end, and the generally higher material absorption rate at the higher end. While the behavior of specific building materials (brick, concrete, glass, etc) has been well characterized in isolation, using this data to predict large-scale building entry loss is not trivial. Even very small gaps in structures can have a large and frequency-dependent impact, and windows with multiple glazing can form band-stop filters as exemplified in Fig. 2.4 [KLHNK$^+$18].

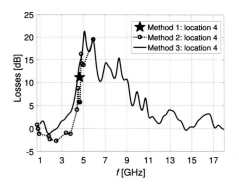

FIGURE 2.4

Double-glazed window as bandstop filter [KLHNK$^+$18].

2.1.5 Train and other vehicular environments

Vehicle connectivity has become an important application area for 5G systems; beyond the obvious opportunities for delivering high bandwidth content to rail passengers, the promised ubiquity and low-latency of such systems will be a key enabling technology for assisted-driving or self-driving cars. This has inspired much recent work to characterize the Vehicle-to-Infrastructure (V2I) and Vehicle-to-Vehicle (V2V) channels, as detailed in Section 3.4. Such channels are very dynamic and characterized by large, and rapidly-changing Doppler shift. The condensed parameters of the railway channel change very rapidly between environments, e.g., from railway cuttings to viaducts, to tunnels and stations, as exemplified in Figs. 2.5 and 2.6.

Much attention has therefore been paid to the modeling of these dynamics within the COST-IRACON action.

FIGURE 2.5

Showing the diversity of railway environments [AHZ+12].

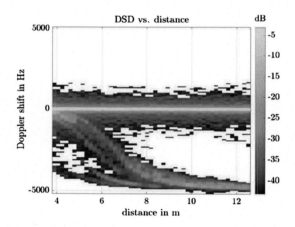

FIGURE 2.6

Rapid change to Doppler spectrum for railway channel [ZHL+18].

2.1.6 Body-centric environments

The on-body channel is very different from most other cases. Interaction between the antenna and the propagation environment is more apparent and may not be separable. On-body propagation at low frequencies is mainly due to creeping diffracted waves and penetration through biological tissue, both of which can be hard to characterize, leading to most works relying on empirical evidence. Body-to-body and body-to-infrastructure channels are similar to those in indoor scenarios, with added variability due to the body motion and installed antenna performance, e.g., [TCB19].

2.2 **Channel model classification**

Among diverse channel models and approaches, this subsection reviews the most popular ones in the scientific community as well as those for industrial standardization. The channel modeling methods and approaches decide their applicability and hence purposes, as detailed in each category of channel model hereinafter. The first approach to review is site-specific channel models which are mainly intended for coverage estimates when wireless network is deployed at a particular site. Stochastic models are then following where the main goal is comparison of physical layer proposals and inter-system interference analysis. The stochastic models do not aim at reproducing channel responses at a particular site but something plausible and realistic given an imaginary environment. They are therefore often called reference models. Different examples of reference models are introduced, including the ITU-R pathloss models, 3GPP and COST models.

2.2.1 **Site-specific channel models**

IRACON activities regarding site-specific model are presented here. In this context, the majority of the studies have been conducted in the field of ray models, which represent a good compromise between accuracy and computation time. Deterministic RF propagation prediction models using ray-optical approximations have been studied and tested with success since the nineties. More recently, due to the advent of MIMO transmission schemes and to the use of higher frequency bands in 5G-and-beyond communication systems, ray models have gained popularity and have been proposed for a variety of uses, including the design and planning of mobile radio systems and services, fingerprinting and multipath-based localization methods, or real-time use for the optimization of wireless systems performance. In serving all these purposes, there are technical challenges in ray-based models as outlined in the following.

- *Computation time:* despite the various approximations of radio wave propagation behavior that are considered, and the always-increasing computation efficiency of modern computers, the required scale of the radio channel simulations can easily go beyond the capability of the presently available computational resources.
- *Description of the environment:* the approach of ray-based models is necessarily deterministic or quasi-deterministic, therefore an accurate knowledge of the environment is needed, usually in terms of 3D digital maps. But there are often practical problems in having all the environment information stated deterministically, leading thus either to simplifications in the modeling, or to the use of stochastic modeling methods. On the other hand, point-cloud maps obtained through laser scanning are nowadays widely available for different kind of environments, and their use can be leveraged to develop new models with a reasonable compromise between complexity and accuracy.
- *Application to complex environments and higher frequency bands:* new channel models need to support 5G-and-beyond technologies such as Massive MIMO and

hybrid beamforming, among others, in a wide variety of settings, including the use of millimeter-wave frequencies and also in new applications and link types, such as Device-to-Device (D2D), vehicle-to-everything (V2X), and air-to-everything (A2X).

The following subsections introduce activities in COST-IRACON to address the mentioned technical challenges, with particular reference to ray models. Latest developments regarding the use of deterministic and physics-based models other than ray-based models, such as full-wave models, are also outlined at the end of the section.

2.2.1.1 Acceleration techniques for ray-based models

In order to cope with the high computation time of for Ray Tracing (RT) and ray launching models, novel efficient algorithms and acceleration techniques have been introduced.

2.2.1.1.1 Reduction of image trees

An efficient image-based RT algorithm is presented in [HB16]. The efficient implementation is thanks to reduction of the size of an image tree for tracing rays to a given mobile location, in addition to a fast ray-object intersection test method. The visibility region of an image can be represented by so-called lit polygons, and the buildings within this lit region will block the rays to form shadow regions as well as the higher order images, as shown in Fig. 2.7. Lit polygons can also be exploited to accelerate finding rays when a mobile is on a linear route. If entry and exit points of a mobile into the lit region can be identified, they make it unnecessary to perform geometrical check of meaningful polygons during the mobile is in the lit region.

2.2.1.1.2 Acceleration based on graph representation

Propagation environments and their interaction with radio waves are represented as a graph in RT according to [ALU16]. The use of the graph accelerates and simplifies the ray tracing process.

2.2.1.1.3 Acceleration based on preprocessing

Analyzing visibility between all surfaces inside a scenario allows acceleration of ray-optical multipath propagation simulations for path loss predictions, according to [DK19]. The visibility preprocessing is formulated through a Surface Relation Tree (SRT), which stores the relation between all surfaces in the scenario, and a Location Surface Tree (LST), which is used by the Raytracer to determine visible surfaces for any given location in the scenario, as outline in Fig. 2.8. The visibility relations between surfaces are represented as polygons similarly to Fig. 2.7. This allows to transform most needed calculations of LOS-check into a point-in-polygon-problem that can be solved by an efficient algorithm. This approach also applies to scenarios with moving devices.

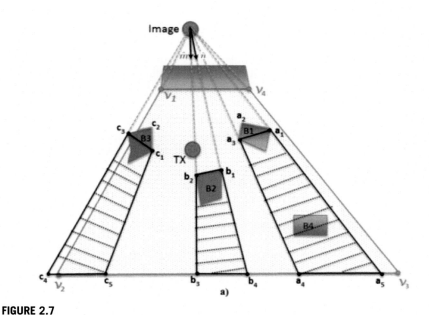

FIGURE 2.7

Lit and shadow polygon of a reflection image [HB16].

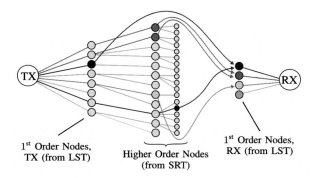

FIGURE 2.8

A set of surfaces represented by the first-order nodes looked up in the LST is visible by each Tx and Rx. The SRT is then used to find other visible surfaces leading to higher order nodes and to set up a visibility graph. It is finally used to determine a connection from the Tx-node to the Rx-node [DK19].

2.2.1.1.4 Discrete ray launching with visibility preprocessing and GPU parallelization

The preprocessing of environment visibility is also shown to facilitate faster geometric computations for both specular and diffuse interactions, according to Lu and Degli-Esposti [LVD+19,DLV+18]. They call it a fully Discrete, Environment-

Driven, Ray Launching (DED-RL) field prediction algorithm. The environment is discretized into simple regular shapes (tiles), which allows for visibility preprocessing and straightforward parallelization of the algorithm on NVIDIA-compatible Graphics Processing Units (GPUs). All the visibility relations among tiles ares stored in a visibility matrix, then ray tubes are launched toward each tile visible from the Tx and then bounced towards the other tiles with the aid of the pre-computed visibility matrix. Both the visibility matrix creation and bouncing procedure are parallelized. Combining these innovative features boosts the computation speed of up to 4 orders of magnitude compared to a standard image-RT algorithm as evidenced in Fig. 2.9. Despite the significant speed-up, DED-RL retains the same level of prediction accuracy as image-RT tool with respect to RF coverage measurements.

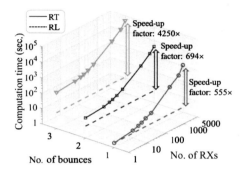

FIGURE 2.9

Computation times and speed up factors of DED-RL with respect to non-parallelized Image-RT in a dense urban scenario (Union Square, San Francisco, USA), for different number of bounces. Computation times of DED-RL are represented with dashed lines [LVD⁺19,DLV⁺18].

2.2.1.1.5 Parallelization with the aid of cloud HPC platforms

Parallelization of ray-path identification has obvious improvements in computational run-time. He et al. [HAG⁺19] implemented a parallel computing technology where High Performance Computing (HPC) in the cloud is used with several computing nodes involving Central Processing Unit (CPU) and Graphics Processing Unit (GPU). Their high-performance cloud-based RT simulation platform (CloudRT) has an architecture illustrated in Fig. 2.10, which is publicly available at http://www.raytracer.cloud/.

2.2.1.2 Applications of ray-optical wave propagation simulation methods to important systems and use cases of 5G cellular

Ray-based wave propagation models are applied to studies of a number of systems and use cases of 5G cellular. The systems include mm-wave and massive MIMO, while use cases usually refer to complex ones of outdoor, indoor, outdoor-to-indoor, vehicular, and air-to-ground scenarios. Ray-based wave propagation simulations

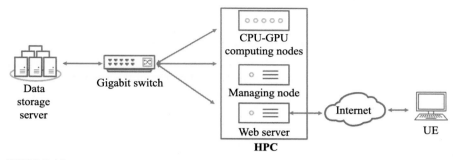

FIGURE 2.10

Network architecture of CloudRT [HAG+19].

sometimes complement measurements where channel sounding is not straightforward due to complexity of the cellular scenarios.

2.2.1.2.1 Application to mmWave frequencies

One of the distinguishing features of 5G cellular is the use of mmWave frequencies. Given the scarcity of understanding and measurements of cellular multipath channels at these frequencies, various efforts have been made in COST-IRACON action to gain more understanding about them through implementing ray-optical wave propagation models. For example, Corre et al. highlighted differences of wave propagation characteristics between mmWave and sub-6 GHz bands [CTS+16] and furthermore studied multipath propagation characteristics at frequencies above 90 GHz [DCB+18]. The former study mentions the importance of considering vegetation into the simulations, while the latter covering indoor and outdoor radio propagation shows that not only the frequency and antenna type, but also the properties of objects such as wall and furniture have significant impact on the channel features. Aslam et al. report a ray-based model implementation for the assessment of a 60 GHz outdoor small-cell networks [CACL17]. Multipaths are predicted from interactions with the static environment, but also with randomly-positioned vehicles and user-bodies causing ray-path blockage and new propagation paths. Importance of considering small physical objects that may cause shadowing and scattering, e.g., parked cars and lampposts, was also pointed out in [MVB+18] according to their outdoor RT model run at 26 GHz and 38 GHz. The results show that, in non-line-of-sight conditions, the contribution from non-specular components, even at these frequencies, is determinant to have a good prediction accuracy of measured pathloss.

2.2.1.2.2 Application to a wave propagation study in complex environments

Ray-based wave propagation models also help studying multipath channel characteristics where measurements may not always be feasible. Degli-Esposti et al. study for example power-Doppler-profiles of a highly dynamic vehicular radio communication link. Their dynamic ray tracing model provides multidimensional channel prediction in any instant within a channel's coherence time. Other typical dynamic scenario of interest is air-to-ground (A2G) radio communications for Unmanned Aerial Vehicles (UAV). Vitucci et al. investigate the A2G channel in a reference urban scenario by means of the powerful DED-RL tool introduced earlier in this section [AVB+19]. Due to its computational efficiency, it is possible for the tool to cover a quite wide urban area, providing a large data-set for statistical assessment of A2G propagation and the optimization of drone trajectories.

Large-scale simulations of wave propagation, e.g., across a city for cellular coverage study, are another typical example where measurement-based analysis may practically be infeasible. A research group of Degli-Esposti and Vitucci made pioneering efforts in large-scale wave propagation simulations. For example, a hybrid method of deterministic 3D outdoor multipath prediction and its indoor extension using a radiosity-based method allows efficient computation of indoor coverage as demonstrated in [LVD+19]. The ray-tracer calculates received field strength on building surfaces, while the radiosity method extends the analysis to indoor using volume information of buildings. The hybrid method works well for indoor coverage estimation in high-rise buildings, where a 3D outdoor map of buildings is readily available but the detailed indoor map of the building is unknown. In a separate work [VFB+18], Over-Roof-Top (ORT) propagation in dense urban environment is investigated. ORT propagation is known to provide energy to mobiles when a base station is elevated above rooftop to cover a wide area of a built environment. Standard ORT models of wave propagation based on multiple diffracting knife-edge modeling of a series of building are considered, in addition to possible reduction of diffracting edges and consideration of heuristic correction factors. Results show that standard ORT models generally overestimate the attenuation, while their combination with edge reduction and correction factors improve coverage prediction accuracy.

2.2.1.2.3 Elaborated diffuse scattering models

In recent years, several works have shown the importance of DiS to achieve accurate prediction of the radio channel characteristics in real propagation environments. The importance is verified through a series of analyses of measured multipath channels where specular wave propagation such as reflection and diffraction does not account for all the power delivered to the receiver from the transmitter. DiS models are often embedded into ray tracing in order to consider the effect of surface roughness and other irregularities which cannot be modeled in a fully deterministic way. The models can either be built by highlighting their micro- or macroscopic nature. Microscopic models to represent DiS include Lambertian, directive, and backscattering models [EFVG07]. As depicted in Fig. 2.11, the scattered field is directed to the normal direction of a surface in the Lambertian model, while it is directed to the direction of specular reflection in the directive model. Examples of DiS models from macroscopic perspectives are elaborated in Section 2.2.5.

Improved mathematical models of microscopic DiS were proposed in the COST-IRACON action. For example, Wagen [Wag19b] introduces a formulation of scat-

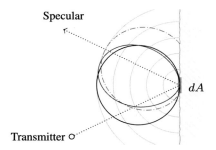

Specular

dA

Transmitter ○

FIGURE 2.11

Polar plot of scattered power at one tile for an impinging path at 30°: the Lambertian model is in black, the Directive model is in red (mid gray dash-dotted line in print version), and the dotted arrow denotes a specular reflection pathway [KWG⁺18].

tered fields accounting for both specular reflection and DiS. It satisfies the Helmholtz reciprocity principle, has dependency on distances between a scatterer and field transmit/receive points and finally takes into account the facet size from smooth to rough surfaces. Freire et al. [FPK⁺18] test if the Kirchhoff theory can provide reasonable estimation of scattered fields from a rough surface when surface shadowing may not be negligible. Such shadowing matters when a surface roughness is greater than a wavelength of radio signals. According to measured DiS from a rough brick wall at 60 GHz, surface shadowing effects were found to have moderate influence and the Kirchhoff theory brought good prediction accuracy of scattered fields from the rough wall.

2.2.1.2.4 Application to massive MIMO channel modeling

Another important aspect of future communication systems is the use of very large antenna arrays to take advantage of beamforming technology in multi-user scenarios, commonly known as massive MIMO. Ray models are particularly suitable to evaluate the performance of MIMO systems for various channel conditions and antenna configurations. However, one peculiarity to apply the ray models to a very large array is the validity of a plane wave incident model of a ray when a scatterer is close to the array. Zentner et al. [ZMM17] therefore studied the validity of the plane wave model for polarization, radio frequency and the sizes of antenna arrays at both sides of communication channel. Another peculiarity of massive MIMO channels is the fact that different parts of a single large antenna array may see different scattering environment and hence rays. This peculiarity was addressed by Sayer et al. [SBHN19]. They found that properties of ray-paths traced from a single transmit antenna element to a single receive antenna element, both in a large antenna array, can be different from those traced between the center of each antenna array. More accurate MIMO channel matrices can be obtained by the RT between individual antenna elements in arrays at link ends, though computational load increases. A possible compromise to keep both accuracy and computational load reasonable may be to trace rays for a vertical column of antenna elements only once using its center coordinate, while rays are traced for each element across horizontal row of antennas. Considering these peculiarities allows for realistic evaluation of radio network performance using RT. For example, Aslam et al. [ACBL18a] reported the system-level performance of a 16-macro-cell MIMO network in outdoor environment. Array configurations at the massive base station array showed significant impacts on the system-level performance.

2.2.1.3 Use of maps and point-clouds for ray-optical propagation modeling

Digital maps of the environment are the natural input for most site-specific propagation modeling, like ray tracing. However, there are also challenges of using digital maps for demanding air interface evaluations because there are currently no open source digital 3D map material that includes all features required for successful radio propagation simulations [HHS18]. The same work showed that a database containing all the needed information can be built from multiple sources for simulation use. In the following work, a novel method to join maps with different coordinate systems together is introduced [SHP+18]. The combined map is useful for site-specific propagation modeling and their visualization on a three-dimensional environment.

Describing the environment using point clouds, an example shown in Fig. 2.12, is another promising method for wave propagation simulations. They are obtained through laser scanning or photogrammetric survey of the physical environment. Proper modeling of propagation mechanisms for point clouds allows accurate simulation of multipath channels, as summarized in the following.

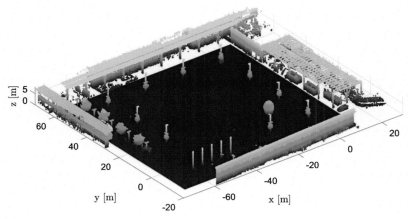

FIGURE 2.12

A sample point cloud of an open square [KSH19]. Points are colored according to their heights above the floor.

2.2.1.3.1 Wave-object interaction analysis based on point clouds

Wagen et al. [WVH16,Wag17,Wag18,WK18a,WK18b,Wag19a] made several contributions to radio propagation simulations using point clouds:

- *Simulation of specular reflection, scattering, and diffraction:* [WVH16] proposed a formulation of the specular reflection for point clouds. The model is verified with a reflection measurement from a small metal surface in an anechoic chamber at 15 GHz RF. The measurement shows that the specular reflected field was observed mainly from the illumination of about a third of the metal surface; [Wag17] illustrates simple mathematical models for simulating scattering and diffraction. Thereafter, [Wag19a] computes the scattering and specular reflection components using a single formulation, instead of accounting for them separately.
- *Application to large-area simulations:* [Wag17] uses detailed topographical data available in Switzerland in the form of point cloud for propagation simulations. Comparisons between the simulations and the wideband impulse response measurements at 69 and 254 MHz show that the dominant channel responses can be reproduced; [Wag18], [WK18a], and [WK18b] compare simulated and measured multipath channels using delay spread and a new metric called channel rise time.
- *Application to small-cell scenarios:* Comparisons of simulated and measured channels were reported in [WK18a] for 5 GHz WiFi connectivity in indoor and outdoor picocells. The work [WK18b] provides more comparisons in terms of spatial fading and data loss rate.
- The mentioned works highlight two potential challenges for propagation modeling using point clouds: (1) identifying long-delayed multipaths with up to a hundred of microseconds or more and (2) defining objective measures to rank the accuracy of different propagation models.

2.2.1.3.2 Use of point clouds for above-6 GHz radio channel modeling

As point cloud description of physical environments preserves small details such as lampposts and roughness in facade in outdoor environments and tables and chairs in indoor scenarios, it is particularly useful in site-specific propagation modeling at above-6 GHz RF where their wavelength is comparable to the details. A ray-tracer implementing all relevant propagation mechanisms including specular reflections, DiS, diffraction and shadowing is applied to an indoor point cloud, showing its validity at 60 GHz band in [JHK16]. In applying the point cloud data to above-6 GHz propagation channel modeling, a particular attention goes to pre-processing techniques of point clouds as they may not be applicable to RT as such. For example, Pascual-Garcia et al. [PGMGPMI+18] show a novel algorithm to extract a geometrical model from point clouds. Flat rectangular surfaces are extracted from a point cloud to produce a complete surface model of the environment, which is used in a ray tracing tool to conveniently estimate the specular and the diffuse components. The technique applied to an indoor environment demonstrates that the obtained surface model contains the needed detail to yield a good level of accuracy in wireless channel simulations. Koivumäki et al. [KSH19] compare multipath channels simulated

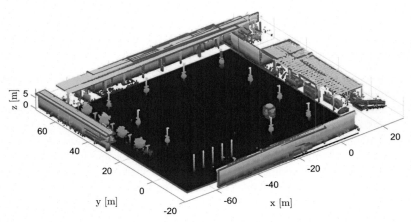

FIGURE 2.13

Processed point cloud after the de-noising, grouping, and filling that preserve details; original one is shown in Fig. 2.12. Wedges are detected for diffraction simulations as presented in red [KSH19].

from ray tracing using point clouds with different pre-processing techniques applied. Their results indicate that preserving small details of the environment is important for accurate ray-based radio propagation simulations. Proper processing of point clouds, which includes de-noising and wedges detection as shown in Fig. 2.13, led to the best accuracy in simulating physical propagation paths of the measured channels.

2.2.1.4 Full-wave models and other physics-based models

Steady increase of available computational power in CPU and GPU makes it feasible to solve electromagnetic fields of a radio environment using full-wave method. Despite being much more computationally demanding than ray-based method, it has a potential to provide very accurate results, and furthermore can also complement ray models. Analyses of scattering due to cluttering and vegetation are good examples where full-wave analysis can provide better understanding, and hence provide better mathematical models that can be incorporated into ray models.

2.2.1.4.1 Volume electric field integral equation

Kavanagh et al. solve the volume electric field integral equation (VEFIE) for indoor environments [KB19]. Comparison between two- and three-dimensional solutions of the equation reveals tradeoff between computational load and accuracy of the solutions. For a wideband channel simulation, it is necessary to solve the integral equation over a range of frequencies. Lin et al. [LAG$^+$18] made an interesting comparison between solutions of the VEFIE and RT along with measured channels in an indoor environment. Estimated received power levels are shown in Fig. 2.14. The plot indicates that optimized RT achieved the highest accuracy in estimating received power level, while a good compromise between computation time and accuracy is found by the 2D VEFIE with a heuristic path loss correction factor. The two-dimensional VEFIE solver was faster in field calculation than RT.

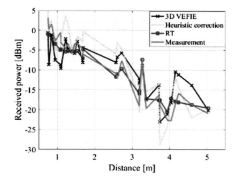

FIGURE 2.14

Received power estimated from 3D VEFIE, its heuristic correction and RT, compared against measurements [LAG+18].

2.2.1.4.2 Method of moments

Another popular full-wave method solver applied to electromagnetic scattering problems and wave propagation studies is the method of moments (MoM). For example, the method was applied to compute EM scattering from rough surfaces with Gaussian or exponential roughness profiles [PTCB15]; field distributions in indoor two-dimensional scenarios were solved using the method in [POC19] in order to study human blockage effects in beamformed 60 GHz radio channels.

2.2.1.4.3 Physical optics

The last method introduced in this subsection is physical optics (PO). PO [And96] is the high frequency approximation of the full-wave model in the lit region of a source, and has wide applicability for scatterers with different shapes and materials. It was originally developed for analyzing scattering from perfect electric conductor, but the concept of currents approximation is general and applicable to magnetic conductors, dielectric materials, and bodies with surface impedance [GMMLLH11]. PO takes less computation time compared to other rigorous numerical approaches where the induced currents on scatterer surface are determined by a large set of linear equations and could be extremely time-consuming to solve. The major source of errors in PO is at the edge of surface, or when the surface curvature is large and multiple reflection occurs. Gueuning et al. [GCO17,GCO19] used PO for fast computation of the near-field radiation from a planar object. The computation took advantage of an non-equispaced Fast Fourier Transform to convert from spatial to complex spectral domains.

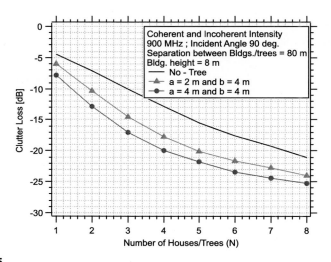

FIGURE 2.15

Clutter loss for vertically polarized fields at the top of the houses/trees for different tree widths a; b is a height of the elliptic model of leaves in a tree [TL18].

2.2.1.4.4 Physics-based model for path loss prediction in vegetated areas

Finally, for outdoor macrocellular wireless links, it is important to consider diffraction and penetration through vegetation and houses. Physics-based models to analyze path loss for such links have been studied for a long time, but still require further improvement. One of existing models is a Torrico-Bertoni-Lang model [TBL98]. It considers a row of houses or buildings, each of which is modeled by an absorbing screen. Each screen is topped with a canopy of trees modeled by a partially absorbing phase screen. Properties of the phase screens are determined to simulate the mean field in the canopy of the tree. PO is then used to evaluate the field at the receiver by using a multiple Kirchhoff-Huygens integration for each screen. The model is extended in [TL19] to reproduce measurements of the propagation loss for point-to-point systems at 3.5 GHz and 5.8 GHz in a vegetated residential area. The attenuation and phase delay of the mean field propagating through the tree canopy are evaluated using a random media model. Another extension is a physics-based model of clutter loss in a vegetated residential environment [TL18]. The improved model of the clutter loss adds the incoherent intensity of the received field produced by multiple trees/houses to the coherent intensity of the same. Estimated clutter losses in Fig. 2.15 show that fields scattered by the first few houses/trees seen from the base station led to both coherent and incoherent components and increased the losses. This observation is important for 5G mobile wireless systems since the size of the cells is few hundred meters; underestimating the clutter loss may result in an overestimation of cell size. Finally, the work [SUL19] is dedicated to accurate prediction propagation loss in a trunk dominated forest as an important application of Wireless Sensor Network

(WSN) to forested areas. A two-dimensional Radiative Transfer Equation (RTE) is solved for a plane wave incident on a trunk-dominated forest. The solutions of RTE are used to compute attenuation constants of both the coherent and the incoherent components. It is noted that at low frequencies, only the coherent intensity needs to be considered. As the frequency increases, the incoherent intensity becomes more important than the coherent intensity. Furthermore, as the transmit-receive distance increases, the coherent intensity decreases.

2.2.2 Geometry-based stochastic channel models (GSCM)
2.2.2.1 GSCM and their features

The family of Geometry-based Stochastic Channel Model (GSCM) has been widely used for physical layer standardization. It has a long history of development throughout the COST models, the 3GPP Spatial Channel Model (SCM), Wireless World Initiative New Radio (WINNER), IMT-Advanced (IMT-A) to recent 3GPP Three-Dimensional (3-D) and the latest 3GPP 5G channel models. The characteristics of GSCM can be defined from multiple features. Generally, the model is a GSCM if it has three characteristic features. Firstly, multipath propagation parameters contain *geometrical* definition of the environment, either in the angular domain or in the Cartesian coordinate system. Secondly, propagation parameters are at least partially *stochastic* and specified by probability distributions. Thirdly, the model enables consideration of realistic antenna characteristics by embedding of antenna radiation patterns sampled in the angular domain.

GSCMs in wireless standards have additional features. They all share close to identical mathematical frameworks and an algorithmic description for generating time-variant MIMO channel matrices containing transfer functions. The first feature is mostly limited to the angular domain, i.e., in the mentioned list of GSCMs the Multipath Component (MPC) or so-called clusters do not have locations but only directions (except for the COST2100 model where clusters are defined in the coordinate system). As a fourth feature, the path loss and shadowing are modeled by separate functions, where path loss model is an empirical function of at least the carrier frequency and the link distance, and the shadowing is a log-Normal zero-mean random process. Finally, these standard GSCMs contain the drop or the channel segment concept. Its essential feature is the so-called virtual motion, that provides Doppler shifts and temporal fading, but keeps all propagation parameters quasi-static and ensures Wide-Sense Stationary (WSS) for a small-scale movement of a mobile device.

2.2.2.2 Requirements for 5G channel models are still not fulfilled in the 3GPP model

Several channel modeling activities targeting 5G evaluations were completed, like European METIS [RKKe15] and mmMagic [Pet17] projects, before 3GPP started specifying its 5G channel model. In these projects, the first step was to envision functionalities of 5G networks and air interfaces, and to identify the new requirements set to channel modeling. These requirements are discussed in several articles, e.g., in [MKK+16]. The key drivers for 5G channel models are: large antenna arrays, new frequency bands, and new deployment scenarios of transceivers. A comprehensive gap analysis for upgrading 3GPP 3-D [36.15] to the 5G model (now [3GP17b]), based on the identified requirements, is performed in [JS16]. In total 17 categories of inaccuracies were found in the 3GPP 3-D model. Many problems were identified, such as missing scenarios, limited frequency band coverage, lack of spatial consistency with many associated defects to it, and inappropriate cluster definition for higher frequency bands and large antenna arrays. For example, large arrays may provide sufficient aperture to resolve individual sub-paths of clusters, which necessitates a new careful modeling of intra-cluster characteristics. The 5G channel model of 3GPP specified in [3GP17b] does still not fill all the gaps or meet all the requirements. As [JS16] summarizes, a significant research effort is needed to address all problems.

In addition to communications, radio can also be used for other purposes, such as positioning and radar. They are both applications that would require extension of the current GSCMs. The peculiarities and modeling requirements of vehicular multipath channel for radar are discussed in [JY19]. Intended and interfering path types in both urban and highway scenarios are thoroughly described. The automotive radar channel model must capture the physical locations of the radar, targets to detect and possible wave reflection and scattering objects. Thus, [JY19] proposes to study a hybrid of deterministic and stochastic model for radar evaluations.

2.2.2.3 Antenna modeling is a part of channel modeling

A realistic modeling of antennas, both individual elements and arrays, has been essential since the advent of MIMO communications and it is even more important now with 5G where antenna arrays may be very large (even massive) and their angular resolution very high. Antenna modeling, with polarimetric angular radiation patters, is a strength of GSCMs and deterministic channel models. Antenna patterns can be imported from antenna measurements or electromagnetic simulation tools. Alternatively patterns can be specified by mathematical functions. Arrays are specified in standard GSCMs by individual element patterns and array geometries, i.e. relative positions of elements within the array. The modeling principle also supports definition of antenna elements within an array into a common phase reference. In this case the array geometry is redundant as its effect is included in phases of element patterns.

2.2.2.3.1 Rotating antennas is a not trivial task

There is often a need to turn and tilt antennas to other orientations than the original measured/simulated orientation. The operation is not trivial, as the radiation pattern has vector coefficients with, e.g., E_ϕ and E_θ unit vectors. [ZMK18] gives a practical procedural description of rotating a complex polarimetric antenna radiation pattern to any direction specified by an arbitrary unit vector. First the antenna pattern, specified presumably in the spherical coordinate system, is transformed to the Cartesian coordinates. Target rotations along the three axes of Cartesian coordinate system are specified by 3×3 rotation matrices and the actual rotation operation is the matrix product of the rotation matrix and the radiation pattern in Cartesian coordinates. Finally the rotated pattern is transformed back to spherical coordinates. The power normalization has to be considered carefully when transforming between coordinate systems and performing the rotation matrix product.

2.2.2.4 Generality of models

Another benefit of GSCMs is their capability to construct different propagation environments and radio channel conditions boundlessly. A new "environment" is constructed each time a new model parametrization is drawn from probability distributions that specify a canonical propagation scenario. On the other hand, deterministic channel models, discussed in Section 2.2.1, are usually site specific. They are assumed to be capable of constructing only limited statistical variability of propagation conditions. Validity of this assumption is briefly touched in [HK16], where a simulation study using a map-based model [KLMLa17] is performed. The map-based channel model is a simplified site-specific model that also contains a random element, as originally proposed in [MSA02]. The question is, whether a deterministic model can provide similar statistical variations of propagation parameters as a GSCM specifies. Strictly and mathematically the answer is yes, if the map area is large and versatile enough and sufficiently many transceiver locations can be deployed. In [HK16] the so-called Madrid grid is 400×550 meters and 24 Transmitter (Tx) (Base Station (BS)) sites and 501 Receiver (Rx) (User Equipment (UE)) sites were simulated and resulting delay spreads, azimuth (UE side) spreads and Ricean K-factors were collected to derive empirical probability distributions. Rather high variation of parameters was found and the "satisfactory generality" of the map-based model was concluded. This is an interesting insight as deterministic channel model have become more attractive, and further work is required to determine to which extent this can be generalized to other scenarios and situations.

A related study is performed in [SR19] where a ray tracing, map-based, and a hybrid channel model output parameters and their second moments are compared. The first two models characterized specific streets of the city of Beijing and the hybrid model is built on identified dominant propagation paths of the map-based model. The hybrid model adds wave scattering points around those of dominant propagation paths randomly using an angular spread specific to the interaction type. A significant variation of resulting propagation parameters along the modeled route was observed.

2.2.2.5 Deterministic modeling of ground reflections in mmWave cellular GSCM

A characteristic of the mmWave band radio channel in cellular scenarios is the dominance of specular reflections over other propagation interaction mechanisms. In particular the ground reflection has a strong effect on the received power level. This is illustrated by a 60 GHz channel measurement in Fig. 2.16 [PWK+15]. The measured channel gain shows a similar dependence over link distance, up to 25 dB, as the well known two-ray model of Line of Sight (LOS) and ground reflection. The mean of both curves would follow approximately the free space path loss, but with a significant fluctuation that depends on the link distance and antenna heights. As a

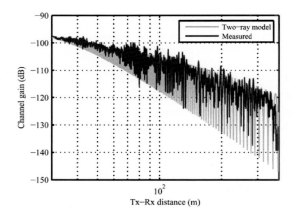

FIGURE 2.16

Comparison of measured channel gain vs. the two-ray model of LOS and ground reflection [PWK+15].

consequence, it is essential to model the ground reflection as a deterministic geometry dependent component, not as a random path. Introducing this deterministic component improves spatial consistency of a GSCM. A detailed derivation of a ground reflection model is given in [PWK+15]. It is based on Fresnel reflection coefficient, deterministic path lengths, and material permittivity.

2.2.2.6 Probability of LOS and reflected paths is derived

A GSCM does not contain a map, but typically only transceiver locations and antenna heights in a Cartesian coordinate system. Consequently, the availability of the LOS path is typically drawn randomly based on a LOS probability function. Furthermore, GSCMs usually specify propagation parameter distributions and path loss models, which may differ substantially, for LOS and Non Line of Sight (NLOS) conditions. LOS probabilities as well as ground and single wall reflection probabilities are investigated by a simulation study in [SJG16]. The map-based model assuming the so-called Madrid grid map is used together with numerous random transceiver locations to determine the probabilities as a function of link distance. The four studied scenarios are mobile-to-mobile, urban macro and micro, and urban micro open square. The study derives empirical cumulative distribution functions based on the binary existence of the LOS or reflected paths. Some interesting remarks were made, partly with respect to the LOS probabilities of the 3GPP model [36.15]. Firstly, the ground reflection does not always follow the LOS path, since in macro scenarios the roof-edge may block the path. Secondly, in the open square scenario the current 3GPP LOS probability formula is inadequate and needs a simple modification. Thirdly, most of the time the probability of finding a single wall reflection is higher than finding the LOS or ground reflected paths. Further simulations with various city layouts are planned as future work.

2.2.2.7 Map assisted LOS determination

The same question of LOS probability, but now with the aim of spatial consistency is discussed in [Ale17b] where a spatial consistent LOS/NLOS state model for GSCMs is proposed. The spatially-consistent model provides areas in a Cartesian coordinate system that indicates whether a UE connected to a particular BS site has LOS. The proposed model is map-assisted and accounts for building heights and densities, dividing the map area to two categories depending on this building information. This work is extended to support for time-varying MIMO channels in [ACS18]. There the so-called large scale parameters are generated on a map according to the determined LOS/NLOS regions. Cluster centers (wave interaction points) are drawn randomly in the coordinate system. Overall, the proposal aims to combine elements from stochastic and map-based models.

2.2.2.8 GSCM for non-terrestrial networks is under development

Though many satellite, atmospheric, and satellite-to-ground channel models have been defined in the past decades, 3GPP has not specified a MIMO channel model for Non-Terrestrial Networks (NTN). NTNs contain communications to ground level from different satellite orbits and from so-called High Altitude Platform Stations (HAPS), operating typically in altitudes between 8 and 50 km. A unified channel model concept for both terrestrial and NTN is described in [JBGS18]. The proposal is to keep the [3GP17b] model for fast fading and to extend it by adding components for long propagation delays, high Doppler shifts, and atmospheric effects like molecular absorption and scintillation. This is sketched in Fig. 2.17. The elevation angle is an essential parameter determining characteristics of MPCs and also molecular absorption. Large scale parameters, i.e., delay and angular spreads are extracted from ray-tracing simulations for various elevation angles and reported in [JBGS18]. These parameters are directly applicable to the GSCM of [3GP17b].

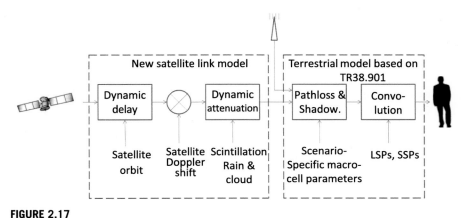

FIGURE 2.17

Block diagram of concatenated satellite and terrestrial models [JBGS18].

2.2.2.9 Clustered delay line model is a degenerated reference model of GSCM

Along the course of standardized GSCMs, there has been a need for limited randomness models, typically for uses like calibration of different implementation of the same channel model. For that purpose the concept of Clustered Delay Line (CDL) model was original developed. In CDL models the stochastic parameters are fixed and each time statistically identical MIMO fading channels are generated. This target is not reached, however, with the current procedure of, e.g., [3GP17b] in all conditions, as described in [KKH18]. Namely, with non-isotropic antennas and varying wave polarizations, the power angular distribution may vary because of a few random components of CDL. This may cause pessimistic performances in Time Division Duplex (TDD) communication simulations and prevent comparison of different uses of CDL models in link performance evaluations; [KKH18] proposes a few simple and non-disruptive modifications to guarantee wide sense stationarity over the ensemble of model runs by removing the unwanted randomness.

2.2.2.10 Analytical SINR model in urban microcells is derived by considering environmental randomness in BS deployment

Stochastic geometry has gained interest in many applications; [WVO17] is using it for urban coverage estimation by introducing random environments and physically motivated shadowing models. A Manhattan grid with straight perpendicular streets together with randomly drawn BS sites is generated by two one-dimensional homogeneous Poisson Point Processes. A UE is located in the center of the coordinate system. Different path types can be easily identified; LOS path is modeled with the free space path loss, penetration loss depends on the number of walls, and corner diffracted paths are modeled with the Berg's recursive model [Ber95]. For the case of identical transmit powers of BS, a mathematically tractable model has been determined for the Signal-to-Interference-plus-Noise Ratio (SINR) experienced by the UE and consequently for the coverage probability.

2.2.3 Enhanced COST2100 model

The COST 2100 MIMO channel model [LOP+12] is a spatially consistent GSCM that uses the concepts of clusters and Visibility Regions (VRs) to capture correlation effects and control the extent of scatterers' contribution to the channel. The model is a measurement based model for frequencies below 6 GHz and has been extended to cope with massive MIMO channels [FLET20,GFD+15] to cover the specific properties of the radio channel that are important in such systems during the COST-IRACON action. A MATLAB® implementation of the extension as well as the legacy model is freely available [IRA18]. Transmit arrays, receive arrays, clusters and scatterers are placed in a simulation area according to the routes of interest and prescribed distributions. It is the positions of the scatterers in the map, rather than directions and angles, that form the basis for the calculations of the contribution of each MPC to the total received signal. The model is inherently spatially consistent and capable of handling spherical wavefronts. The COST IRACON massive MIMO extension [FLET20] includes:

1. introduction of VR at the BS,
2. introduction of a gain function of individual MPCs,
3. generalization to full 3D geometries, and
4. full parameterization in indoor and outdoor scenarios at 2.6 GHz.

2.2.3.1 Visibility regions on base station sides

The introduction of VR at the BS is motivated by the fact that when the antenna arrays get physically larger, the radio channel cannot be seen as WSS over the array, but the statistics change over the array. Measurements in [PT12,GTE13] have shown that clusters appear and disappear along physically large arrays, which means that both the angular spread as well as the delay spread change over the array. This effect is typically not captured by conventional MIMO channel models, but could be important both when determining beamforming strategies as well as for realistic performance assessment and system simulations.

The appearance and disappearance of new clusters are modeled by a Poisson process along the array with an intensity of λ new or dead clusters per meter. For an array spanning the interval x_1 to x_2, the number of observed BS VRs (and clusters) is thus given by

$$N(x_1, x_2) \in Poi(\lambda \cdot (x_2 - x_1) + \lambda \cdot E(Y)), \tag{2.1}$$

where $E(Y)$ is the scenario dependent mean length of the visibility area at the BS, further details are described in [FLET20]. Fig. 2.18 shows an example from a measurement with a 7.5 m long uniform linear array in a LOS scenario at 2.6 GHz together with the modeled parameters. Over the whole array the median value of clusters seen is 23, but not all of them are visible at the same spot of the array. Six of the clusters can in median be seen over the whole array, and 17 clusters are in the median observable only at some parts of the array.

2.2.3.2 Gain functions of multipath components

The gain function of individual MPCs is introduced as measurements have shown [LLO+19] that individual MPCs have a limited lifetime within the cluster when the UE moves and that different MPCs of a cluster are active at different locations within a VR as illustrated in Fig. 2.19. To model this phenomenon, a gain function, with a Gaussian shape in the spatial domain and with its peak randomly located within the cluster, is connected to each MPC. These gain functions are used as weighting functions for the MPCs so that depending on where the UE is located in the VR, it sees different weighted combinations of them. This retains the spatial consistency of the model at the same time as it captures the fine details of the channel that are important for realistic assessment of user separability, especially when they are closely located, and for more advanced forms of radio based localization and navigation such as multipath aided navigation and tracking.

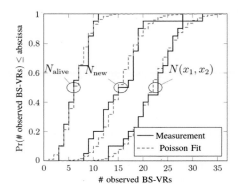

FIGURE 2.18

Observed visibility regions over a physically large array. $N(x_1, x_2)$ is the number of observed visibility regions (or clusters), N_{new} is the number of appearing visibility regions and N_{alive} is the number of already existing visibility regions [FLET20].

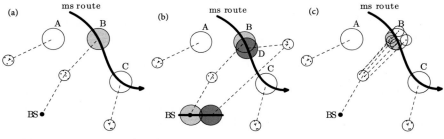

FIGURE 2.19

A MS moves in and out of various VRs, here named A, B, C, and D. (a) When the MS enters a VR, the associated cluster becomes active (VR B). (b) When a physically large array is used, different parts can observe different sets of clusters. (c) The relative gain of individual MPCs is controlled by MPC VRs, one for each MPC.

The lifetime, or rather the length of the spatial region within the VR where an MPC has a somewhat significant contribution to the observed impulse response, is a random parameter determining the radius of the MPC VR and the corresponding width of the gain function. Some MPCs have long lifetimes whereas the majority of MPCs actually can be observed only in a limited area and thus have short lifetimes and small MPC VRs. Measurements in an indoor sports hall [LLO+19] show that MPC lifetimes are best described by a log-normal distribution for the radii of the MPC VR. To have a smooth onset of the activation of a specific MPC we model the relative contribution of each MPC to the total channel gain by means of the gain function [GFD$^+$15,GERT15]. A Gaussian profile

$$g_{\text{MPC},l}(\vec{r}_{\text{MS}}) = \exp\left(-\frac{(\vec{r}_{\text{MS}} - \vec{r}_{g,l})^2}{2\sigma_{g,l}^2}\right) \tag{2.2}$$

multiplies the complex amplitude of each MPC. Thus, the weight of a particular MPC depends on the Euclidean distance between the UE position $\vec{r}_{\text{MS}}$ and the center of the l-th MPC VR $\vec{r}_{g,l}$. The centers of the MPC VRs are uniformly distributed within the cluster VR. The lifetime of the MPC is in this way determined by the width of the Gaussian profile. This width is controlled by $\sigma_{g,l}$ and is modeled as a log-normal parameter [FLET20].

2.2.3.3 Support of three-dimensional geometry

The need for generalization to 3D geometries and support for polarimetric channels comes naturally with the more advanced antenna arrangements used for massive MIMO. The BSs, UEs, scatterer locations and VRs can all be described by their 3D coordinates in the simulation area and any array geometry is supported as the individual antenna locations also are described by their individual coordinates. Antenna gain patterns are included at an individual antenna level, and can either be in the form of measured antenna responses, simulated responses or set as an arbitrary mathematical function (including the case of omnidirectional antenna responses). The COST family of channel models inherently captures spherical wave front effects as scatterer locations are defined by their coordinates in the simulation area rather than their directions with respect to BS and UE antennas. As the model output is the transfer function matrix (or equivalently, the impulse response matrix after inverse Fourier transform) between BS antenna array and UE antennas, any kind of digital, analog or hybrid beamforming can be analyzed.

2.2.3.4 Available parameters of the model

Finally, the COST IRACON massive MIMO extension is parameterized and validated based on measurements for physically large outdoor arrays at 2.6 GHz in LOS and NLOS, and with indoor and outdoor measurements for closely located users at 2.6 GHz using a physically smaller circular array. The list of parameters can be found in Table 2.1, and a detailed model description can be found in [FLET20].

2.2.4 Reference ITU-R path loss models

A number of ITU recommendations address path loss models depending on the environment, frequency band and BS-UE range. These include

- Recommendation ITU-R P.1411 [Rec17b] for short range outdoor scenarios up to 1 km and in the frequency range of 300 MHz to 100 GHz;
- Recommendation ITU-R P.1238 [Rec17a] for indoor scenarios in the frequency range of 300 MHz to 100 GHz;

Table 2.1 Parameterization of the COST 2100 model extension for closely-located users with physically-large arrays and CLA at 2.6 GHz [FLET20].

Parameter	Outdoor (NLOS)	Indoor (LOS)
Length of BS-VRs, L_{BS} [m]	3.2	-
Slope of BS-VR gain, μ_{BS} [dB/m]	0	-
Slope of BS-VR gain, σ_{BS} [dB/m]	0.9	-
MPC gain function, $\mu_{R_{MPC}}$ [dB]	-	−19.8
MPC gain function, $\sigma_{R_{MPC}}$ [dB]	-	10.1
Average number of visible far clusters, N_C	$2.9 \times (L_{BS} + L)$	15
Radius of the cluster visibility region, R_C [m]	10	5
Radius of cluster transition region, T_C [m]	2	0.5
Number of MPCs per cluster, N_{MPC}	31	1000
Cluster power decay factor, k_τ [dB/µs]	43	31
Cluster cut-off delay, τ_B [µs]	0.91	0.25
Cluster shadowing, σ_S [dB]	7.6	2.7
Cluster delay spread, m_τ [µs]	0.14	0.005
Cluster delay spread, S_τ [dB]	2.85	1.5
Cluster angular spread in azimuth (at BS), $m_{\psi_{BS}}$ [deg]	7.0	4.6
Cluster angular spread in azimuth (at BS), $S_{\psi_{BS}}$ [dB]	2.4	2.1
Cluster angular spread in elevation (at BS), $m_{\theta_{BS}}$ [deg]	0	3.7
Cluster angular spread in elevation (at BS), $S_{\theta_{BS}}$ [dB]	0	2.6
Cluster angular spread in azimuth (at MS), $m_{\psi_{MS}}$ [deg]	19	3.6
Cluster angular spread in azimuth (at MS), $S_{\psi_{MS}}$ [dB]	2.0	2.1
Cluster angular spread in elevation (at MS), $m_{\theta_{MS}}$ [deg]	0	0.7
Cluster angular spread in elevation (at MS), $S_{\theta_{MS}}$ [dB]	0	3.6
Cluster spread cross-correlation,		
$\rho_{\sigma_S\tau}$	−0.09	−0.45
$\rho_{\sigma_S\psi_{BS}}$	0.04	−0.56
$\rho_{\sigma_S\theta_{BS}}$	0	−0.50
$\rho_{\tau\psi_{BS}}$	0.42	0.70
$\rho_{\tau\theta_{BS}}$	0	0.34
$\rho_{\psi_{BS}\theta_{BS}}$	0	0.50
Radius of LOS visibility region, R_L [m]	-	30
Radius of LOS transition region, T_L [m]	-	0
LOS power factor, $\mu_{K_{LOS}}$ [dB]	-	−5.2
LOS power factor, $\sigma_{K_{LOS}}$ [dB]	-	2.9
XPR, μ_{XPR} [dB]	0	9
XPR, σ_{XPR} [dB]	0	3

- Recommendation ITU-R P.2108 [Rec17c] gives models for the prediction of clutter loss;
- Recommendation ITU-R P.2109 [Rec17d] is for building entry loss;

- Recommendation ITU-R P.1546 which covers point-to-area path loss predictions for terrestrial services in the frequency range 30 MHz to 3000 MHz;
- Recommendation ITU-R P.528 on the prediction of point-to-area path loss for the aeronautical mobile service for the frequency range 125 MHz to 15.5 GHz and distance range up to 1800 km;
- Recommendation ITU-R P.617 for the prediction of point-to-point (P-P) path loss for trans-horizon radio-relay systems for the frequency range above 30 MHz and for distances from 100 to 1000 km;
- Recommendation ITU-R P.530 for the prediction of P-P path loss for terrestrial line-of-sight systems; and
- Recommendation ITU-R P.2001 [Rec19] which provides a wide-range of terrestrial propagation models for the frequency range 30 MHz to 50 GHz including both fading and link enhancement statistics.

In this section we review the ITU path loss models for outdoor short range, indoor environments, clutter loss and building entry loss, which cover the high frequency bands including the millimeter wave bands proposed for 5G cellular networks.

2.2.4.1 Indoor environments

Recommendation ITU-R P.1238-9 [Rec17a] gives transmission loss models for indoor environments which assume that the BS and the UE are located inside the same building. It provides two models: a site-general model and a site specific model.

The site-general model is given by

$$L_{\text{total}} = L(d_o) + N \log_{10}\left(\frac{d}{d_o}\right) + L_f(n) \text{ dB}, \qquad (2.3)$$

where N is a distance power loss coefficient, f is frequency (MHz), d is a separation distance (m) between BS and UE where $d > 1$ m, d_o is a reference distance (m), $L(d_o)$ is the path loss at d_o (dB) for a reference distance d_o at 1 m, and assuming free space propagation $L(d_o) = 20 \log_{10} f - 28$ where f is in MHz; L_f is the floor penetration loss factor (dB) and finally n is the number of floors between BS and UE ($n \geq 0$), with $L_f = 0$ dB for $n = 0$.

The recommendation provides typical parameters, based on various measurement results from 0.8 GHz to 70 GHz in residential, office, commercial, factory and corridor environments with the office environment being the most characterized across the bands, and at 300 GHz for a data center. Tables 2 and 3 in [Rec17a] give the values of the coefficient N based on $d_o = 1$ m, and floor penetration loss factors, L_f (dB), respectively. The shadow fading statistics, standard deviation in dB, is given in Table 4 of [Rec17a].

The site specific model refers to the estimation of path loss or field strength, based on the uniform theory of diffraction and RT techniques which require detailed information of the building structure. It recommends including reflected and diffracted rays to improve the accuracy of the path loss prediction.

2.2.4.2 Outdoor short range environment

Recommendation ITU-R P.1411-9 [Rec17b] also provides site-specific and site-general propagation models across various frequency bands for LOS and NLOS scenarios. It also classifies the environments as urban very high rise, urban high rise, urban low rise/suburban, residential and rural. It also defines three cell types: micro-cell for ranges between 0.05 to 1 km, dense urban micro-cell for ranges between 0.05 to 0.5 km and pico-cells for ranges up to 50 m where the station is mounted below rooftop.

The site-general model applies to two scenarios where the two terminals, i.e., Tx and Rx, are below rooftop heights and where one terminal is above the rooftop height and the second terminal is below the rooftop. The site general model for both scenarios is given by

$$PL(d, f) = L(d_o) + 10\alpha \log_{10}(d) + \beta + 10\gamma \log_{10}(f) + N(0, \sigma) \text{ dB}, \quad (2.4)$$

where d is a 3D direct distance between the transmitting and receiving stations (m); f is the operating frequency (GHz); α is a coefficient associated with the increase of the path loss with distance; β is a coefficient associated with the offset value of the path loss; γ is a coefficient associated with the increase of the path loss with frequency; and finally, $N(0, \sigma)$ is a zero mean Gaussian random variable with standard deviation σ (dB).

This model provides coefficients which cover a wide frequency range. Table 4 in the recommendation provides the coefficients of the model for below the rooftop scenario for urban high rise, urban low rise and suburban for LOS and NLOS for up to 73 GHz and from 5 m to 715 m depending on the environment. Frequencies are covered depending on the environment and with ranges varying. Table 8 in the recommendation provides the coefficients for above the rooftop scenario for urban and suburban environments up to 73 GHz.

The recommendation also provides two site-specific models for LOS and NLOS scenarios where in the LOS case, it adopts a dual slope model with a breakpoint up to 15 GHz. For the millimeter wave band a single slope model is recommended since the breakpoint will occur beyond the expected range of the cell. In this case the path loss model is given by

$$L_{\text{LoS}} = L_0 + 10n \log_{10}\left(\frac{d}{d_0}\right) + L_{\text{gas}} + L_{\text{rain}} \text{ dB}, \quad (2.5)$$

where L_{gas} and L_{rain}, are attenuation by atmospheric gases and by rain which can be calculated from Recommendation ITU-R P.676 and Recommendation ITU-R P.530, respectively. Typical values of the path loss coefficient from LOS measurements with directional antennas are in the range of 1.90 to 2.21.

2.2.4.3 Clutter loss

Clutter refers to objects, such as buildings and vegetation, which are on the surface of the earth but not actually terrain. Clutter loss models in Recommendation ITU-R

P.2108 [Rec17c] are statistical, where clutter loss is defined as the difference in the transmission loss with and without the presence of clutter. It is therefore, the difference between free space loss and the loss in the presence of clutter.

The statistical clutter loss model for terrestrial links can be applied for urban and suburban environments. It gives the clutter loss L_{ctt}, not exceeded for p % of locations for the terrestrial-to-terrestrial link by the following set of equations:

$$L_{ctt} = -5\log(10^{-0.2L_l} + 10^{-0.2L_s}) - 6Q^{-1}(p/100) \text{ dB}, \qquad (2.6)$$

where $Q^{-1}(p/100)$ is the inverse complementary normal distribution function, and

$$L_l = 23.5 + 9.6\log(f) \text{ dB}, \qquad (2.7)$$
$$L_s = 32.98 + 23.9\log(d) + 3\log(f) \text{ dB}, \qquad (2.8)$$

where d is the 3D link distance.

2.2.4.4 Building entry loss

The building entry loss model in Recommendation ITU-R P.2109-0 [Rec17d] is based on measurement data collated in Report ITU-R P.2346 in the range from 80 MHz to 73 GHz. The model is given for two types of buildings: traditional and thermally efficient; where in the case of modern, thermally-efficient buildings (metalized glass, foil-backed panels) building entry loss is generally significantly higher than for "traditional" buildings without such materials. The model therefore gives predictions for these two cases where it is assumed that the indoor antenna is omnidirectional.

The building entry loss distribution is given by a combination of two log-normal distributions. The building entry loss not exceeded for the probability, P, is given by

$$L_{\text{BEL}}(P) = 10\log(10^{0.1C(P)} + 10^{0.1B(P)} + 10^{0.1C}) \text{ dB}, \qquad (2.9)$$

where:

$$A(P) = F^{-1}(P)\sigma_1 + \mu_1, \qquad (2.10)$$
$$B(P) = F^{-1}(P)\sigma_2 + \mu_2, \qquad (2.11)$$
$$C = -3.0, \qquad (2.12)$$
$$\mu_1 = L_h + L_e, \qquad (2.13)$$
$$\mu_2 = w + x\log(f), \qquad (2.14)$$
$$\sigma_1 = u + v\log(f), \qquad (2.15)$$
$$\sigma_2 = y + z\log(f), \qquad (2.16)$$

where $L_h = r + s\log(f) + t(\log(f))^2$ is the median loss for horizontal link paths and $L_e = 0.212|\theta|$ is the correction for elevation angle of the link path at the building facade; f is the frequency (GHz); θ is the elevation angle of the path at the building facade (degrees); P is the probability that the loss is not exceeded ($0.0 < P < 1.0$); finally, $F^{-1}(P)$ is the inverse cumulative normal distribution as a function of probability.

2.2.5 Models for dense multipath components

Dense multipath components (DMCs) represent the part of the radio propagation channel that can not be characterized by a superresolution of distinct plane waves [PSH+15]. In other words, the continuous power spectrum representation is more appropriate for the DMC. In typical indoor scenarios, DMC may include 1) DiS originated from rough surfaces whose height deviation is greater than the wavelength of the carrier frequency, and 2) the reverberation of radiated energy due to a multitude of specular components, e.g. multiple reflections and multiple scatterings, which typically occurs in a closed or partially-closed cavity.

While Section 2.2.1.2.3 covers microscopic modeling of DiS, there also exists macroscopic modeling of the DMC. It can be modeled empirically and holistically as fitting to the residue of the measured channel after excluding Specular Multipath Component (SMC), as detailed in Section 2.3. Empirical evidence of significant DMC is found in a number of papers in literature. It contributes to more than 30% of the total channel power for LOS scenario at 11 GHz in indoor scenarios, and depends on angle and polarization for a frequency range from 3 to 28 GHz [VBJO17,SiTK17,HST+18,SHF+19]. DMC contributes up to 70% of the total channel power at 3 GHz in an industrial environment [TGL+14] and up to 40% and 86% in LOS and NLOS indoor channels [PSH+15]. Polarimetric characterization of DMC is also of great interests, as discussed for an industrial hall at 1.3 GHz and an urban site at 4.5 GHz [GTJ+15,LST+07]. Steinböck et al. [SPF+13] defined a term reverberation ratio to model the DMC contribution to the total channel power that depends on the distance between Tx and Rx antennas.

Despite the potential large power contribution to the total channel power and the impact on the polarization and angular characteristics of multipath channels, DMC have not yet been considered explicitly in standard channel models, except for the COST2100 model detailed in Section 2.2.3. Therefore, measurements, characterization and modeling of DMC for various application scenarios are necessary. In particular, when DMC dictates the total channel power, the comprehensive modeling of DMC in temporal-spatial-frequency domain is crucial for studies of wideband and multiple-antenna radio systems [MGO18,KWG+18].

Statistical and macroscopic modeling of DMC looks at reverberation of electromagnetic fields inside a confined space as in [Ped18,MPG+19b,AP18b,AP18a,Ped19,APB18]. Multipaths due to DiS typically have longer delays than those of SMC and hence forming a "tail" in the power delay profile (PDP) of channels. The tail can be described as an exponentially decaying function with fluctuations; an exemplary indoor PDP is shown in Fig. 2.20. As the DMC represents reverberation of energy, multiple bounce specular reflections or multiple bounce DiS describe the phenomenon.

2.2.5.1 Analysis of diffuse scattering

One of the challenges in proper modeling of DiS is to obtain its observations either from measurements or computation. The latter usually allows more rigorous analysis than measurements as different surface samples can be individually considered. DiS

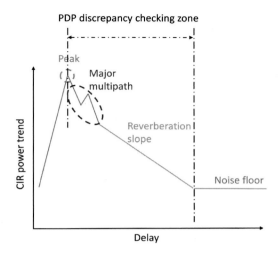

FIGURE 2.20

Typical shape of indoor PDP consisting of SMC, a slope representing DMC and noise floor [MPG+19b]. CIR indicates Channel Impulse Response.

from rough surfaces can be computed, for example, by Physical Optics (PO) introduced in Section 2.2.1.4. The surface roughness (or equivalently irregularity) can be statistically described as an Effective Roughness (ER) model using, e.g., a correlated Gaussian process [Ves91], while the same can also be measured for existing surfaces as point clouds as discussed in Section 2.2.1.3 [JHK16]. Miao et al. [MGO18] investigated the spatial coherence of DiS multipath by applying PO to ER model of a surface. In PO, the rough surface is first divided into triangles whose side length is in the order of 0.1 times the wavelength. The EM field on each triangle is then approximated to equivalent PO current to derive the Channel Transfer Function (CTF). Different from PO, the side length of facet in ER is determined by the far-field distance from Tx and Rx. They derived the spatial correlation of DiS multipath from a rough wall using its ER model. With a fixed Tx antenna, phase correlation of DiS multipaths over Rx antenna locations decreases. The speed of decrease depends on the surface roughness and the relative position between Rx and the wall.

The work by Kulmer et al. [KWG+18] showed that DiS due to rough surfaces can have negative impacts on radio localization accuracy. DiS was generated in their GSCM using the ER models that reflect dielectric properties and locations of a rough surface. The ER models clearly impacted the shape of angular power spectrum and Power Delay Profile (PDP) of channels and hence the radio localization performance. Dense Multipath Component (DMC) behaves like an added Gaussian non-zero-mean noise to SMCs that are used for localization, resulting in biased angular and delay estimates of SMCs.

2.2.5.2 Modeling reverberation by propagation graphs

Propagation graph (PG) [PSF12,PSF14] is a promising approach to model the diffuse "tail" of a PDP due to reverberation of electromagnetic fields in a confined space. An exemplary PG is illustrated in Fig. 2.21, where the vertices represent the transmitters, the receivers and the scatterers presenting in the propagation environment. The edges model wave propagation between vertices including time delay, leading to time-dispersion of the channel. The surface roughness of objects in a propagation environment may impact the choice of the vertices in PG models [TEVY16,MPG+19b]. The major merit of PG modeling of DiS is that it yields a closed-form solution of a transfer function even if infinite bounces of radio waves on objects in the environment are involved.

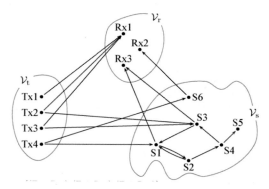

FIGURE 2.21

An exemplary propagation graph with four transmit vertices, three receive vertices, and six scatterer vertices [PSF12]. Signal propagation between vertices is represented by the edges and associated transfer matrices.

2.2.5.2.1 Application of propagation graph models to different scenarios

PG can be either stochastically or deterministically generated. It has been applied to different scenarios and has successfully predicted the radio channel properties, such as scenarios of in-room [PSF12], outdoor-to-indoor [PSF14], high speed railway [ZTS+15] and millimeter wave systems [CYTK17]. In addition, the hybrid of PG with RT has been applied to predict in-room scenarios [TEVY16,SGM+16]; RT and PG are capable of capturing major multipaths and diffuse reverberation "tail", respectively.

2.2.5.2.2 Extension of the propagation graph models

PG modeling of diffuse scattering and the calculation of reverberation components are extended to various application scenarios [Ped18,MPG+19b,AP18b,AP18a, Ped19,APB18]. For example, Adeogun and Pedersen [AP18b] have extended the original PG model in [PSF12] to be applicable to polarized multiple-antenna systems. The polarized PG model contains not only attenuation, delay, phase shifts, but also depolarization effects of wave propagation, scattering and antenna effects. The model was applied to predict the dual-polarized indoor MIMO channel in a cuboid. The simulation results showed that the diffuse reverberation "tail" of PDP of co- and cross-polarized channels has almost the same decay rate. The model is further extended to relate the powers of co- and cross-polarized signal components and their ratio with the number of scatterers, probability of scatterer visibility, reflection gain and polarization coupling parameters [AP18a].

Miao et al. [MPG+19b] show the use of PG for modeling DiS in room-to-room scenarios with the help of RT. The PG vertices were obtained in each room separately by RT with the assumption that Rx (or Tx) virtually locates on the surface of the separating wall between two rooms. The rays transmitted from one room to the other through the separating wall are calculated by Snell's law. Placing vertices of PG in each room separately allows simplest modeling of DiS multipaths in the scenario, based on two assumptions: 1) no propagation from the room containing Rx back to the room containing Tx and 2) RT applied separately to each room only involves propagation mechanisms of LOS and first order specular reflection.

2.2.5.2.3 Propagation graph of reduced complexity

DiS observed in PDPs results from waves bounced on surfaces many times. In the closed-form CTF produced by PGs, there is a fraction of matrix inversion. The inversion represents the infinite interaction loops among scatterers, but may result in a heavy computation load with the increased number of scatterers and hence vertices. To cope with this problem, [APB18] has proposed an equivalent PG model of reduced complexity in computing the CTF when Tx and Rx are located in different rooms. A vector signal flow has been proposed for PG where rooms in the complex large environment are denoted as nodes and the propagation between rooms is as branches. The proposed equivalent model was shown to yield the same accuracy of prediction as the original PG model in [PSF12] but with much lower computational complexity.

2.2.5.2.4 Path arrival rate

Pedersen [Ped18,Ped19] has modeled the path arrival rate of a propagation channel when Tx and Rx equipped with directive antennas are in the same room. PG and the mirror source theory are used for the model. A notion of "mixing time" is defined as the onset of the diffuse reverberation "tail", and as the point in time where the arrival rate exceeds one component per transmit pulse. This definition was inspired by Eyring's empirical model of the reverberation time, which has been experimentally evaluated in [SPF+15].

2.2.5.2.5 Combination of propagation graph and geometry-based channel model

Finally, Vinogradov et al. [VBJO17] have modeled dynamic PDPs of room-to-room channels. The model is a physical-statistical combination of 1) room electromagnetic theory for the deterministic components and of 2) GSCM with time-variant statistics for the stochastic components. The deterministic components are the major multipaths followed by an exponentially decaying slope in PDP, while the stochastic components represent large-scale variations such as shadowing and small-scale variations over time and delays. The work reported non-stationary fading of dynamic indoor channels across time and delays. Their variation has been characterized using the Rician K-factor and the t-location scale distribution parameters for small- and large-scale fading. A Hidden Markov model was used to describe the evolution of the small-scale fading statistics.

2.2.5.3 Experimental characterization of dense multipath components from channel measurements

Observations of DMC from real-world radio channel sounding serve as a basis for developing its realistic models [VBJO17,SiTK17,HST+18,SHF+19,D'E19b]. In practice, DMC can be obtained as residue of measured channels after subtracting SMC.

2.2.5.3.1 Directional and polarimetric characteristics

Saito et al. [SiTK17] have characterized the DMC from the indoor LOS MIMO measurement at 11 GHz band with 400 MHz bandwidth. Statistics of DMC over frequency, angular and polarization domains were estimated jointly with a set of parameters for SMC as detailed in Section 2.3. Estimated DMC has shown directional and polarization dependence. The DMC occupies up to 30% of total channel power according to their measurements. In addition, with the decrease of the room size, the angular spreads of DMC increase, and their diffuse "tails" last longer over the delay, owing to stronger reverberation of energy inside the room. A similar analysis has also been reported by Hanssens et al. [HST+18] for measured radio channels at 11 GHz RF, where the Angular Power Spectrum (APS) of DMC is characterized by a multimodal Von Mises distribution better than the conventional unimodal one.

2.2.5.3.2 Dependence on carrier frequency

Measurements in IRACON also shed lights on frequency dependency of DMC. For example, D'Errico [D'E19b] characterized the DMC in a machinery room full of metallic objects for a band of 2 to 6 GHz. The DMC contributed to 11.8% and 24.7% of the total channel power in the measurements of LOS and NLOS scenarios, respectively. Their trend of exponentially decaying power in PDP, as the delay increases, was found to be similar across the band. Finally, Saito et al. [SHF+19] estimated DMC from MIMO measurements at 3, 10, and 28 GHz with the same bandwidth in a hall and a laboratory room. The work found that 1) the SMCs were similar in all investigated frequency bands and 2) the power and the angular spread of DMC tended to decrease as the carrier frequency increased.

2.3 **Algorithms for estimation of radio channel parameters**

Multipath component estimation for radio channels is limited by the finite apertures of the radio channel sounding equipment in the time, frequency, and spatial domains. Propagation paths that are low in power or are densely packed in the Doppler, delay, or angular domains can often not be fully resolved by state-of-the-art estimation algorithms. These paths are put under the umbrella of *dense multipath components (DMC)* and include diffusely scattered paths and low-power specular reflections. The DMC phases are considered to be random (i.e., are incoherent) and as such DMC can only be described by their power density in the Doppler, delay, and angular domains. This is in contrast to the high-power *specular multipath components (SMC)* that possess well-defined coherent phases.

2.3.1 **Narrowband multipath component estimation**

In classical narrowband multipath estimation, it is commonly assumed that the frequential and spatial domains are uncorrelated. Specifically, the change in wavelength (or equivalently, the electrical dimensions of the environment and the antenna arrays) over the considered bandwidth is deemed insignificant. This means that, in narrowband algorithms, we assume a constant wavelength while still allowing the frequency to vary across the bandwidth. We thus effectively separate space (wavelength) and frequency. This is an approximation to simplify calculations because, of course, frequency and wavelength are still always related by the speed of light. Moreover, the limited bandwidth implies that the reflectivity of the environment is constant and independent of frequency within that band. The narrowband assumption offers a few computational benefits to maximum-likelihood multipath estimation algorithms such as SAGE [FTH+99] and RiMAX [Ric05]. The separation of the frequential and spatial domains means that certain vectors or matrices in the signal model can be simply written as the Kronecker product of their one-dimensional versions across each of the multipath parameter domains separately. This so-called Kronecker separability is applied to the specular multipath steering vector and the dense multipath covariance matrix. The reduction in computational burden then stems from replacing the multidimensional parameter estimation with an iterative one-dimensional initialization and optimization scheme executed sequentially in each of the parameter domains.

2.3.1.1 *An estimator for non-uniform angular dense multipath*

Modern multipath estimation algorithms include DMC estimation based on mathematical models of the DMC power spectral densities with a priori unknown parameters. For example, the original RiMAX maximum-likelihood estimator assumes an exponentially decaying DMC power spectral density in the delay domain and a uniform distribution of DMC power in the angular domains [Ric05]. The uniform DMC angular spectrum in particular is hard to maintain in most realistic non-reverberating indoor and outdoor environments. For this reason, modifications to the mathematical model of the DMC angular spectrum have been proposed.

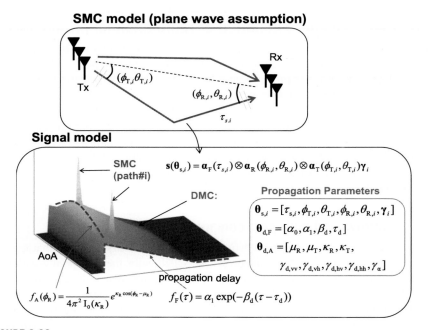

FIGURE 2.22

SMC and DMC data model for their parameter estimation [SiTK17].

A more fitting model for the DMC angular spectrum at transmitter and receiver is the Von Mises distribution, originally proposed in [RRK07]. The work in [SiTK17] incorporates a unimodal Von Mises model at transmitter and at receiver in the RiMAX estimator. Fig. 2.22 shows the used SMC and DMC signal model: the original RiMAX model is extended with a DMC propagation parameter vector $\theta_{d,A}$ that includes the Von Mises mean angle and concentration parameters, and the power gains for the four polarization subchannels. Follow-up work in [HST$^+$18] extends this to a multimodal Von Mises model that allows for multiple angular DMC clusters. The Von Mises extension to the RiMAX DMC data model has been validated with indoor radio channel sounding measurements at 11 GHz in both [SiTK17] and [HST$^+$18]. The agreement between measurements and model was checked by comparing the distribution of the largest eigenvalues calculated from the measured MIMO channel matrix and from the channel matrix reconstructed with the model.

2.3.1.2 Specular multipath estimation from power spectra

The clustered MPCs are widely observed and used in channel characterization. To develop and evaluate advanced wireless systems at high frequency, directional scanning by highly directive antennas is the most popular method to obtain angular channel characteristics. However, it is often insufficient for ray-/cluster-level characterization and channel modeling because the angular resolution of the measured data is actually limited by the angular sampling interval and antenna half power beamwidth for a given scanning angle range. In [KIU⁺20], the subgrid CLEAN algorithm was developed, which is a novel technique for high-resolution MPC extraction from the multi-dimensional power image, the so-called double-directional angular delay power spectrum. By applying the developed technique to measured data taken at 58.5 GHz in indoor environments assuming an in-room 5G hotspot access scenario, scattering processes are identified and multipath clusters are characterized: 8-12 clusters were observed within a dynamic range of approximately 40 dB as illustrated in Fig. 2.3.

2.3.1.3 A Bayesian estimator for specular and dense multipath

In [GLFW18], a sparse Bayesian learning (SBL) algorithm for the estimation of SMC and DMC parameters is introduced. The benefit of the SBL algorithm is that the number of SMC is assumed to be unknown and that number is estimated jointly with the SMC and DMC parameters. This is in contrast to maximum-likelihood or maximum a posteriori estimation for which the number of SMC is a priori fixed. The SBL algorithm is applied to both synthetic and measured data and returns SMC and DMC parameters with good accuracy. Importantly, the algorithm is shown able to detect SMC that have real physical representations in the radio environment and is good at suppressing specular artifacts that cannot be linked back to the environment.

2.3.2 Wideband multipath component estimation

2.3.2.1 A specular multipath estimator for ultra-wideband channels

A new multipath component estimator specifically developed for Ultra WideBand (UWB) radio channels is presented in [HTG⁺18]. The estimation framework is based on the RiMAX narrowband maximum-likelihood estimator [Ric05]. The ultrawide bandwidth is divided into sub-bands wherein the narrowband assumption can reasonably be expected to hold. Parameters of SMC and DMC are classified into parameters that are assumed to be constant over the full UWB bandwidth and parameters that are constant only in each sub-band. The parameters constant over the full bandwidth are specular multipath parameters that only depend on the geometry of the propagation environment, namely AoD, AoA, and ToA. These parameters are estimated by maximizing a *global* cost function that is a weighted sum of the narrowband RiMAX cost functions for each sub-band. The multipath parameters that are only constant in each sub-band are the complex amplitudes of the specular paths and the parameters of the theoretical power spectral densities associated with the dense multipath. These are estimated in each sub-band separately with the classical RiMAX estimator by maximizing *local* cost functions.

2.3.2.2 Joint delay and Doppler estimation for bistatic radar

In [DKS⁺19], a RiMAX-inspired maximum likelihood estimator is presented for the purpose of joint delay and Doppler estimation. The estimator is developed specifically for bistatic radar measurements that typically have transmitting sequences that are relatively long. To take this into consideration, Doppler is modeled as a phase that changes linearly with time within one transmit symbol period. Delay and Doppler are estimated with an iterative maximum-likelihood scheme similar to RiMAX, i.e., a coarse grid-based initialization of both parameters followed by non-linear gradient-based optimization with the Gauss-Newton method. Model order selection is performed by discarding paths based on a criterion that compares the estimation error variance of path power to the estimated path power itself. The proposed estimator is applied to measurements with a carousel-type bistatic radar setup in a semi-anechoic chamber.

2.3.2.3 Comparison of multipath estimators for millimeter-wave channels

The work in [FJZP] presents frequency-swept millimeter-wave (28-30 GHz) indoor radio channel sounding measurements in both LOS and obstructed LOS scenarios. The goal of the work was to compare multipath estimates obtained from the channel sounding data with various multipath estimation algorithms. A single biconical antenna was used as Tx, while for the Rx a high-density virtual uniform circular array was created with a biconical antenna. Multipath component delay and AoA were estimated with these estimation algorithms: classical beamforming, frequency-invariant beamforming, space-alternating generalized expectation-maximization (SAGE) with either delay or AoA as the first search space, and maximum-likelihood estimation (MLE) with either a plane-wave or spherical wave signal model. The estimation output is compared with the true angular and delay distribution of multipath power, obtained by rotating an Rx horn antenna in all directions and measuring the channel impulse response. The comparative study of multipath estimation algorithms revealed a number of observations on their relative performance. For example, the multipath estimates of the MLE algorithm with spherical wavefronts are closest to the true angular and delay power distribution. This highlights the necessity of using the spherical-wave model for short-range millimeter-wave communication links.

2.3.3 Multipath component clustering

A large body of MIMO measurements has shown that MPCs occur in groups, also known as clusters, such that the MPCs within one group have similar characteristics, but have significantly different characteristics from the MPCs in other clusters. Separately characterizing the intra-cluster and inter-cluster properties can allow to significantly simplify channel models without major loss of accuracy. Therefore, many channel models have been proposed and developed based on the concept of clusters. Recently, automated algorithms of MPC clustering have gained popularity, but still suffer from the use of arbitrary thresholds and/or a priori assumption of the number of clusters [HAM+18].

2.3.3.1 Clustering based on kernel power density

In [HLA+17], a novel kernel power density (KPD) based algorithm is proposed for MPC clustering in MIMO channels, where the kernel density of MPCs is adopted to

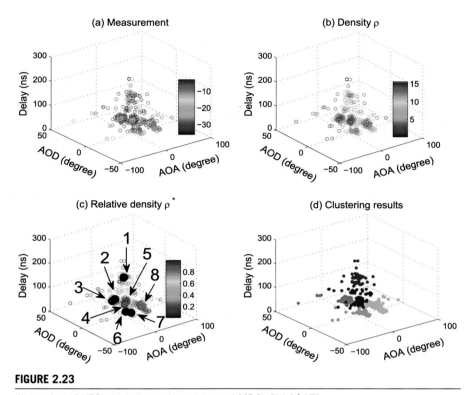

FIGURE 2.23

Illustration of KPD clustering using measured MPCs [HLA$^+$17].

characterize the modeled behavior of MPCs. The KPD uses the kernel density and only considers the neighboring MPCs when computing the density. It also uses the relative density normalized within a local region, and a threshold is used to determine whether two clusters are density-reachable and are to be merged. To illustrate, an example is given in Fig. 2.23, where (a) shows the measured MPCs, (b) shows the estimated density of MPCs, (c) shows the estimated relative density, and (d) gives the final clustering results by using the KPD algorithm.

2.3.3.2 Sparsity-based clustering

The work in [HCA$^+$16] introduces a novel algorithm to identify clusters in the time-delay domain. The intra- and inter-cluster power decay behavior is assumed to follow the well-known Saleh-Valenzuela (SV) double-exponential formulation [SV87]. The SV model parameters are estimated by fitting its mathematical model to the measured PDP in the least-squares sense, with an added weighted l_1-norm constraint. The l_1-norm constraint ensures that the intra-cluster power decay follows the exponential SV model and that the algorithm favors a small number of clusters (i.e., that the solution is sparse). A separate clustering enhancement approach is used, which also forces the inter-cluster power decay to follow the expected SV model behavior. The result is an algorithm of fairly low computation complexity.

2.3.3.3 Clustering based on the Hough transform

In [CPYY18], a new MPC clustering algorithm is presented for vehicular channels. The algorithm is founded in the observation that, in time-variant power delay profiles $|h(t, \tau)|^2$ of vehicular channels, the power follows continuous curves with smoothly varying slopes. The MPC clustering detects and tracks these curves based on a modified Hough transform. The Hough transform is originally an image processing technique to find straight lines in black-and-white images [CPYY18]. The modified Hough transform for cluster detection includes a procedure to convert time-variant power delay profiles to black-and-white images, the removal of clutter points that are not part of a cluster of significant lifetime, and cluster tracking accounting for the smooth change of the cluster velocity relative to the receiver motion.

2.3.4 Large-scale parameter estimation with limited dynamic range

The work in [KH19a] deals with the estimation of Large-Scale Parameter (LSP) of radio channels when there is insufficient dynamic range. Large path loss for example can cause *outage*, meaning that the received signal power is smaller than the measurement noise floor. Often outage samples are not considered for LSP estimation. A more truthful approach would be to include outage samples in the LSP estimation, but with the notion that the sample value is unknown and *censored* by the measurement noise floor. The mean and standard deviation of the LSPs (path loss, shadow fading, and delay and angular spread) are estimated with a modified log-likelihood function that accounts for interval censoring of received power due to limited dynamic range. The estimation method is applied to outdoor-to-indoor measurements in the 14-14.5 GHz range.

IRACON channel measurements and models

3

Sana Salous[a,b], Davy Gaillot[c], Jose-Maria Molina-Garcia-Pardo[d], Ruisi He[e], Kentaro Saito[f], Ke Guan[e], Marina Barbiroli[g], Juan Pascual-Garcia[d], and Katsuyuki Haneda[h]

[a]*University of Durham, Durham, United Kingdom*
[c]*Université des Sciences et Technologies de Lille, Lille, France*
[d]*Universidad Politécnica de Cartagena, Cartagena (Murcia), Spain*
[e]*Beijing Jiaotong University, Beijing, China*
[f]*Tokyo Institute of Technology, Tokyo, Japan*
[g]*University of Bologna, Bologna, Italy*
[h]*Aalto University, Espoo, Finland*

Following Chapter 2 where the latest radio channel modeling methods, tools, and frameworks are reviewed along with scenario definitions, Chapter 3 presents the activities of the IRACON action to derive a radio channel model through extensive measurements. The measurements are in line with recent trends of wireless communications development, covering 1) *above-6 GHz radios in the millimeter-wave and Terahertz bands*, 2) *multiple-antenna radios* including massive and distributed MIMO, and 3) *high-mobility radios* such as trains, cars, and aircrafts.

3.1 Measurement scenarios

3.1.1 Summary

With the continuous development of communication technology, the demand for wireless communications has expanded from traditional cellular networks to other scenarios, such as vehicular including unmanned aerial vehicle (UAV), device-to-device (D2D), and body-centric scenarios. Table 3.1 summarizes the channel measurements carried out by COST-IRACON Action including conventional scenarios [ODV17][VZL+17], massive MIMO [LAH+19], vehicular [YLB16][YAH+19], D2D [MMEO16][FAC+17][KHH+19], and railway scenarios [AHZ+12]. A series

---

[b] Chapter editor.

Inclusive Radio Communications for 5G and Beyond. https://doi.org/10.1016/B978-0-12-820581-5.00009-2
Copyright © 2021 Elsevier Ltd. All rights reserved.

of measurements and research on emerging communication scenarios, such as body-to-body (B2B) [ACT16a], industrial [NE19], long-range, back- and front-haul, and ground-to-air scenarios [Lal16] have also been reported. In addition, millimeter wave (mm-Wave) radio channels have also been characterized [GPdP17][SCR18], primarily in the 60-80 GHz band, and a few measurements above 80 GHz, e.g., at 100 and 300 GHz [MIMGPG+17]. Due to the limitation of hardware equipment, most of the mm-Wave measurements are single input single output (SISO).

3.1.1.1 Cellular scenarios

[ODV17] reports 8×8 MIMO channel measurements at 3.8 GHz, in urban microcell environment in Louvain-la-Neuve. Fig. 3.1 shows an example of the results. Large-scale estimated channel parameters of shadow fading, delay and angular spreads, are compared with values from COST 273, COST 2100, and 3GPP-3D models. The measurements were performed, at a base station (BS) height of 18 m with a dual polarized antenna array with orthogonal slanted polarization and a 3-dB beamwidth of 95 degrees. At the mobile terminal (MT), a uniform circular array of vertically polarized dipoles spaced by approximately 4 cm was placed on a moving trolley at 2 m height.

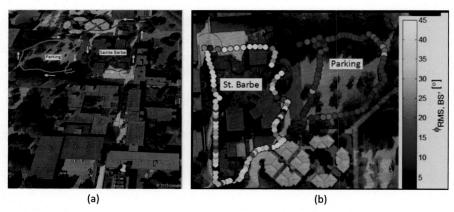

(a) (b)

FIGURE 3.1

An example of results in [ODV17]. (a) Measurement routes and BS location. (b) BS azimuth-spread for both routes.

3.1.1.2 Vehicular scenarios

Vehicle-to-vehicle (V2V) communications can significantly improve efficiency and safety of road traffic. To model the V2V channel, several channel measurements have been conducted such as those reported in [YAH+19] in typical vehicular scenarios, e.g. urban, suburban, crossroads, vehicle-to-infrastructure (V2I), underground garages, and vehicle obstructions. The measurements used a 16-element dual-polarized cylindrical antenna array at the receiver as shown in Fig. 3.2(a), us-

Table 3.1 Configurations of measurement system.

Reference	Category	Carrier Frequency	Antenna	Scenarios
[ODV17]	Cellular	3.8 GHz	MIMO	Parking, Urban
[VZL+17]	Cellular	2.6 GHz	SISO	Urban
[YL17]	Vehicular	900 MHz	SISO	Anechoic chamber
[YLB16]	Vehicular	900 MHz	SISO	Anechoic chamber
[YAH+19]	Vehicular	5.9 GHz	MIMO	Urban, Suburban, Crossroads, V2I, Underground garages, Vehicle obstructions
[AHZ+12]	HSR	930 MHz	SISO	Viaducts, Cuttings, Tunnels, Stations, In-carriage
[MMEO16]	D2D	2.48 GHz	SISO	Indoor
[FAC+17]	D2D	0.5-4.0 GHz	SISO	Maritime Container Terminal
[KHH+19]	D2D	2.45 GHz, 5.8 GHz	SISO	Elevator Shaft
[BS19]	D2D	470 MHz, 580 MHz, 670 MHz	SISO	Indoor-to-outdoor
[NE19]	D2D	2-6 GHz	MIMO	Indoor
[LAH+19]	massive MIMO	2-26 GHz	MIMO	Campus, theater, railway station, lobby
[GPdP17]	mm-Wave	5-67 GHz	SISO	Indoor
[MIMGPG+17]	mm-Wave	100 GHz, 300 GHz	SISO	Indoor
[SCR18]	mm-Wave	26 GHz, 78 GHz	SISO	Outdoor
[ACT16b]	B2B	2.45 GHz	SISO	Indoor and outdoor
[NE19]	Industrial	2-6 GHz	SISO	Indoor
[Lai16]	Ground-to-X	1.35 GHz	MIMO	Ground-to-ground, Ground-to-air
[CDG+18]	Maritime	1.39 GHz, 4.5 GHz	MIMO	Shore-to-shore, Ship-to-shore
[AO18]	Frequency Modulation	88-108 MHz	MIMO	Outdoor

B2B: Body-to-Body; D2D: Device-to-Device; HSR: High speed railway; V2I: Vehicle-to-infrastructure.

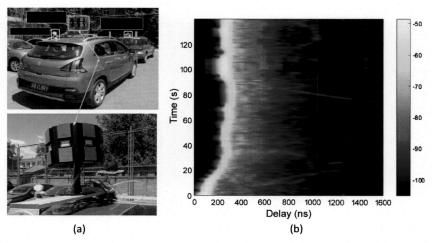

(a)

(b)

FIGURE 3.2

Measurements in [YAH+19]. (a) Measurement System. (b) PDP for Urban Scenario.

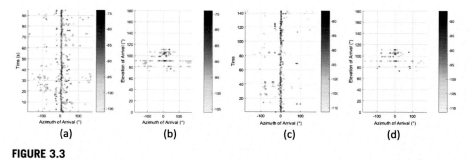

(a) (b) (c) (d)

FIGURE 3.3

Results in [YAH+19]. (a-b) AoA and EoA for urban environment. (c-d) AoA and EoA for suburban environment.

ing a single RF module with an electronic switch to realize MIMO measurements. Using the SAGE algorithm, the measurements were used to estimate the spatial distribution of multipath components (MPCs) as shown in Fig. 3.3. Based on the measurement data in [YAH+19], a cluster-based three-dimensional (3D) channel model is proposed. The clusters are divided into two categories: global-clusters and scatterer-clusters, and the spatio-temporal distributions of the two types of clusters are characterized by inter- and intra-cluster parameters.

[AHZ+12] proposes a scheme based on channel measurements to categorize high-speed railway (HSR) scenes into twelve scenarios. Narrowband measurements at 930 MHz along "Zhengzhou-Xian" HSR in China, using GSM-R base stations (BS), at different heights ranging from 20-60 m were conducted at train speeds up to 350 km/h. The results enable the characterization of HSR channels and highlight their differences from for example cellular measurements.

3.1.1.3 Massive MIMO scenarios

As one of the key technologies of 5G, massive MIMO is promising to be deployed in both outdoor and indoor scenarios. In [LAH+19], a customized 3D positioner with a position accuracy of 0.1 mm is used to form up to a 256 virtual massive MIMO array by moving the Tx antenna in both the horizontal and vertical directions. The measurements were performed with a sounder which transmits a multicarrier spread spectrum signal at frequencies ranging from 2 GHz to 26 GHz with a bandwidth up to 160 MHz. Measurements were conducted in campus, theater, railway station, and lobby environments. Based on these measurements, a novel cluster-based stochastic channel model is proposed for massive MIMO communications in hotspot scenarios. Fig. 3.4 shows two photos of the measurement scenarios, together with some insightful results. The clusters which can only be observed by parts of the large-scale array

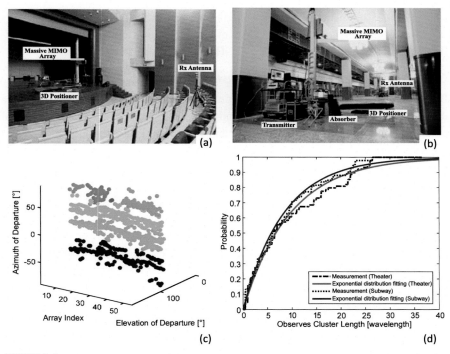

FIGURE 3.4

Example results in [LAH+19]. (a-b) Two measurement scenarios. The virtual Tx array (4 × 64) is marked with a yellow block as a schematic drawing. (c) An example of clustering result in the theater scenario. (d)'Cumulative distribution functions of the observed cluster length over the array, with the corresponding exponential distribution fittings.

are identified by the hybrid clustering method described in the TD. The observed cluster length, modeled by an exponential distribution, is introduced to statistically model the size of the cluster visibility region.

3.1.1.4 Long-range scenarios

Millimeter-wave and Terahertz frequencies offer unique characteristics to simultaneously obtain high spatial resolution, and high penetration loss. To study the impact of precipitation on mm-Waves for back-haul and front-haul radio links, [SCR18] gives an overview of the experimental set up in the 25.84 GHz and 77.52 GHz bands. The transmission can be set up either as CW for rain attenuation or wideband FMCW (frequency modulated continuous wave) for the estimation of narrow pulse distortion. Preliminary dual polarized measurements cover effects of rain on the quality of mm-Wave radio links with the help of a disdrometer which gives rain-fall rate and drop size distribution.

[CDG+18] studies the effect of different ducting conditions on long range maritime broad-band radio links based on one week of extensive maritime propagation tests. The measurement campaign covered both static shore-to-shore overwater links and mobile ship-to-shore overwater links of various ranges, and included dawn, daylight, night and sunset tests at both 1.39 GHz (Band 3) and 4.5 GHz (Band 4).

3.1.1.5 D2D, body-centric, and industrial scenarios

D2D scenarios are applicable to many communication systems, as they can realize direct communication between devices and have numerous applications in industrial environments. [FAC+17], gives detailed analysis of slow and fast fading effects in container terminal environments including fading distribution parameters and proposes analytical models.

[KHH+19] reports results of dynamic polarimetric wideband channel sounding in an elevator shaft using a USRP-based channel sounder capable of dual-frequency dynamic wideband channel sounding, with 9.77 kHz of maximal resolvable Doppler shift. The results show significantly higher path loss at 2.45 GHz for electric fields parallel to the longer side of the shaft than to the shorter side due to a break point.

[NE19] presents channel characterization in the 2-6 GHz band of an industrial environment with high clutter density. Using a Vector Network Analyzer, the measurements were performed in a machinery room, with virtual arrays at both the transmitting and receiving side using two omni-directional antennas, placed on automatic positioners. Using the amplitude and phase information of the measured channel, fading statistics of the MPCs were estimated. Characterizations of path loss, delay spread, and angular spread are presented.

Finally, [ACT16a] presents body-to-body radio channel measurements with initial analysis of results at the 2.45 GHz band, for different body motion scenarios and on-body antenna placements, in indoor and outdoor environments.

3.2 **mm-Wave and Terahertz channels**

This sections presents channel modeling above 6 GHz, including mm-Wave and THz channels, from both an experimental and theoretical point of view. These include results of path loss and RMS delay spread, with emphasis on specific scenarios, outdoor-to-indoor propagation, cross-polar discrimination/polarimetric, clustering, and massive MIMO. Simulations of these channels are also presented.

3.2.1 **Path loss and RMS delay spread**

The most fundamental and important characteristics of radio channels for wireless communications are path loss and RMS delay spread. This section highlights various IRACON measurements reporting those characteristics.

3.2.1.1 Path loss

Path loss, which measures the loss of energy of a wave propagating between the transmitter and the receiver, is the main parameter in the design of wireless networks. In the mm-Wave band the wavelength is of the order of millimeter, so the interaction with the environment becomes challenging. In particular, IRACON measurements highlight

- the importance of reflected and scattered energy in mm-Wave path loss,
- a significant path loss increase with frequency, specially in outdoor scenarios, along with a periodic behavior in frequency caused by structure of windows,
- the good matching of ITU-R Recommendation P.2040 model regarding the transmission through single panes of glass,
- the specific influence of roughness on the specular reflection.

Many propagation mechanisms contribute to multipath propagation in the mm-Wave band, but their importance changes dramatically compared to the frequency bands below 6 GHz. While strong reflections are present at frequencies below 2 GHz, their impact on 60 GHz can be negligible. For example, in [Rud17], results from both indoor and outdoor measurements at 28 GHz highlight the importance of reflected and scattered energy in determining the overall isolation that can be expected between terminals sharing the spectrum in an urban environment. Deterministic models are unlikely to be sufficient to estimate the intra- and inter-system interference in a reliable and efficient manner.

To study the frequency dependence, a VNA was used to carry out measurements at 2.44, 5.8, 14.8, and 58.68 GHz in different urban scenarios such as an indoor office, outdoor-to-indoor, and outdoor street canyon [MSA$^+$17]. The results show that the path loss increases significantly with frequency in all the measured scenarios, particularly in the outdoor to indoor scenario. Fig. 3.5 shows the path loss, in excess of free space loss, versus distance for the indoor office environment. A periodic varying attenuation as a function of frequency was detected which was attributed to the structure of the glass of the window panes. For the outdoor street canyon scenario, it

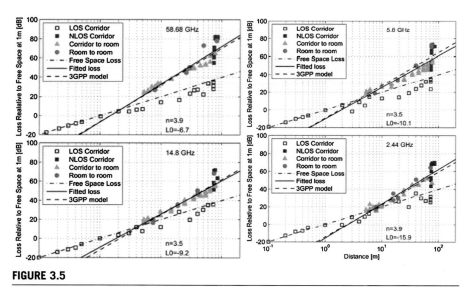

FIGURE 3.5

Path loss, in excess of free space loss, versus distance for 2.44, 5.8, 14.8, and 58.68 GHz, measured in the indoor office environment.

was shown that this increase may not be explained by diffraction. This indicates that it is more likely that the main propagation mechanisms, in NLOS, are due to scattering by rough surfaces or small objects as well as specular reflection. This would also explain why the measured increase in path loss with frequency being substantially smaller than what is expected by diffraction.

To study the transmission loss through a single pane glass, composed of different types of glass, measurements were performed using a VNA with omnidirectional antennas at 5, 15, 28, and 58.7 GHz using a maximum bandwidth between 2 and 16 GHz [DM18]. The results of the measurements were compared with the multi-layer model in ITU-R Recommendation P.2040 which provides an expression for the attenuation based on the number of layers, dimensions, and electromagnetic properties of the material model. A gating operation is applied in the time domain to select only the transmission component; and the transmission loss (attenuation) is then estimated using a Fourier transform. The results showed that the P.2040 model for complex permeability including parameter values for glass matched the measurements very well. Only slight adjustments of the corresponding parameter values were required in some cases. (See Fig. 3.6.)

Rough surfaces, or surfaces with millimetric irregularities of typical building materials cause a reduction of the specular reflected power in the mmWave band due to the comparable size of the wavelength with the size of the irregularities of the surface. In [LFBH19] this effect was studied at 26 GHz. The results indicate that the reduction of reflected power is not constant in a specific material; but is rather random according to the irregularities of the surface. Rayleigh, Nakagami, and Rician

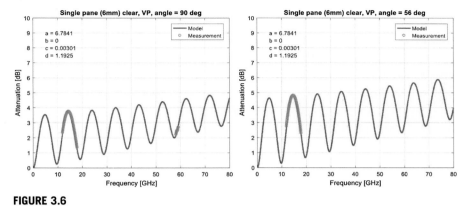

FIGURE 3.6

Measured transmission attenuation versus frequency for non-coated 2, 4, 6 mm glass panes at different angles of incidence relative to the surface of the planes [DM18].

distributions were used to model the fluctuations of the specular reflection, and it was observed that the Nakagami distribution has the flexibility to characterize the specular reflection from smooth surfaces like glass, and from rough surfaces like stones. Conversely, the Rayleigh distribution adequately models the specular reflection from stones but fails to characterize glass. The Rician distribution was found to provide a good fit for the specular reflection generated by different surfaces.

Another interesting result deals with measuring and modeling specular reflection at mm-Wave frequencies. In [LFGB+16] using an arbitrary waveform generator with 2 GHz bandwidth at 60 GHz; and vertical horn directional antennas, the measured diffuse scattering of the specular path for three building materials of different roughness was analyzed. The results show very large signal strength fluctuations in the first order scattered components as the user moves over very short distances (on the order of centimeters). This effect is thought to be produced by diffuse components that add constructively and destructively, creating an effect similar to the multipath small-scale fading. The result also shows that reflection from rough materials may suffer from high depolarization. (See Fig. 3.7.)

Clutter loss (CL) is another important factor to consider in the mmWave band. In [MCB+18], clutter loss is estimated from measurements in a suburban environment at 26 GHz and 40 GHz and the results in Table 3.2 compare the median values obtained from the measurements with the ITU-R model for different elevation angles. Initial results showed some agreements and some discrepancies for the limited number of cases that were possible to compare. The discrepancies could be attributed to the ITU-R model being valid for urban and suburban environments. The results indicate that more measurements are needed to investigate whether the height of the receiver relative to the height of the clutter, and the distance from the receiver to the clutter are needed to be added to the parameters of the current ITU-R model.

In [MBC19], CL was also investigated with ray-tracing simulations at 26 GHz for a 22 m transmitter antenna height to improve and validate CL propagation models in

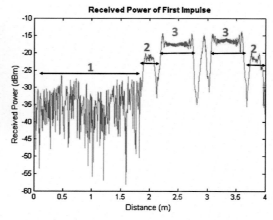

- Area 1: **Rough wall**
- Area 2: **Smooth transition**
- Area 3: **Window (smooth surface)**

FIGURE 3.7

Measured profile for dressed stone wall and window scattering [LFGB⁺16].

Table 3.2 Median CL from measurements and ITU-R model for 0, 2, and 3.

Elevation angle [°]	Measurements Median CL [dB]		ITU-R Model 50%	
	26 GHz	**40 GHz**	**26 GHz**	**40 GHz**
0	39.3	47	46.6	48.56
2	28.7	41.5	33	34.3
3	22.3	31.6	28.8	30

Table 3.3 Clutter loss mean comparison.

	Clutter loss mean [dB]			
	26 GHz		**40 GHz**	
Elev [°]	**Meas**	**Sim**	**Meas**	**Sim**
[0,1]	34.7	34.3	45.7	43.6
[1,3]	26.8	25.8	35.2	32.9
[3,4,5]	23.4	25.2	32.7	28.4

the mmWave band. The CDF results indicate that CL obtained from the simulations and the measurements are within 3 dB. The difference could be attributed to the error introduced by the parameters considered in the simulation, such as characteristics of the materials, and that the ray tracing only considered two different types of building facades, and the exact size of buildings and positions of trees and cars are not identical to the actual measured environment. (See Table 3.3.)

3.2.1.2 Delay spread

The delay spread (DS) is one of the important wideband channel parameters. In [DCCS17a], DS was estimated at 3, 17, and 60 GHz for different urban outdoor scenarios. Overall, a slight decrease of DS values with increasing frequency was observed from the measurements although the different specific outdoor environments indicated different trends. As the frequency increases, the DS decreased in the Open Square (OS) and Street Intersection (SI) environments while it increased in the Street Canyon (SC). The main conclusions were that the DS was not impacted by the different threshold values used in the estimation, and atmospheric oxygen absorption had a very small impact at 60 GHz.

The frequency dependence of the DS was also studied in the mmMAGIC project [NMP+18] in indoor, outdoor, and outdoor-to-indoor (O2I) scenarios over a wide frequency range from 2 to 86 GHz. Several requirements have been identified that define the parameters which need to be aligned in order to make a reasonable comparison among the different channel sounders that were employed in the study. After careful analysis, the conclusion is that any frequency trend of the delay spread is small considering its confidence intervals. The results were significantly different from the 3GPP New Radio (NR) model 3GPP38.901, except for the O2I scenario (see Table 3.4).

The angular variation of DS is studied in [LFGB17], based on channel sounding measurements at 60 GHz at Bristol University. The angular and spatial consistency are analyzed by correlating the dispersion for different azimuth and elevation angles, and comparing them for different distances. A very small DS value was computed for the Optical Line Of Sight when using a highly directive transmitter antenna with respect to an omni-directional antenna. Significant variations were observed in the measured scenario which includes a large reflector, and foliage.

Table 3.4 Frequency-dependent linear regression for the DS in each campaign.

Scenario			Model: $\mu_{\log DS/1s} = \alpha \log(1 + f_c/1\text{GHz}) + \beta$		
			α (95% confidence bound)	β	p-value
Indoor	LOS	**CEA Office**	**-0.98 (-1.17, -0.78)**	**-6.31**	**0.000**
		CEA Conference Room	**-0.24 (-0.46, -0.03)**	**-7.79**	**0.025**
		HWDU Lecture Room	**-0.05 (-0.08, -0.03)**	**-8.08**	**0.000**
		Aalto Airport	-0.18 (-0.47, 0.11)	-7.11	0.207
		EAB office	-0.07 (-0.27, 0.12)	-7.83	0.134
	NLOS	EAB office	-0.01 (-0.06, 0.04)	-7.58	0.735
O2I		Orange Low-loss	-0.11 (-0.29, 0.07)	-7.71	0.231
		Orange High-loss	-0.05 (-0.14, 0.04)	-7.66	0.286
		EAB Traditional building	**0.27 (0.16, 0.39)**	**-7.58**	**0.000**
Outdoor	LOS	**HHI Street Canyon**	**-0.12 (-0.226, -0.005)**	**-7.45**	**0.041**
		Aalto Open Square	-0.23 (-2.00, 1.53)	-7.12	0.768
		Aalto Street Canyon	-0.12 (-0.50, 0.26)	-7.73	0.521
		Orange Open Square	-0.05 (-0.32, 0.22)	-7.53	0.773
		Orange Street Canyon	0.10 (-0.28, 0.49)	-8.14	0.549
	NLOS	**Orange Open Square**	**-0.11 (-0.20, -0.02)**	**-7.13**	**0.014**
		Orange Street Canyon	0.13 (-0.20, 0.47)	-7.57	0.294
		EAB Street Canyon	0.01 (-0.09, 0.12)	-6.82	0.803

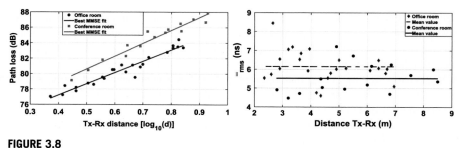

FIGURE 3.8

Path Loss model and RMS delay spread results for the E band [BMD16].

3.2.1.3 Specific scenarios: Indoor and confined environments

Channel characteristics in an office and a conference room in the E-Band (80.5-86.5 GHz) are reported in ([BMD16]), from measurements performed with mechanical steering of directive antennas at both the transmitter and receiver, allowing a double-directional angular characterization in the azimuth plane. In each antenna position a 4-port Vector Network Analyzer (VNA) and frequency converters are used to record the frequency response in the desired bandwidth. An inverse Fourier transform is applied to obtain the channel impulse response. A peak detection algorithm is then applied to the synthetic omni-directional PDP to extract the angles of departure and arrival and the delay spread. Path loss exponents less than 2 were estimated due to the waveguide effect. The mean DS spread values were 6.1 ns for the office and 5.5 ns for the conference room with the highest arrival angular spread occurring when the Rx location is close to the scatterers and corners whereas locations in the middle of the room exhibited lower dispersion of the AoA. (See Fig. 3.8.)

In [MPG+19a], MIMO indoor channel transfer functions were obtained using a VNA, Optical-Radio transceivers, and omnidirectional antennas at 1-40 GHz. The frequency response was measured over all the bandwidth, and an inverse Fourier transform was applied to find the PDP. The DS and relative received power were estimated for the 2 GHz, 10 GHz, 20 GHz, 30 GHz, and 40 GHz sub-bands. The Rician K-factor was also estimated over the full bandwidth. The results showed a path loss exponent below 2; which is attributed to the reverberation effects in a rich multipath environment. The DS as a function of distance was also observed (see Fig. 3.9). Although all measured locations were LOS, negative values of the K-factor were found due to the large number of received contributions. The time-variant MPCs observed during a wideband measurement campaign performed at 28 GHz in two realistic indoor environments were also discussed in [DBKU+17]. The receiver antenna was fixed and the transmitter was moving along a route with slow speed (0.05 m/s). The observed channel was confirmed to be rich, exhibiting significant spatially consistent contributions. The MPCs were observed to be highly correlated and having almost the same slopes while in NLOS situation, the direct visibility was observed by transmission mechanism.

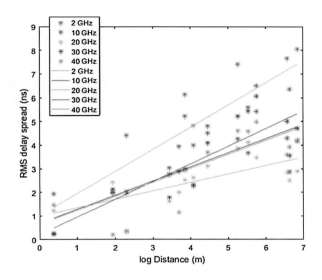

FIGURE 3.9

RMS Delay Spread at 2, 10, 20, 30, and 40 GHz with 5 GHz bandwidth.

Measurements in an intra-wagon, underground convoy were carried out in the frequency range of 25-40 GHz using a VNA with omnidirectional antennas to estimate path loss, DS and the coherence bandwidth [RPRR+18]. Two different scenarios corresponding to two different locations of the receiver were measured. The results were similar to the indoor environment with one location exhibiting the waveguide effect with respect to the path loss exponent. The DS increased with distance and then at a certain limit distance it started to decrease. In addition, measurements in an engine and steering control room of a 200 m long bulk carrier vessel were carried out at 60.48 GHz using the commercial Terragraph sounder which uses 36 x 8 transmit/receive antennas and measures directional path loss and the impulse response of the channel [BTY+ed]. The measured path loss at 1.5 m was equal to 74.6 dB, which is higher than free space path loss, whereas the estimated path loss exponent of 1.7 is lower than free space due to the guiding effect, which enables high data rate communication. Reflection in the highly metallic environment seen in Fig. 3.10 enable non line of sight communication but no clear distance relationship was detected. The inherent vibration caused by train locomotives can affect the performance as studied in [SDS+18], where the wireless channel variations were recorded with vibration data. The results show a large coherence time which indicates a static behavior for short packets used in wagon-to-wagon wireless train control and management systems (TCMS). The results also show that the Doppler frequency shift ranges up to 50 Hz where the reflected signal had a higher Doppler frequency shift. The authors suggested that further analysis is needed to get more conclusive results regarding the effects on path-loss, antenna misalignment, and MPCs reflections from the surroundings.

(a) Floorplan engine room   (b) Upper level engine room   (c) Steering room

FIGURE 3.10

Measurement locations on the vessel.

Sub-THz measurements in the frequency band 126-156 GHz were performed up to 10 m distances in different indoor scenarios which include office, conference room, and laboratory [D'E19a]. The double directional measurements enable the development of a full geometry based stochastic channel model. The estimated path loss exponents were 1.45, 1.93, and 1.91 for the three scenarios, while the DS values were below 10 ns, and a maximum of five clusters were detected (see Fig. 3.11).

3.2.2 Outdoor-to-indoor propagation

Building entry loss (BEL) refers to the attenuation of the signal as it propagates from outside to inside a building. This is particularly important in the mmWave band which exhibits significantly higher attenuation than below 6 GHz bands.

To study BEL as a function of frequency O2I measurements were conducted at 3, 10, 17, and 60 GHz [DCCS17b]. Both BEL and DS were estimated for various Rx positions in different rooms. The results reveal a strong variation in signal attenuation from 5 to 40 dB depending on the composition of the window material. Relatively low and non-frequency-dependent attenuation (around 5 dB) is observed for non-coated glass windows while a high attenuation (30 dB), increasing with frequency, is recorded for coated glass windows (see Fig. 3.12.) The estimated DS values are relatively low (below 30 ns for all the measurements) and uniformly distributed across the different frequency bands.

In [SMVB20], data sets at 26 GHz from three different measurement campaigns in the same site but with different experimental setups were used to characterize CL, BEL in isolated building and building entry loss over a cluttered path. The measurements were performed to estimate BEL by placing the transmitter: in front of the building on a street canyon, in front of the building on an open field, or at a high level in a cluttered environment with NLOS. The data were processed to estimate median values of the BEL. From the three sets of data the difference between the estimation of BEL from a transmitter placed outside in LOS to the facade of the building or from a transmitter above a cluttered environment with NLOS gave

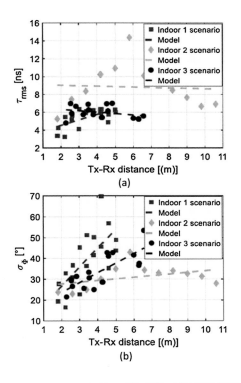

FIGURE 3.11

(a) Delay spread and (b) Angular spread for all three scenarios.

values which differ by less than 1.8 dB which indicates that either method can be used for estimating the median value of BEL. The median of clutter loss estimated from two different data sets was found to be 18 dB, which confirms that both experimental setups and the methodology are valid for this kind of measurements. With these results, it can be concluded that at 26 GHz, the median values of BEL and CL can be treated as multiplicative, i.e., the overall excess path loss is the sum in dB of the individual losses. A summary of the obtained results is shown in Table 3.6 [MVS19].

Measurements of BEL for a "traditional" building (non-coated window glass) for radio frequencies in the range 2-60 GHz and for different elevation angles by locating the transmitter and receiver at different floors at two orthogonal sections of a building are reported in [DMS18]. It is proposed to use a linear model to account for the angle of elevation dependence of BEL. $L_{ef} = k|\theta|$ (dB). The results give a value of k greater than 24 dB/90° which is in contrast to the k value in the ITU and mmMagic models of 20 dB/90°.

Simulations and measurements using a VNA and directional horn antennas were used to study infra-red reflector glass over a frequency range 11-67 GHz [BCX17].

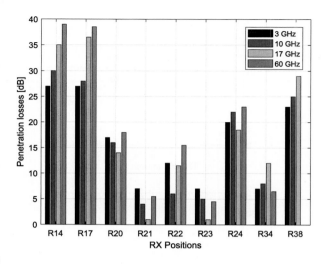

FIGURE 3.12

Building Entry Losses across the four bands [DCCS17b].

Different situations are analyzed specifically to assess the losses due to diffraction, penetration and the possible exploitation of gaps in the window pane by placing the antennas on either side of the window. The results showed that the penetration loss through the glass is in excess of 30 dB which is further exacerbated by the problem of angle of incidence. The opportunity to exploit gaps in the building infrastructure was shown to reduce the penetration loss in excess of 10 dB. This benefit is greater as frequency increases, though there is more vulnerability to scattering where it is more beneficial to rely on reflection from the far end of a window as opposed to the diffraction off the near end. (See Fig. 3.13.)

To measure BEL in a typical thermally efficient building, and a traditional building measurements were performed between 0.4 to 73 GHz using two custom channel sounders based on the frequency swept continuous wave (FMCW) architecture [TSRC18]. BEL was then estimated by taking the difference between the received power inside the house and the mean power of the locations measured outside the house. Directional antennas were used outside the building as transmitters and omnidirectional antennas as receivers inside the two types of houses. The CDF of the BEL did not display significant frequency dependence between 25 and 73 GHz. Comparative measurements indicated that opening external windows reduces the penetration loss by 10 dB in the millimeter wave bands between 25-73 GHz. A difference of 2-16 dB between the two types of build is measured across the entire frequency range. (See Fig. 3.14 and Table 3.5.)

- *Strong variations between 10 and 40 dB depending on the room*

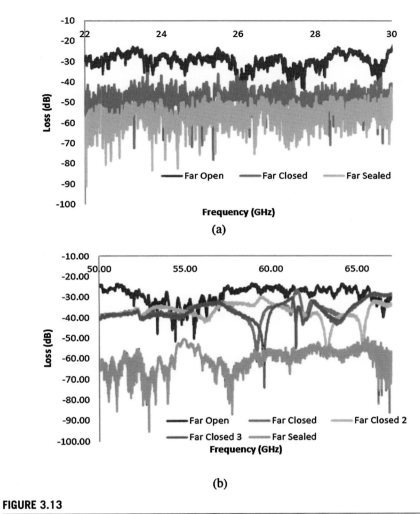

FIGURE 3.13

Analysis of gap effects in the range a) 22-30 GHz and b) 50-67 GHz.

3.2.3 Cross-polar discrimination, clustering, and massive MIMO

State-of-the-art MIMO GSCM including the COST 2100 model, 3GPP model, and the WINNER model have been widely used to describe the multidimensional characteristics of the polarimetric MIMO radio channel with low computational cost. Since polarization brings an additional degree of freedom to the radio channel, a deep understanding of the complex polarization effects is important. Similarly, the spatial diversity brought by massive MIMO as a key enabler for 5G communication systems. Massive Tx arrays are capable of multiplexing many UE in the same time-frequency resource. In addition, the high gain antennas or beamforming techniques, allow com-

FIGURE 3.14

CDF for the traditional house for the higher frequency bands [TSRC18].

Table 3.5 Measured BELs values.

Penetration Loss [dB] Average	Omni. Dir.			Omni. Dir.
	3 GHz	10 GHz	17 GHz	17 GHz
Corridor	20	20	21	21
Flat	27	30	33	35
Rest room	11	9	8	9
Large Office 1	14	17	18	18
Delay Spread [ns] Average	**3 GHz**	**10 GHz**	**17 GHz**	**17 GHz**
Corridor	37	34	37	34
Flat	41	38	39	35
Rest room	33	30	36	38
Large Office 1	29	31	31	31
Large Office 2	49	42	42	40

Table 3.6 Measurement campaigns, experimental setups and results.

Campaign	Experimental setup	Losses characterized	Measured values [dB]	
			No GTx correction	GTx correction
Measurements 2019	Durham University	BEL over cluttered path, NLOS	44.03	42.6
	Channel Sounder	BEL isolated building, LOS	45.9	42.9
		Clutter loss	18.07	18.07
Measurements 2018	CW + downconverter	BEL isolated building, LOS	44.27	
Measurements 2017	CW at 26 GHz	Clutter loss		18.21

pensating the additional path-loss at mmWave frequencies, with reduced fading and delay spread. Hence, dual-polarized communications become more relevant to provide spatial and polarization diversity at mmWave frequencies. Several studies have shown that a more deterministic modeling approach is needed to address polarization with a more realistic treatment but that diffuse scattering must be included. In general, the tuning of model parameters requires a rich set of statistics from measurements and/or simulations obtained for a wide range of indoor and outdoor propagation scenarios and frequency bands.

3.2.3.1 Cross-polar discrimination

With regard to the polarimetric properties, the main results can be summarized as follows:

- a novel excess cross-polarization ratio (XPR) model per path has been developed and no frequency or environment dependence was observed;
- scatterers are spatially consistent across the whole radio spectrum for vehicle-to-vehicle scenarios;
- a polarimetric description of diffuse scattering must be included in mmWave radio channels, and a Lambertian model should be used over a Single-lobe directive model.

The authors in [KJNH17] have proposed a novel excess loss-based XPR model per path in LOS and NLOS indoor and outdoor environments using a modified lognormal distribution introduced in WINNER and 3GPP models between 15-86 GHz, with the mean value and standard deviation changing linearly as a function of the MPC excess loss. The results for LOS in an open-square scenario indicate that the XPR of deterministic paths is not strongly frequency or environment dependent. Furthermore, the XPR distributions derived from measurements with different bandwidths are quite similar such that the same model and parameters can be used at least within the studied frequency range in both indoor and outdoor environments. From this analysis, it is proposed to use the mean values of XPR per path for all frequencies. In addition, double-directional dual-polarized ultra-wideband measurements were performed in a conference room scenario at 190 GHz and compared with RT simulations [Duped]. Both confirm that the propagation channel includes multiple components from reflections in window frames, walls, tables, and other furniture. Table 3.7 presents the measured large-scale parameters of the extracted MPCs. The strongest MPCs have a large polarization discrimination that depends on the orientation of the scattering surface.

A vehicle-to-vehicle (V2V) dual-polarized double-directional radio channel was also investigated at 6.75 GHz, 30 GHz, and 60 GHz in a corner scenario at a "T" intersection in an urban environment [DDR19]. It was concluded that similar scattering was observed across the three bands; but with different relative powers to the LOS component. This allows the implementation of inter-band algorithms as dynamic assisted beam-forming at the mmWave with a priori information from sub-6 GHz.

Diffuse scattering, which can be strong in indoor scenarios and specifically below 6 GHz, tends to decrease as the frequency increases and exhibits polarization dependence. The large presence of scatterers in indoor scenarios even at mmWave frequencies can create a partial reverberating field with complex depolarizing characteristics. Scattering models can be used in tools such as ray tracers to provide useful

Table 3.7 Table of parameters with 15° HPBW at Tx and Rx [Duped].

Link	DS$_{SO}$ [ns]	ASA [°]	ESA [°]	ASD [°]	ESD [°]
TX1-RX1	2.27	48.16	11.35	49.65	11.53
TX1-RX2	2.6	48.16	12.15	58.23	12.52

Table 3.8 Median average and std of XPR within the channel bandwidth (V-POL-Tx / H-POL-Tx).

	Median average XPR (dB)	Median std of XPR (dB)
60 GHz short range wall	12 / 10	6 / 7
26 GHz short range wall	10 / 9	8 / 7
60 GHz short range window	18 / 20	6 / 5
26 GHz short range window	10 / 9	6 / 6
60 GHz long range wall	10 / 10	8 / 8
26 GHz long range wall	9 / 8	8 / 8
60 GHz long range window	10 / 10	7 / 7
26 GHz long range window	10 / 11	7 / 6

information about angle of arrival, power delay profile, and Doppler shifts, by using data sets from measurements. For example, representative electromagnetic parameters such as permittivity and roughness must be obtained. Measurement data also helps determine the level of detail that is necessary to include in the geospatial and building databases such as non-uniform wall structure, windows, and window frames, for reliable predictions. Extensive polarimetric measurements at 26 GHz and 60 GHz further reported that diffuse scattering can be a dominant propagation mechanism for mmWave links especially when Tx and Rx are located near a surface with rough or with multiple small scale features and discontinuities [LPK+18]. The XPR of the channel was found to vary significantly over the channel bandwidth, especially in the non-specular areas where diffuse scattering is the main propagation mechanism as in Table 3.8.

In [KNHS18] two polarimetric diffuse scattering models of electromagnetic fields at 28 GHz were investigated, from a facade of modern shopping center. The models were integrated into an in-house ray-based wave propagation simulation tool with a point cloud of the measurement site. The Lambertian model was found to reproduce the measured reality for all investigated polarimetric links better than the single-lobe directive model of scattering (SLD). The parameter values giving the best fit of the scattering model to measurements are in the lower end of the previously reported values in the literature for below-6 GHz radio frequencies (Table 3.9), possibly reflecting the fact that a large portion of the studied facade consists of glass walls.

3.2.3.2 Clustering

The concept of clusters is now almost ubiquitous in stochastic physical multipath channel models. It can be traced back to the COST 207 models of PDP, and the

Table 3.9 Summary of parameters reported in the literature.

Paper	RF [GHz]	S	K_{xpol}	wall properties
[36.15]	26	0.6	n/a	university campus
[3GP08b]	1.296	0.05	n/a	airport hangar
[3GP08b]	1.296	0.2	n/a	warehouse brick wall
[3GP08b]	1.296	0.4	n/a	brick wall with windows and other details
[3GP09]	3.8	0.6	0.2	office building glass facade
[3GP09]	3.8	0.45	0	simple rural building
[3GP16]	5.3	0.2	0 ... 0.01	urban street canyons, smooth walls
Present study	28	0.08	0.05	urban open square wall, complex building facade

Saleh-Valenzuela (S-V) indoor propagation model. These models, being SISO models, involved the delay domain only, but the concept was incorporated later into SIMO and MIMO models such as COST 259 and the 3GPP spatial channel model. Since 5G models must be extended at mmWave frequencies, the main features of the existing radio channel models, originally defined for frequencies below 6 GHz, ought to be revisited. In particular, it was found that the mmWave radio channel can indeed be described by a small number of dominant clusters, whose parameters strongly depend on the antenna beamwidth characteristics.

The validity of multipath clustering, and the influence of the antenna aperture on 5G mmWave modeling, is investigated in [BHK18], which describes a statistical technique to determine whether a measured multipath distribution shows evidence of clustering of multipath components beyond what might be expected by chance in a Poisson random process, in which the occurrence of multipath components is independent. The results obtained from a set of indoor 60 GHz measurements confirm that the data are indeed clustered. Radio channel sounding followed by high-resolution parametric estimation of the MPC or identification from the APDP were also conducted in an atrium entrance hall environment at 58.5 GHz with 400 MHz bandwidth [KIU$^+$17]. The path parameters were directly estimated from the double directional channel impulse response using novel HRE from SIC-SAGE and clustered using simple successive clustering algorithm from which the characteristics were computed for a LOS and OLOS scenario. An example of a LOS position is illustrated in Fig. 3.24. The results of 8 dominant clusters for the LOS and 3 for the OLOS (see Table 3.10) are in agreement with other studies with respect to the low number of clusters and MPC in this frequency range. (See Fig. 3.15.)

In [BHB18] a cluster parameterization for the COST 2100 channel model using mobile channel simulations at 61 GHz in Helsinki Airport is proposed. Using a point-cloud model of the airport (Fig. 3.16), a ray-tracer which has been optimized to match measurements was considered to obtain double-directional replica channels. The results confirm that the joint clustering (K-PowerMeans)-and-tracking (Kalman filter) is a suitable tool for cluster identification and tracking for the ray-tracer results. The simulation results show that almost 80% of clusters disappear after a single snapshot.

Table 3.10 Identified propagation mechanisms of dominant clusters.

Rx Pos.	CL#	Cluster centroid [°]				Delay [ns]	Cluster Gain [dB]	Excess Loss [dB]	Mechanism
		TxEl	TxAz	RxEl	RxAz				
Rx1 (LOS)	1	101.3	-32.2	77.9	137.2	19.6	-87.4	0.0	LOS path
	2	95.7	62.0	82.1	106.6	33.8	-90.6	3.2	wall reflection
	3	94.0	-91.0	81.5	142.1	59.7	-93.5	6.0	scattered by metallic pillar
	4	91.2	2.5	79.5	30.3	48.7	-101.9	14.4	scattered by column
	5	100.0	-95.0	114.0	142.0	61.8	-105.5	18.1	metallic pillar and ground
	6	84.5	-21.5	74.1	-9.8	48.0	-105.9	18.5	scattered by column
	7	88.6	-6.1	89.2	-3.3	132.1	-106.7	19.2	wall reflection
	8	114.8	-32.0	114.0	137.6	21.8	-107.5	20.1	ground reflection
Rx3 (OLOS)	1	96.2	40.0	89.3	130.9	55.5	-94.2	3.9	wall reflection
	2	94.4	-37.5	85.4	-137.6	53.5	-106.3	16.0	scattered by column
	3	99.4	-5.8	114.0	131.1	53.0	-108.1	17.7	column and ground

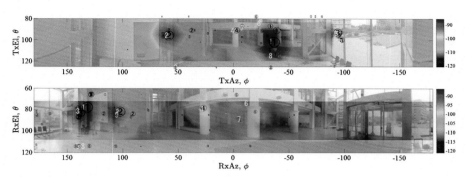

FIGURE 3.15

Example of extracted MPCs clusters in the Angular Power Spectrum (APS) at Rx1.

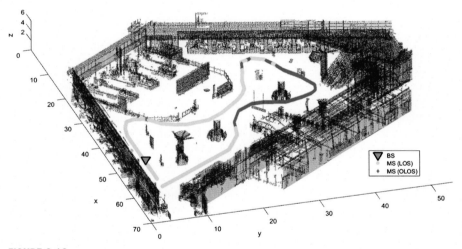

FIGURE 3.16

Floor plan of the small-cell site in Helsinki airport. For this simulation set-up $f_c = 6.1$ GHz, BW = 2 GHz refer to the carrier frequency and bandwidth, respectively. Moreover, the position of BS is fixed (the green triangle), while we investigate 2639 positions for MS (the yellow and red points demonstrate the LOS and OLOS, respectively). The total MS route is 132 m, and channels simulated at every 5 cm.

Directional-scan-sounding (DSS) was used in [LYWT17] to capture the spatial characteristics of propagation channels from 13 to 17 GHz in a lecture hall scenario. The MPC cluster statistics without the antenna effect were compared with those obtained from the raw data with the antenna effect. Results show that using the 10°-HPBW antenna, more objects located farther away from the transceivers are involved in the propagation, and the 30°-HPBW antenna enables observing more local scatterers in the vicinity of the Tx and Rx. Using wide-HPBW antennas with dense scanning of the environment enables higher precision of the estimated values

Table 3.11 Cluster delay and angular dispersion σ_τ, σ_ϕ, σ_θ and their mutual dependencies $\sigma_{c,\tau,\phi}$, $\sigma_{c,\tau,\theta}$, $\sigma_{c,\phi,\theta}$ obtained with 10°- and 30°-HPBW antennas.

HPBW		10°		30°			
γ		1	2	1	2	3	6
$\sigma_{c,\tau}$ [ns]	μ	0.59	0.49	0.56	0.35	0.25	0.27
	σ	0.49	0.37	0.34	0.30	0.21	0.26
$\sigma_{c,\phi}$ [°]	μ	5.25	2.51	6.92	6.03	6.76	6.45
	σ	4.07	4.37	2.57	2.4	2.29	2.29
$\sigma_{c,\theta}$ [°]	μ	4.91	3.47	4.68	5.75	6.92	6.46
	σ	1.91	2.01	2.14	1.82	1.91	1.91
$\rho_{c,\tau\varphi}$	μ	0.14	0.06	-0.01	0.07	-0.04	-0.03
	σ	0.62	0.56	0.59	0.55	0.51	0.53
$\rho_{c,\tau\theta}$	μ	0.05	0.07	0.13	0.14	0.18	0.10
	σ	0.55	0.47	0.51	0.47	0.42	0.48
$\rho_{c,\varphi\theta}$	μ	0.07	0.04	-0.03	0.06	0.03	0.03
	σ	0.51	0.48	0.50	0.47	0.45	0.55

and detailed reflection of the propagation channel. The modeling results demonstrate that by using the proposed HRPE-based parametric approach, channel characteristics obtained with different HPBW antennas are similar, and thus, stochastic models independent of the underlying antenna and the rotation scheme adopted during the measurements can be established. (See Table 3.11.)

3.2.3.3 Massive MIMO

As will be discussed in Section 3.3.1, channel hardening and favorable propagation conditions are two main pillars of massive MIMO. Massive MIMO in the mmWave band is well suited for 5G systems to improve throughput performance, wireless access, and reduced latency among other features. However, measurements, simulations and performance evaluation of the mmWave Massive MIMO channel are scarce. Hence, a thorough analysis and understanding of the massive MIMO channel in the mmWave band through a depiction of the propagation mechanisms and related parameters retrieved from the measurements is missing. This can be explained by the fact that characterizing or testing a full physical massive MIMO system is a complex and costly approach, specifically at mmWave frequencies. Hence, the most popular solution is virtual measurements of massive MIMO where the use of one antenna moved over several positions simulates a full system. Results show that

- massive MIMO provides full spatial diversity at mmWave frequencies;
- the massive radio channel is deterministic with few specular MPCs;
- linear precoders are optimal even with closely spaced users.

In [CML+18], fading and correlation characteristics of virtual LOS 4 x 2500 massive MIMO channels were investigated at 94 GHz in an indoor office environment for three Tx - Rx scenarios (close, center, and far). The results suggest that the receivers

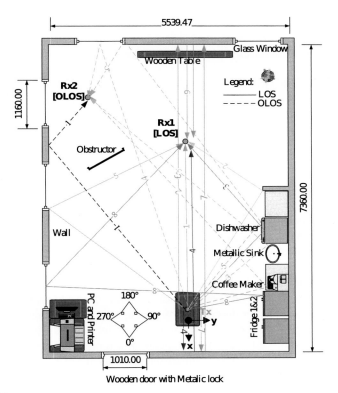

FIGURE 3.17

Trajectory of the MPCs in relation to the room geometry for the LOS and OLOS scenario. The dashed lines represent the MPCs corresponding to the OLOS scenario. Dimensions are in millimeters.

of the users can be spatially uncorrelated with marginal user interference and high power focusing. It is concluded that the diversity brought by the indoor radio channel can be harvested even with a smaller array.

The propagation characteristics of an indoor kitchen/meeting room were also evaluated in [MFJP18] with a 30 x 30 x 10 uniform cubic array (UCuA) at 28 GHz center frequency with 4 GHz bandwidth. The APDPs under LOS and OLOS scenarios extracted from the measurement data were found in good agreement with a ray-traced version of the estimated paths as shown in Fig. 3.17. In particular, the channel is shown to be sparse and deterministic with few specular MPCs.

The analysis was extended to a multi-user scenario using a virtual 30 x 10 uniform rectangular array (URA) which acts as the BS and with users under LOS and OLOS conditions. The effects of the propagation channel impairments on the achievable interference suppression are then demonstrated for critical multi-user scenarios. Results showed interference suppression with zero-forcing beamforming (ZF-BF) performs well even in critical multi-user scenarios based on the measured channels, i.e. a scenario with closely spaced users in LOS conditions, and a scenario with a strong interfering user and a weak desired user (see Fig. 3.18(a)). In addition, the narrow beam pattern illustrated in Fig. 3.18(b) due to the higher number of antennas implies that there is improved spatial separation of the users. However, at this location, the non-orthogonality of the user channels implies that there is a power penalty due to the transmit power constraint.

3.2.4 Millimeter-Wave and Terahertz channel simulations

Ray tracing is one of the most practical techniques in the simulation of the wireless channel at the mmWave and Terahertz bands as the ray-optical approximation of wave propagation is better held than in the lower frequency bands. High speed computing and the use of accelerating techniques, led to both commercial and own-programmed ray tracing tools to become a widespread tool in the analysis of the wireless channel.

Ray tracing tools traditionally used direct rays, specular reflections, diffracted rays, and their transmissions and combinations. However, the addition of diffuse scattering models allows a better description of propagation effects. Thus, in [FHZ^{+}17] diffuse scattering rays and their combinations with reflected rays were included. The simulations were validated with measurements in a room using an ultra-wideband channel sounder at 70 GHz. The comparison showed that in LOS conditions, the direct LOS path carried less than 40% of the total power and in NLOS conditions, reflections carried at least 60% of the total power, whereas scattering was evaluated to be responsible for 20-30% of the total power. Therefore, a relevant amount of the total power was produced by the diffuse components.

The integration of diffuse models in ray tracing tools in the simulation of outdoor scenarios is less common than in indoor environments. Nevertheless, the traditional components are still enough to extract interesting results from simulations as in [KIU^{+}16], [MVBSR17], and [SAJG17]. Thus, in [KIU^{+}16] up to two order reflections and single diffraction were considered in the analysis at 58.5 GHz of an urban micro-cellular, open square outdoor-to-outdoor, and outdoor-to-indoor scenarios. In these three outdoor environments the full polarimetric double directional channel transfer functions, in particular the Angular Delay Power Spectrum (ADPS) functions, were measured using a 2x2 MIMO configuration. The comparison between the measured and simulated APDS allowed the identification of the specular components as seen in Fig. 3.19. The rest of the propagation mechanisms were classified as random rays and their angular probability density functions were obtained.

Outdoor scenarios can have a large number of multipath components due to the presence of metallic elements, trees, and buildings with different construction materials. In this case a detailed description, using a geometrical model, is needed to achieve a good accuracy as suggested in [MVBSR17]. In this work, Power Delay Profile (PDP) simulations at 52 GHz carried out with a commercial ray tracing tool

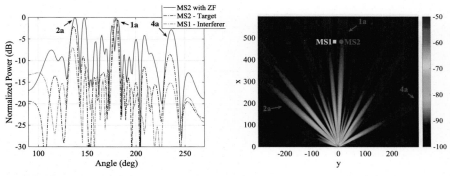

FIGURE 3.18

(a) PAP with ZF-BF at 28 GHz for MS2 in comparison to the PAP for MS2 (target at location A) and MS1 (interferer) without ZF-BF. The MPCs labels correspond to paths identified in Figure 9 of [MFJP18]. (b) Field Pattern Distribution for LOS case. MS2 is the targeted user at position A while MS1 is the interfering user. x and y denote the location coordinates with units λ and the color bar units in dB.

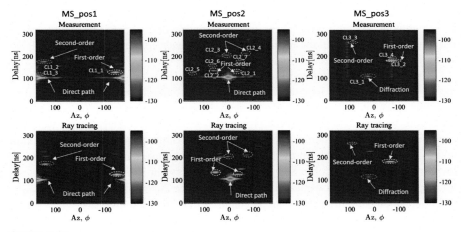

FIGURE 3.19

Measured and simulated ADPS seen from BS and MS in one outdoor to indoor scenario [KIU+16].

were compared to measurements conducted using the chirp (FMCW) technique with heterodyne detection. The comparisons, supported with graphical view of the paths as seen in Fig. 3.20, showed that long range reflections from metallic structures, even 100 m far from the transmitter, contributed significantly to the received multipath components. While the simulator predicted the main features of the channel, the relative magnitudes of the multipath components were inaccurate. The simulation did not predict the cluster of rays accompanying the strong multipath components, that

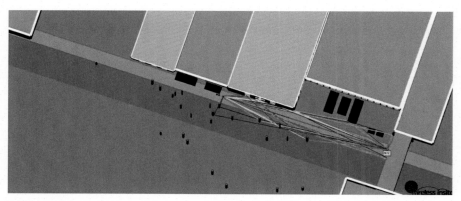

FIGURE 3.20

Simulated rays at 39 m distance in an outdoor scenario [MVBSR17].

sometimes appear in the measurements. These unpredicted clusters were attributed to the diffused multipath components arising from rough surfaces. The impact of these components can be significant in the estimation of channel parameters such as the time delay spread and the received power. However, refinement of the simulated scenario improved the accuracy of the results.

In [SAJG17], the polarization behavior at 28 GHz is studied for a Wireless to the x (WTTx) urban outdoor scenario using two commercial ray tracing tools. Ideal isotropic antennas were considered and the main specular components (except ground reflection) were included. As in the previous case, diffused scattering rays were not considered. One of the main results was the analysis of the cross polar discrimination which showed that the two tools differed greatly and appeared to have at least 9 dB lower median values as compared to the largest settings in 3GPP UMa. Thus, the ray tracing tools simulations would advocate utilizing the different polarization domains for independent streams rather than for the use of polarization diversity.

Ray tracing tools can be also used in different vehicular scenarios. In [HWZ+19] the mmWave channel is characterized and modeled for intra-wagon communication in the metro environment. The measurement was conducted in Madrid Metro in the frequency range from 26.5-40 GHz. By minimizing the error of the dominant multipath components (MPCs) between the ray-tracing simulation (RT) and the measurement, the environment model and electromagnetic (EM) parameters of the main objects are calibrated. The mean calibration errors of the delay and power of the tracked rays are 0.08 ns and 2.76 dB, respectively. It is found that the poles and the train body have significant contribution to the propagation channel. Although up to second order reflection and first order scattering propagation are found, the dense MPCs and the avalanche effect were missing. A hybrid modeling approach is proposed in this work by combining propagation graph model and RT model. The proposed approach is evaluated and the accuracy is found to be higher than

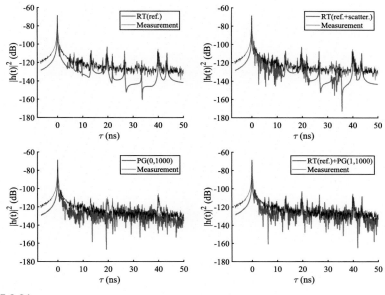

FIGURE 3.21

Comparison of models of [HWZ+19]. a) Measurement VS RT only (without scatter MPCs). b) Measurement VS RT only (with all MPCs). c) Measurement VS PG only. d) Measurement VS Proposed.

RT or PG models. The advantage of the hybrid modeling approach can be seen in Fig. 3.21.

3.2.5 Other effects

In this section, several other phenomena that may influence the performance of the wireless link are discussed. In particular, the effect produced by human blockage, the importance of the alignment between directional antennas, the influence of vibration in trains, and the effect of precipitations for outdoor point-to-point links. Furthermore, analysis of drone detection, MPC, and mobile antenna design for 5G is also presented.

3.2.5.1 Human blockage

High gain directional antennas or beamforming are proposed for mmWave radio links to reduce the impact of high path-loss. This has the advantage of less fading and reduced delay spread. However, moving human bodies often shadow some plane waves, leading to time-varying radio channel responses even if the transmitter and receiver are static. Measurements of human blockage at different frequency bands have shown that human shadowing loss increases with frequency. A novel IRACON model based on Quadruple Knife-Edge Diffraction (QKED) has been proposed to describe hu-

FIGURE 3.22

Illustration of shield edge diffraction around a body [BK19].

man shadowing at mmWave frequencies. One solution to overcome this effect is the utilization of simultaneous multiple-beams which could also be used to achieve diversity gains or increase Rx power. A thorough literature survey of existing human blockage models and their quantitative comparison can be found in [VH20].

Typically, the human body model in simulations is based on diffraction which can be implemented in ray-tracing tools to improve their accuracy (Section 3.2.4). In [BK19], the impact of the far field distance on the diffraction loss was analyzed for Rayleigh distances which are significant compared to the distance from the transmitter or receiver to the diffracting object. This is an additional factor to include in addition to the phase and antenna pattern. The model fitted well to structures that have some regularity in their shape and position. However, the model fails for more irregular shapes where positioning is difficult to ascertain such as a body (see Fig. 3.22).

Hence, in [RPaC19], the feasibility of using a double Knife-Edge (DKE) diffraction model to compute the human blockage in indoor environments was evaluated. The recommended basic rules for this model at mmWave frequencies were: (1) the minimum electrical dimension of the cross section (cube width and cube height) of the blocking element should be greater than 10λ; (2) all geometrical shapes can be simplified to an absorbing screen infinitesimally thin; (3) the antenna polarization in the diffracted field is only relevant when the cube width and cube height are lower than 10λ; and (4) the minimum distance from the source to the absorbing screen model is 15λ, but will strongly depend on the antenna electrical size.

The human blockage was further characterized in [VH20] via measurements in an anechoic chamber at 15, 28, and 60 GHz employing 15 human subjects of different sizes from which a 3D physical human blockage model based on Quadruple Knife-

Edge diffraction (QKED) was developed. The results showed that the blockage loss increases with frequency and is on average 7-10 dB more at 60 GHz than at 15 GHz. It was observed that the body width and height are the key parameters which affect the human blockage significantly. The oblique orientation of human body, results in relatively more blockage loss compared to the positions where the frontal plane of the human body is either parallel or perpendicular to the direct-wave path. Furthermore, the blockage loss decreases as the height of transmitter (Tx) increases as seen in Fig. 3.23.

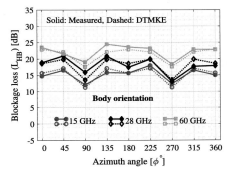 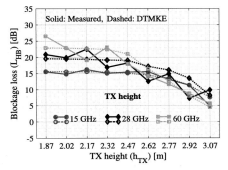

FIGURE 3.23

Measured and DTMKE-predicted median human blockage loss (LHB) over 15 human subjects at 15, 28, and 60 GHz frequency bands for different (a) body orientations and (b) Tx heights [VH20].

In [ZACCL17] an urban ray-based model was used to study the impact of human body blockage and vehicles in an outdoor small-cell network at 60 GHz. The results indicate that they generate ray-path blockage and new MPCs, which can affect the cell selection or beam orientation, and significantly change the received signal strength and inter-cell interference. The blockage effect was introduced into a whole mmWave small-cell network simulation via a stochastic process. The results show that the user-body can lead to stronger inter-cell isolation. In addition, the degradation of the useful signal strength does not significantly impact the global network performance, since the victim users can switch to another cell without body blockage. The obstruction by the user-body and vehicles can lead to a limited reduction of the SINR. Significant degradation occurs only on the very worst-served users: 1.1 dB on the 10% percentile; and 0.2 dB on the median percentile. The results also indicate that the impact of large vehicles can be reduced by physically placing the small-cell base stations on opposite sides of the streets in an asymmetrical manner.

3.2.5.2 Precipitations

The rainfall and other meteorological phenomena attenuate the received signal in particular for frequencies above 6 GHz. To estimate the impact of precipitation on the mmWave band to generate statistics for an appropriate model, fixed links were set up in Durham (UK) and Gothenburg (Sweden) with single and dual polarization configurations ranging from a few 10's of meters to a few kilometers in the 18/19 GHz, 25.84 GHz, and Eband (71-76/81-86 GHz), [SCR19] and later in [RPaC19] and [LB20]. To achieve the objective, a high performance disdrometer was used to measure the drop size distribution and velocity of falling hydrometeors. The antennas were shielded to reduce the impact of antenna wetness on the measurements. An example of received signal strength versus rain rate is shown in Fig. 3.24(a). It was also reported that gas attenuation and fading impact the Eband link more than the 18 GHz link, giving rise to considerable changes in path attenuation also during dry weather compared with the ITU recommendation as shown in Fig. 3.24(b).

3.2.5.3 Drones

Drones have become a very successful entertainment product and their applications to transport, surveillance, security and many other uses are being investigated. Drones can be also used in illegal operations and mmWave radars can be useful to detect malignant drones. In [SHP+20] and [SVC19], a quasi-monostatic radar was used to measure the Radar Cross Section of different Unmanned Aerial Vehicles in the 26-40 GHz band. It was concluded that mmWave radars with machine learning algorithms could be used to detect UAV using their signatures (example in Fig. 3.25) which depend on the material type and batteries.

3.2.5.4 Impact of antenna aperture and misalignment

The introduction of mmWave antennas in 5G mobile terminals is a challenging task and must be evaluated at the system level for inter-operability between worldwide mobile providers across different frequency bands. The antennas must compensate for the additional path-loss either using a higher array gain or dynamic beamforming. Coupled with a smaller aperture size, the radio channel is more directive with the advantage of less fading and reduced delay spread but can be subject to larger blocking effects, and therefore, communication loss. A Teaching-Learning-Optimization (TLBO) algorithm was applied in [GTB+17] in order to design a dual-band E-shaped patch antenna where the geometrical parameters of the aperture-coupled antenna formed the inputs of the optimization algorithm. The method gave acceptable design solutions achieving simultaneously S_{11} minimization and low VSWR at the frequencies of interest (25 GHz and 37 GHz).

Moreover, the pros and cons of selecting either omnidirectional or directive antennas must be evaluated by taking into account the characteristics of the radio channel. The impact of the antenna aperture on the channel DS at 17 and 60 GHz in an urban outdoor environment was analyzed in [DCCS18]. The results at 17 GHz were considered to be applicable to the 60 GHz frequency band, given the similarities observed between the azimuth delay power profiles (ADPPs) at these two frequencies. The results show that the channel DS can be reduced by the use of small antenna aperture (such as 30° HPBW) even though obstructions represent a limiting factor. Nonetheless, when the strongest beam is highly attenuated, the wireless system may switch

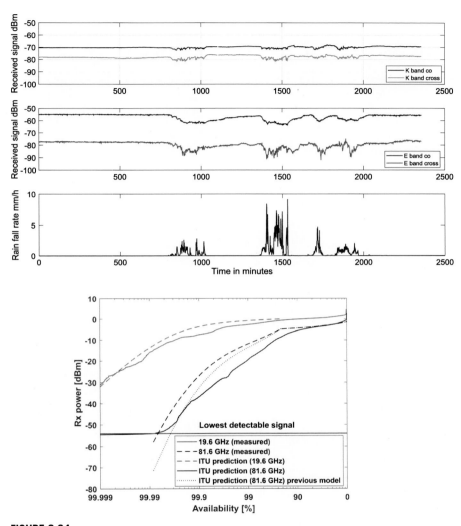

FIGURE 3.24

(a) Received signal strength versus rain rate in mm/hr as a function of time (minutes) at 25 GHz (top) and 77 GHz (bottom) extracted from the data shown in [SCR19].
(b) Measured path attenuation compared with rain attenuation according to the ITU-R P.530-15 recommendation [LB20].

to an alternative beam configuration resulting in an increased channel DS compared to that of omnidirectional antennas.

Finally, the received signal in mmWave point-to-point links is strongly dependent to the alignment of the directional antennas. In [MIGMGP⁺17], it was shown that

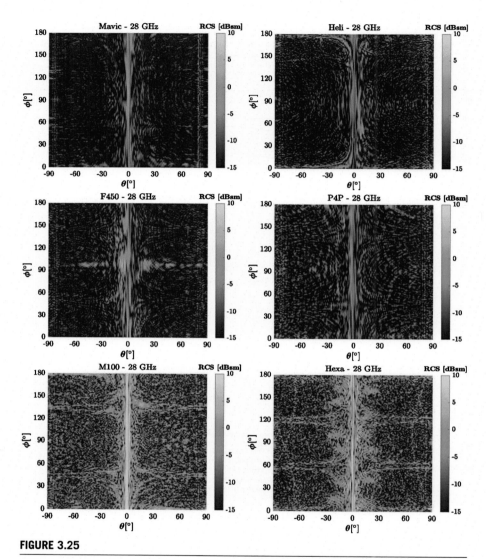

FIGURE 3.25

RCS signatures of the studied objects at 28 GHz extracted from [SHP+20]. a) "Mavic" - Quadcopter DJI Mavic, b) "Heli" - Helicopter Kyosho, c) "F450" - Quadcopter F450, d) "P4P" - Quadcopter DJI Phantom 4 Pro, e) "M100" - Quadcopter DJI Matrice M100, f) "Hexa" - custom built hexacopter. a)-d) drones were mostly made up of plastic whereas e)-f) drones were mostly made up of carbon.

a misalignment of 2° at 1 m resulted in a 3 dB loss between two horn antennas at 300 GHz. In addition, multiple specular reflections were found between both antennas with DS and mean excess values of 2.7 ns and 18.5 ns, respectively.

3.3 **MIMO and massive MIMO channels**

This section presents the results of research into MIMO channels to realize next-generation wireless communication. Results of the various channel measurements from 2 GHz to 90 GHz are analyzed. New research topics such as spatial correlation characteristics and frequency dependence of the propagation channel with results of channel parameters in various environments are discussed. The section also outlines massive MIMO system evaluation technologies through measured channel data and simulation.

3.3.1 **MIMO channel measurement and analysis**

3.3.1.1 Channel sounder development

The evaluation and deep understanding of the mmWave massive MIMO propagation channel through exhaustive channel measurements are indispensable to justify the feasibility of the system. Massive MIMO channel measurement systems below the 6 GHz band were developed using the commercially available software-defined radio (SDR) platform [BBM16][LHB+19]. In [LHB+19], 2 × 64 real-time MIMO channels were measured by synchronizing the 32 commercial SDRs (URSP 2954R) as shown in Fig. 3.26. The system was then used to measure the dynamic propagation channel in a street environment.

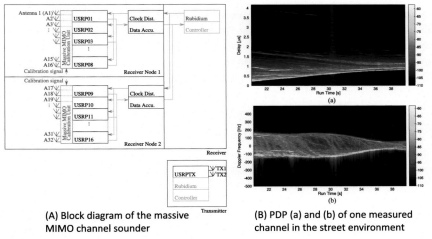

(A) Block diagram of the massive MIMO channel sounder

(B) PDP (a) and (b) of one measured channel in the street environment

FIGURE 3.26

The block diagram of the sounder and measurement result [LHB+19].

3.3.1.2 Channel parameter characteristics

The propagation channel in the 92 GHz band was measured using a vector network analyzer (VNA) based virtual array system [MGG+16][CML+18]. [MGG+16] reported that the specular component could be estimated accurately by 3D ray-tracing simulation as shown in Fig. 3.27. The propagation parameters of the channel were estimated from the measured data by the RiMAX algorithm. Although the result showed that the propagation channel was dominated by the line of sight (LOS) wave in the short distance region, the reflection and scattering waves were also not negligible when the distance increased to several meters in this frequency band. [JFK+19] analyzed the influence of BS and MS side angular spread of the channel on the antenna correlation characteristics.

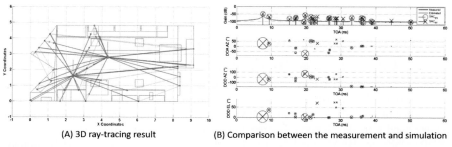

(A) 3D ray-tracing result   (B) Comparison between the measurement and simulation

FIGURE 3.27

The comparison between the measurement and simulation in 92 GHz band indoor environment [MGG+16].

The channel characteristics in industrial environments were investigated in [SZZ19][D19]. In [D19], the angular and delay characteristics of the channel were measured with a virtual array. Because of a large number of metallic objects in the environment, the delay spread tended to be large. The room electromagnetic approach was applied to quantify the channel characteristics. [SZZ19] analyzed the influence of the channel characteristics on a 2-by-2 MIMO channel capacity in an industrial environment.

3.3.1.3 Spatial correlation characteristics

When large antenna arrays are considered, it is not guaranteed that the propagation cluster characteristics are consistent within the array. The spatial stationarity (SS) of the MPC clusters concept which is the cluster transition characteristics among the sub-arrays was introduced in [CYCW17] and [LAH+19]. The 13 and 17 GHz band propagation channels were measured in an outdoor roof-top environment [CYCW17]. The propagation clusters were estimated from the measured data by the SAGE, and K-Power Means algorithms. The SS characteristics of the K factor, DS, and angular spread were investigated. The spatial stationarity of the cluster concept was also investigated in [LAH+19]. The clusters were classified into "common cluster" which is observed by the entire array and "non-common cluster" which is not observed by the array. The measurements were conducted in a theater and a subway environment in the 2.6 GHz band. The results showed that 40% of the clusters were identified as non-common clusters. In [GO19], the spatial correlations between the propagation paths were analyzed. The correlation was modeled according to the similarity of the physical propagation paths which were identified from the channel sounding data.

3.3.1.4 Frequency dependency characteristics

It is envisaged that a very wide range of frequency bands will be utilized in heterogeneous 5G systems to realize large capacity and high reliability. Therefore, comparative investigation of the millimeter-wave band channel to the microwave band is also essential. [MSA16] pointed out that differences in the signal bandwidth and the antenna beamwidth in each frequency band can mislead the discussion of channel frequency characteristics. Therefore, channel measurements in the same location, bandwidth, and beamwidth were conducted from 5 GHz to 60 GHz. As shown in Fig. 3.28, the angular power spectra (APS) were similar in all frequency bands. The frequency characteristics of the delay and angular spreads were also similar except for some conditions where the window attenuation dominated the characteristics. The frequency characteristics of the propagation channel from 3 GHz to 28 GHz were also investigated in an indoor environment [SHiT+18]. [SHiT+18] reported that although the specular wave characteristics were similar in all frequency bands, the scattered wave was frequency dependent; where scattering tended to become "directional" as the carrier frequency increased.

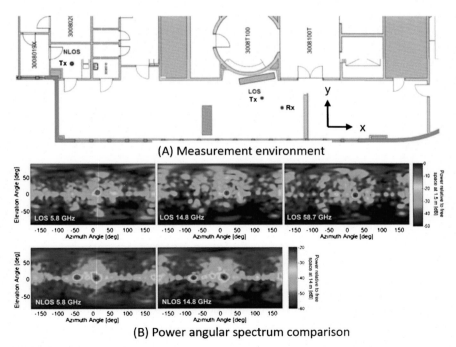

(A) Measurement environment

(B) Power angular spectrum comparison

FIGURE 3.28

The angular power spectrum comparison from 5 GHz to 60 GHz band [MSA16].

3.3.1.5 Propagation mechanism identification

The 28 GHz band massive MIMO channel was measured in an indoor environment [MFJP18]. The multipath components (MPCs) and those interactive objects in the environment were identified from the power angular delay profile (PADP) of the channel data. As shown in Fig. 3.29, although the MSs at each location had common scatterers, the channels of each MS were almost orthogonal using the ZF algorithm. The result indicated the wide applicability of the massive MIMO communication system in the environment. The 11 GHz band propagation channel was measured in an indoor hall environment [HST+17]. The MPCs were estimated by the SAGE algorithm, and the propagation trajectories were also estimated by the measurement based ray-tracing (MBRT) algorithm. It was reported that the receiving power and the cross-polarization ratio (XPR) of the double bounce clusters were lower than the single bounce clusters.

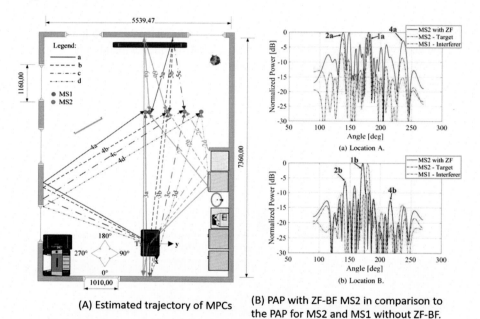

(A) Estimated trajectory of MPCs

(B) PAP with ZF-BF MS2 in comparison to the PAP for MS2 and MS1 without ZF-BF.

FIGURE 3.29

Channel measurement results in the 28 GHz band indoor environment [MFJP18].

3.3.2 Wireless simulation techniques

Wireless simulation and data analysis techniques for massive MIMO communication system evaluation are useful to reduce the workload of channel measurements. MIMO channel reconstruction method from low dimensional measurement data was introduced by assuming the Kronecker-type MIMO channel model [Ale17a]. The

2×4 MIMO channel was reconstructed from the 2×2 MIMO land mobile satellite measurement data in the 2.5 GHz band.

MIMO channel characteristics were also investigated from ray-tracing simulations [ZMM17][ACBL18b]. In [ACBL18b], the performance of a massive MIMO communication system in an urban environment was investigated using real 3D geographic map and considering the appropriate BS and user distribution. The downlink user equipment (UE) throughput was investigated, and the results showed that the system performance was affected by not only the outdoor users but also the indoor user distribution, especially in the case where the indoor users were on multiple floors.

A map-based time-varying MIMO channel model was proposed in [ACS18]. In the model, the LOS and the NLOS regions were determined by assuming the Manhattan grid map model. The LOS probability was determined from the distance between the MS position and the boundary between two regions as shown in Fig. 3.30. The probability distribution of cluster parameters were also estimated from the LOS probability to generate a time-varying MIMO channel. In addition, Doppler power spectral density and the time-varying behavior of the antenna correlation characteristics were investigated by the proposed modeling approach.

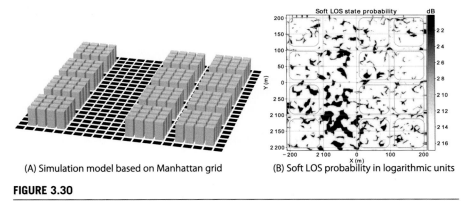

(A) Simulation model based on Manhattan grid   (B) Soft LOS probability in logarithmic units

FIGURE 3.30

The LOS probability characteristics based on the Manhattan grid model [ACS18].

3.3.3 Antenna development

In the design of embedded antennas, it is necessary to consider the influence of other components such as touch screen, cameras, microphones, batteries, and the body of the user. The theory of characteristic modes (CMs) enables the design of high-quality antennas through providing the characteristic modes inherent to structures. For example, embedded antennas for the sub 1 GHz band were developed by considering the influence of the internal components of the UE [MSE+16] and the hands of the user [ML17]. Also, the CMs theory was used to deign a dual-band MIMO terminal antenna with low-correlation [LL17].

Due to the widespread concept of the connected car to realize intelligent transportation systems (ITS), antennas that are optimized for the automotive urban communication are needed. In [KMZ16][KMMZ19] a reconfigurable antenna using phase shift between the elements and a linear tapered microstrip balun to switch the radiation pattern to follow the dynamic environment was developed as shown in Fig. 3.31.

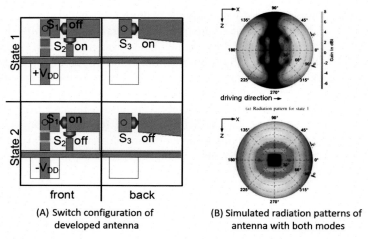

(A) Switch configuration of developed antenna

(B) Simulated radiation patterns of antenna with both modes

FIGURE 3.31

The developed pattern reconfigurable antenna [KMZ16].

3.3.4 Electro-magnetic compatibility analysis

Another aspect to consider is the impact of human exposure to radio transmission such as massive MIMO communication systems in realistic environments. In [STV+19] both the propagation environment and the specific absorption rate (SAR) of the human body, to exposure was estimated by combining ray-tracing and finite-difference-time-domain (FDTD) simulations. The maximum allowed power of the massive MIMO BS was obtained for both LOS and NLOS scenarios as shown in Fig. 3.32.

3.3.5 Massive MIMO system evaluation

3.3.5.1 Channel hardening concept

Massive MIMO is widely considered to be a crucial enabling technology for 5G wireless systems. It has the potential to improve capacity, reliability and spectral efficiency. Channel hardening is the phenomenon where the channel gain becomes more concentrated on its mean and the gain variations due to multipath fading are mitigated by utilizing a massive number of antennas at the BS. The relationship between chan-

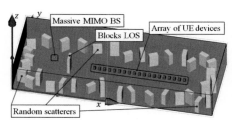

(A) Ray-tracing simulation scenario

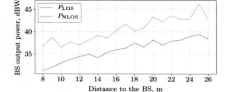

(B) BS output power violating ICNIRP guidelines

FIGURE 3.32

The human exposure assessment in 3.5 GHz band [STV+19].

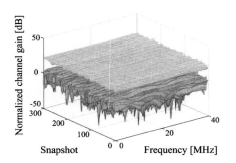

(A) Normalized channel gain for user 1 showing the single antenna with the average mean channel gain (lower) and the channel gain for all 128 base station antennas (upper).

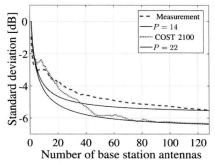

(B) Standard deviation of channel gain as a function of number of base station antennas as a comparison between measurements and COST 2100 for the indoor scenario. (when using all 128 antennas)

FIGURE 3.33

The effect of channel hardening [GFdPT18].

nel hardening and the number of BS antennas has been investigated in [GFdPT18] using indoor and outdoor channel measurements in the 2.6 GHz band. As can be seen in Fig. 3.33, the channel gain becomes stable when 128 BS antennas are used. The measurement results were also compared with the simulation results based on the COST 2100 channel model.

3.3.5.2 Inter-user interference evaluation

The inter-user interference characteristics are also an essential factor for system evaluation. [CLL+19][CLMG18] investigated the effectiveness of the polarization diversity scheme in the 1.3, 3.5, and 6 GHz bands in an industrial hall and conference room environment. The performance of the zero forcing (ZF) and the MRT algorithms were evaluated in the experiment. The result showed that polarization diversity suppressed spatial correlation in all frequency bands. The result also showed that the MRT was suitable as a whole if polarization diversity was utilized.

The spatial correlation characteristics were also investigated through indoor channel measurement in the 5 GHz band [LHZ17b] and with ray-tracing simulations in the 2.4 GHz band [LHZ17a]. The singular value spread which is the ratio between the highest singular value and the lowest singular value of the multi-user MIMO channel matrix was introduced. The authors also pointed out that the channel gains between several BSs and the MS were different by the differences of the distances. In addition, the characteristics of the BS channel with the highest gain tended to dominate the entire multi-user MIMO channel characteristics as shown in Fig. 3.34.

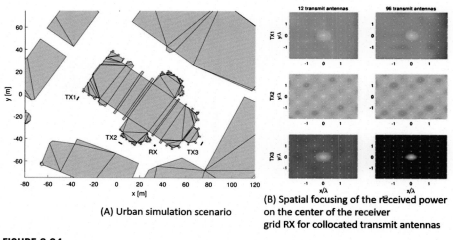

(A) Urban simulation scenario

(B) Spatial focusing of the received power on the center of the receiver grid RX for collocated transmit antennas

FIGURE 3.34

Interference evaluation in urban scenario [LHZ17a].

3.3.5.3 Signal processing techniques

A signal processing algorithm for channel hardening was investigated in [AGE19]. The received signal component was concentrated in a delay bin by combining the maximum ratio transmission (MRT) and the time-reversal filtering, where the channel hardening effect was more easily achieved by the proposed algorithm. Channel measurements were conducted in an indoor environment at 1.3 GHz and 6 GHz [GAEE18]. The results showed both the variations of the channel gain and the delay spread were suppressed by the algorithm as shown in Fig. 3.35.

3.4 Fast time-varying channels

To meet the goals of efficiency, safety, and convenience, both road transport and rail traffic, are expected to evolve into a new era of land intelligent transportation systems where infrastructure, vehicles or connected vehicles (train), travelers and goods will be increasingly interconnected. To realize this vision, a seamless wireless connectiv-

(A) Indoor measurement campaign

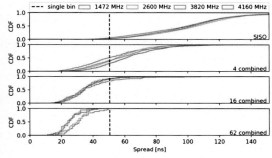

(B) Cumulative distribution functions of the rms delay spread for different numbers of combined SISO channels

FIGURE 3.35

The indoor measurement result [GAEE18].

ity integrating ultra-reliable low latency and high-data rate communications both for vehicles and trains is required. To effectively support the design, simulation, and development of this envisaged system, it is necessary to acquire fundamental knowledge of the propagation channel characteristics, such as path loss, shadow fading, delay spread, Rician K-factor, Azimuth angular Spread of Arrival (ASA), Azimuth angular Spread of Departure (ASD), Elevation angular Spread of Arrival (ESA), Elevation angular Spread of Departure (ESD), cross-polarization ratio, and the cross-correlation of these parameters. Due to the high mobility of the Tx and the Rx of the vehicles or the high-speed train, as well as the scatterers which can be vehicles, bicycles or pedestrians, the channel is fast time-varying, within a rather short period of time. To counteract this effect, in [Art19a][Art19b][Art19c], it is proposed to perform a counter movement of the vehicular antenna to keep it in its original position relative to the outside observer. However, considerable work still needs to be undertaken to transform this concept to reality. Thus, currently for land intelligent transportation systems, the widely used assumption of Wide-Sense Stationary Uncorrelated Scattering (WSSUS) channel is no longer satisfied. Therefore, characteristics such as Doppler shift, Doppler spread, coherence time, and stationary time, are highly important. In general, there are two major challenges for characterizing fast time-varying channels.

1. Diversity of V2X communications: unlike cellular communications where the link is between the base station and the user equipment, the links of V2X communications are very diverse and include V2I, V2V, and Vehicle-to-Pedestrian (V2P). Thus, it is necessary to characterize each of these V2X channels exclusively rather than presenting an averaged channel which may differ from the actual scenario.
2. High-Speed Train (HST) communications: 5G mobile communication systems are expected to support high mobility up to 500 km/h for high-speed trains. However, such high speeds present a challenge for channel sounding especially in the

mmWave band. Thus, a new paradigm exclusively to break the bottleneck of acquiring realistic 5G mmWave HST channels is required.

In this section, measurements and characterization for fast time-varying channels are presented.

3.4.1 Channel characterization of V2X scenarios

The Tapped Delay Line (TDL) model which is based on the WSSUS model, was used in cellular communication systems. A survey of the TDL model can be found in [HKS+18] where the classical TDL model is extended using a Markov chain to implement non-WSSUS for V2X channels. For V2I channels, apart from the studies focusing on the Intelligent Transportation System (ITS) licensed band around 5.9 GHz, the channel at the LTE-Vehicle (LTE-V) radio interface band that supports V2I communications (named Uu-interface) is also characterized. In Yusuf et al. [YTC+19], results of experimental V2I radio channel sounding at 1.35 GHz performed in a suburban environment in Lille, France are presented. Based on channel measurements acquired in vertical and horizontal polarizations, a multitaper estimator is used to estimate the Local Scattering Function (LSF) for sequential regions in time, from which Doppler and delay power profiles are deduced. It is found that a lognormal model best fits the Root Mean Squared (RMS) delay and Doppler spreads for both polarizations. Small-scale fading is also investigated per delay tap. The parameters of the Rician fading for the first tap and the Nakagami fading for the delayed taps are estimated and statistically modeled. It was found that the best fit is the lognormal model. Similar discussions of the delay-Doppler domain for the V2I channels on a highway at 2.53 GHz can be found in Sommerkorn et al. [SCK+18]. Taking into account the higher carrier frequency, the question whether the channel is sparse and how to describe it is posed for V2I communications as well. To address these questions, in Groll et al. [GZP+19], a series of mmWave V2I channel measurements were conducted in an urban street environment, in down-town Vienna, Austria. Frequency-division multiplexed multiple-input single-output channel measurements were acquired with a time-domain channel sounder at 60 GHz with a bandwidth of 100 MHz and a frequency resolution of 5 MHz within urban speed limits. Two horn antennas were used on a moving transmitter vehicle: one horn emitted a beam towards the horizon and the second horn emitted an elevated beam at 15-degrees up-tilt. After applying a multi-taper estimator, the results show high sparsity in the delay-Doppler domain. It was also found that within the region from 28 to 16 meters, the Tx antennas had a clear wide-angle view to the mounted Rx, and therefore, the Probability Density Function (PDF) of the signal magnitudes was bimodal for two-wave interference and for Two-Wave with Diffuse Power (TWDP) interference with two strong distinct specular components [GZH+19]. Al Mallak et al. [ABLH19] investigated the influence of Doppler shift, Doppler spread and signal attenuation on a V2I communication channel employing a 26 GHz phased array antenna from Anokiwave. This antenna facet acts as a spatial filter, eliminating unwanted MPCs and allows the main beam to be electronically steered. The results show a successful LOS communication up to

150 m distance for a transmitted power of 25 dBm. A stable channel gain was observed during the measurements and related to the proper direction of the beamwidth. Further results indicate that there is attenuation between 17 and 27 dB when a vehicle blocks the LOS connection between the Tx and the Rx. However, by driving the test vehicle in an almost full car park at a speed of 5 m/h, a very small Doppler shift was observed due to the suppression of the scattered and reflected waves by the spatial filtering of the array.

The major body of V2X study is V2V channels. For the single-link sub-6 GHz V2V channels, the stationarity distance and time is investigated in Czaniera et al. [CKS+19], using channel sounding measurements taken in a highway scenario at 2.35 GHz. The influence of the antenna positioning on both the receiving and transmitting car is evaluated. The findings indicate in general a larger stationarity time when the respective antennas are mounted on the roof of the cars (as opposed to the bumper or front of the car). However, the results are still inconclusive as in the studied cases mostly LOS or quasi-LOS conditions dominated. It can be expected that investigations for NLOS conditions yield different results (e.g. regarding the antenna placement). For the single-link mmWave V2V channels, the maximum Doppler shifts are much higher than their centimeter wave counterparts. Zöchmann et al. [ZML+18] measured the wireless channel of overtaking vehicles in an urban street environment to obtain delay and Doppler spread values based on ensemble averages. Since the Tx antenna is a horn with 18 degree half power beamwidth, the spatial filtering by the narrow beam decreases the delay and Doppler spread values. Thus, in Zöchmann et al. [ZHL+18], it was observed that the RMS Doppler spread was effectively below one-tenth of the maximum Doppler shift. For example, at 60 GHz a relative speed of 30 km/h yields a maximum Doppler shift of approximately 3.3 kHz and a median RMS Doppler spread of 220 Hz. The highest median RMS delay spread observed was 5.3 ns. When switching the focus from the single-link to the multilink systems, e.g., Vehicular Ad-hoc NETworks (VANETs), shadowing from vehicles becomes critical and therefore, needs to be characterized and modeled. In Nilsson et al. [NGAT17], a measurement based analysis of multilink shadowing effects in a V2V communication system at 5.9 GHz with cars as blocking objects in a highway convoy scenario is presented. The results show that it is essential to separate the instantaneous propagation condition into LOS and Obstructed Line of Sight (OLOS), by other cars, and then apply an appropriate path loss model for each of the two cases. Moreover, the choice of the path loss model not only influences the coherence time but also changes the cross-correlation of shadow fading between different links. It is concluded that it is important that VANET simulators use a geometry based model, that can distinguish between LOS and OLOS communication, otherwise the VANET simulators need to consider the cross-correlation between different communication links to achieve realistic results, e.g. as reported in Nelson et al. [NLVT17]. Nilsson et al. [NGAT18] present an alternative NLOS path loss model based on V2V channel measurements at 5.9 GHz in an urban environment between six vehicles. The model is reciprocal and takes into account whether communicating vehicles are obstructed by other vehicles. Schieler et al. [SSA+19] demonstrated via field measurements at

5.2 GHz that Orthogonal Frequency Division Multiplexing (OFDM) waveform radar sensing approach can be used to detect cars and pedestrians in a bistatic setup. Based on measurements conducted at 5.9 GHz in urban and suburban scenarios, Yang et al. [YAH+19] proposed a cluster-based 3-D channel model taking into account the distribution of MPCs clusters in both the horizontal and vertical dimensions. It was found that both the azimuth and the elevation spreads follow the lognormal distribution. In addition, the power of MPCs within a cluster has a truncated-Gaussian distribution, whereas the angle of MPCs within a cluster has a Laplacian distribution. In [CPYA17], based on V2V measurements at 5.9 GHz, a novel algorithm based on the Hough transform is proposed to identify multiple clusters from the evolution of their trajectories in the concatenated power delay files. Based on the cluster identification results, parameters that can be directly applied in the standard clustered delay line model are investigated. It is interesting to find that the power of a cluster with longer delay can be higher than that of a cluster with a shorter delay due to reflections by vehicles or the facade of buildings. Moreover, Czaniera et al. [CSTD19] discuss the possibilities of utilizing the stationarity of the V2X radio channel for applications in communications as well as passive radar. The stationarity can be estimated in the temporal domain using the Generalized Local Scattering Function (GLSF) and its collinearity. For communications, the necessary pilot spacing can be adapted whereas in radar, a feasible temporal integration window size can be set accordingly. Moving the focus from V2V to aircraft-to-aircraft channels, Walter et al. [WSM+19] derived a 3-D time-variant single-bounce scattering model to provide a description of non-stationary aircraft-to-aircraft channels. Time-variant limiting frequencies of the delay-dependent Doppler probability function are derived by an algebraic analysis of the Doppler frequency in prolate spheroidal coordinates.

V2P communication promises to prevent accidents by enabling collision avoidance applications. In order to develop and test a V2P communications system, accurate knowledge of the propagation channel is essential. However, only limited analysis of V2P channel has been reported in the literature. To fill this gap, the German Aerospace Center conducted extensive channel sounding measurements in a controlled environment for four different scenarios. The measurements were performed at 5.2 GHz with a bandwidth of 120 MHz in Rashdan et al. [RdW+18]. The results indicate that the two-ray path loss model fits the measured path loss at distances between 40 m and 100 m better than the log-distance path loss model. It was found that reflections due to pedestrians lead to rapid fluctuations of the path loss around its mean value. Pedestrians that surround the receiver cause path loss variations and degradation on the order of 5 dB to 10 dB due to shadowing. LOS obstruction by parked vehicles is also investigated where an additional path loss of about 10 to 15 dB by each individual parked vehicle is distinguished. Extensive V2P channel measurements were also conducted by Makhoul et al. [MEO18], in different propagation conditions and different mobility patterns. A Geometry-based Stochastic Channel Model (GSCM) is proposed consisting of the strongest path, the discrete components and the diffuse components. For the strongest path and secondary discrete components the path loss, large-scale and small-scale fading are presented. The results are reported for the LOS scenario with future work to model the diffuse scatterers and the extension of the characterization towards NLOS V2P links.

3.4.2 COST-IRACON V2V channel model for urban intersections

To develop next generation wireless systems for vehicles, detailed propagation models are needed. Particularly, there is a need for a spatially consistent model that can reflect the non-stationarities of the vehicular channel and that also supports multiantenna operation at both ends, either for beamforming purposes, for spatial multiplexing, or just for diversity. Previously, member institutions of the COST 2100 and IC1004 actions have derived a GSCM for V2V communication in highway scenarios [KTC+09] and a non-stationary model for vehicular communication [HRK+15], but a detailed measurement based propagation model supporting multiple antenna configurations with realistic spatio-temporal characteristics in urban intersections has been lacking until now.

The COST IRACON channel model for urban intersections [GMBT20] is a GSCM based on a street geometry defined by a map. It can thus represent typical intersection scenarios or specific intersections. Specular scatterers are randomly dropped, with a given density (number of specular scatterers per area unit), in the simulation area in bands along the walls according to the geometry of the intersection. These scatterers are then labeled as first, second, and third order reflection points. Similarly, diffuse scatterers are also randomly placed along with the walls, but in wider bands. For simulations of specific intersections, especially for simulations of wider or more open intersections, scatterers can also be dropped in areas that are not aligned with the walls. These scatterers typically represent contributions from lampposts or larger street signs, which typically can be observed in measurements. Fig. 3.36 shows an example of a drop of scatterers from a real intersection in Berlin, Germany. Here, the blue, red, yellow, and gray dots represent first- order, second-order, third-order, and diffuse scatterers, respectively. Table 3.12 summarizes the intensity χ, of different types of scatterers as well as the width, W, of the bands of scatterers along the walls.

Table 3.12 Scatterer location parameters.

Type	Order	χ (m^{-2})	W (m)
Wall	1st	0.044	3
Wall	2nd	0.044	3
Wall	3rd	0.044	3
Non-wall	1st	0.034	User defined
Diffuse, wall	1st	0.61	12
Diffuse, non-wall	1st	0.61	User defined

Once the scatterers are dropped in the simulation area the contribution of each scatterer to the impulse response is determined by expressions for 1) distance dependence, 2) losses due to interactions with scattering objects, 3) obstructions by

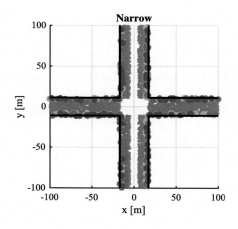

FIGURE 3.36

Example of scatterers dropped in a simulation area. Colored dots are specular components, gray dots represent diffuse scattering [GMBT20]. XXX to be modified to avoid copyright XXX.

buildings, foliage and other objects, 4) diffraction around corners, 5) angular dependence of the scattering interaction, and 6) random, but spatially correlated, large scale fading. The transfer function of the channel between any two antennas is calculated as a superposition of all power weighted contributions from the scatterers as in (3.1)

$$H(f,t) = \sum_{l=1}^{L} g_l e^{-j2\pi f \tau_l} G_{\text{Tx}}(\Omega_{\text{Tx}}) G_{\text{Rx}}(\Omega_{\text{Rx}}), \qquad (3.1)$$

where G is the complex antenna amplitude gain in direction Ω, g_l is the complex amplitude of path l, L is the total number of paths, f is the frequency, and τ_l is the propagation delay, $\tau_l = d_l/c$. For each MPC the average path power gain is modeled as in (3.2)

$$\bar{g}(d)^2 = \left(\frac{g_0 g_a g_b}{d}\right)^2 10^{-\frac{L_p}{10}}, \qquad (3.2)$$

where d is the path propagation distance and g_0^2 is the path power gain at a reference distance of 1 m. The term g_a^2 describes the path angular power gain, which is a function of the incoming and outgoing angles for each scatterer, and g_b is a gain describing the effects of obstruction and blockage by buildings. Each multipath component then undergoes spatially correlated Gamma distributed large scale fading (with parameters k and θ) so that the instantaneous amplitude of each MPC varies around the mean according to a Nakagami distribution. The autocorrelation of the fading process is described by a conventional Gudmundson model [Gud91] with coherence distance d_c. Table 3.13 summarizes the path gain parameters for the different contributions, where $G_0 = 20\log_{10}(g_0)$.

Table 3.13 Path gain parameters.

Type	Order	G_0 (dB)	d_c (m)	k	θ
Wall	1st	$U(-65, -48)$	$U(1, 2)$	$U(2, 8)$	$1/k$
Wall	2nd	$U(-70, -59)$	$U(0, 1.5)$	$U(1, 6)$	$1/k$
Wall	3rd	$U(-75, -65)$	$U(0, 1)$	$U(1, 4)$	$1/k$
Non-wall	1st	$U(-68, -52)$	$U(0, 1)$	$U(1, 6)$	$1/k$
Diffuse	1st	$U(-80, -68)$	$U(0, 1)$	$U(1, 1)$	$1/k$

The model was parameterized and verified against measurements in various types of street intersections in Berlin, Germany. Fig. 3.37 shows an example of simulated and measured impulse responses from the intersection outlined in Fig. 3.36. The y-axis shows the measurement time and the x-axis shows the propagation distance of the multipath components. The measurements are performed with an ultra-wideband channel sounder at a center frequency of 5.7 GHz with a bandwidth of 1 GHz. The white stripes in the left plot are due to the data storage process where no measurements could be taken.

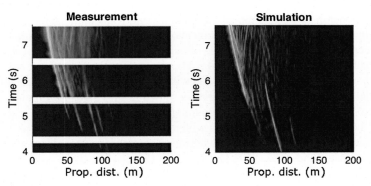

FIGURE 3.37

Measured and simulated impulse responses over time in the intersection in Fig. 3.36. The white stripes are due to data download, with no measurements [GMBT20].

As seen in the figure there is a very good correspondence between the outputs of the channel model and the real life measurements. Some detailed behavior differs, but the overall structure of the channel evolution and variation of the impulse response is well captured by the GSCM.

3.4.3 Wide band propagation in railway scenarios

In this subsection we focus on railway wireless communication and the corresponding propagation environment which include Train-to-Train (T2T) and Train-to-Infrastructure (T2I). To realize autonomously driving trains and virtual coupled trains via wireless communication links instead of mechanical couplers and electrical hard wired connectors, an Ultra Reliable Low Latency Communication (URLLC) system between two or more trains is required. Two important questions emerge. First, how the railway environment influences the T2T link for frequencies above 1 GHz or even at mmWave and Terahertz (THz) bands for 5G and beyond. Second, how the high absolute speed (higher than 300 km/h) or high relative speed (higher than 600 km/h) influences T2T communication.

To address these questions, the first reported high speed train HST channel sounding measurements took place on a HST track with two trains at 5.2 GHz with 120 MHz bandwidth between Naples and Rome in Italy [USS+17]. The distance was estimated from the absolute delay of the LOS of each Channel Impulse Response (CIR), which is only feasible when the LOS path can be accurately tracked by the Kalman Enhanced Super Resolution Tracking (KEST) algorithm [UJWK18]. In these measurements the LOS was continuously tracked starting with a delay of 4.27 μs which was related to a link distance of 1.28 km. Additional MPCs tracked in parallel to the LOS might have been caused by obstacles on the trains roofs and the noise barriers in parallel to the track. While a link distance of 66.1 m was measured as the short distance of T2T communication link, the LOS was tracked well from a delay of 0.22 μs, and more multipath components were also tracked. Doppler estimation is generally based on the velocities of Tx and Rx, and the angle of departure and angle of arrival. In these measurements, due to the Single-Input Single-Output (SISO) configuration, neither the angle of departure nor the angle of arrival were directly available. An alternative method is to estimate Doppler from the delay changes as shown in (3.3), with the delay $\tau_i(\mathbf{x_S}, t)$ with respect to the scatterer's position vector $\mathbf{x_S}$ and time t. The resulting Doppler is shown in Fig. 3.38. Two regions can be separated. First, the Doppler values around $f_d = 288$ Hz are caused by scatterers in front of the Tx and the Rx. For MPCs with a delay of less than 2.6 μs the Tx passes by the scatterers within the measurement time. Therefore, the Doppler changes to the second region around $f_d = -192$ Hz. The high Doppler spread of the measurement data might be caused by the vibrations and the relative movement of the trains. By comparison of the delay behavior and Doppler behavior, the locations of corresponding scatterers, namely the overhead line poles, could be estimated.

$$f_d(\mathbf{x_S}, t) = -\frac{\partial}{\partial t}\tau_i(\mathbf{x_S}, t)f_c. \qquad (3.3)$$

In addition to the channel sounding measurements, ITS-G5 measurements at 5.9 GHz with an Effective Isotropic Radiated Power (EIRP) of 31 dBm were performed. The results of ITS for T2T links indicated that the density and the depth of the fades increased due to high speed. In a rural environment, a stable connection can be built up to 1.2 km. Yet, for tunnels and big buildings as in urban environments a higher link distance up to 2.2 km could be achieved because of the wave guide effect. In the T2T measurements reported in [USS+17], different topographical environments were found to have great influence on radio propagation. In [URW19], T2T Path Loss (PL) models for three typical railway environments are derived and fitted with Free Space Propagation Loss (FSPL) and log-distance PL models. For the open

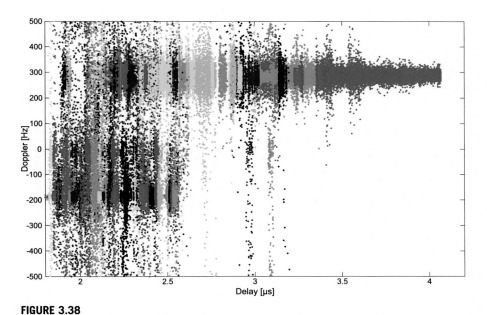

FIGURE 3.38

Doppler over delay for the estimated MPCs.

field environment, a two-ray PL model could also be applied. The new open field PL model is compared with a previous presented model based on ITS-G5 measurements. The new two-stage log-distance PL model sharpens the model for short distances and especially in combination with the two-ray PL model, the new PL model extends the previous one for longer distances.

To verify new communication regimes in railway environments, it is critical to also define T2I scenarios. In the 3GPP R1-163887, it was agreed to include the Macro+relay nodes for HST scenarios in the evaluation. The most influential materials and objects in HST scenarios were determined through measurements of EM and scattering parameters. The classification of HST environments has been widely studied. For example, viaduct, tunnel, cutting, urban, crossing bridge, train station, cut and cover tunnel, and rural are the main HST environments. In Guan et al. [GLH+17], 3-D models of a set of comprehensive railway scenarios are classified and reconstructed, including six modules: "Module 1-Tunnel entrance on steep wall connecting cutting with crossing bridges", "Module 2-Viaduct with open train station", "Module 3-Urban with semi-closed train station", "Module 4-Rural with cut and cover tunnel", "Module 5-Rural connecting double-track tunnel", and "Module 6-Single-track viaduct". These modules can be independently used for site-specific verification of new communication regimes, or combined in various ways for obtaining statistics.

Objects, their components, and their material determine the properties of the propagation channel. In Guan et al. [GAP+18a], the EM and scattering parameters of common HST materials such as metal, concrete, resin, tempered glass, marble, and brick, were estimated from 26.5 GHz to 40 GHz using an in house-built testbed. Fig. 3.39 shows a typical outdoor HST and tunnel scenarios with the most important objects marked with different colors.

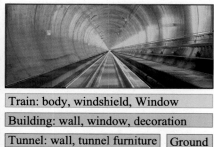

FIGURE 3.39

Main objects and their components in HST scenarios.

The scenarios were divided into outdoor HST and tunnel environments. Among the most important objects in the outdoor HST scenarios are the overhead line poles which affect the LOS communication path periodically, and also act as static and periodic scatterers along the railway lines. In Zhou et al. [ZYL+18], a wideband channel sounder with center frequency of 2.4 GHz is utilized to conduct channel measurements on the high-speed train line in China. From the measurement results, a dynamic channel model consisting of a LOS part with the shortest delay for each distance, and a pole scattering part caused by the poles along the rail is proposed. Fig. 3.40 shows the distance-delay-power profile of the measured HST channel and the influence of overhead line poles.

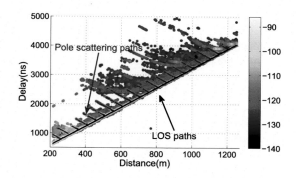

FIGURE 3.40

The distance-delay-power profile of the measured HST channel on the influence of overhead line poles.

Compared to outdoor HST environments, the wireless channel in tunnels is unique due to the confined space and the waveguide effect. Channel characterization for tunnel environments was carried out either theoretically or experimentally. Analyzes show that RMS delay spread, Doppler spread, and the maximum Doppler dispersion are large when both trains are inside the tunnel compared to the "open-air" situation. It is also observed that the signals at lower carrier frequency exhibit higher path loss, which is somehow contradictory for path loss assuming free space channel model. However, the tunnel is a waveguide, where signals with higher carrier frequencies exhibit lower path loss compared to signals with lower frequencies. In addition, with the same carrier frequency but different bandwidth, the wider signal bandwidth leads to lower path loss [HBJ18]. Different from the tunnel measurements with no train or a train with no passengers on board, in Cai et al. [CYCP16], passive channel sounding measurements are conducted when trains are in normal operation. The characteristics in tunnels are presented including path loss which exhibited dual-slope power fluctuations with the break distance of 100 m in most cases, delay spread similar to both the near and far regions with a mean value of 20 ns, Ricean K-factor with mean value of 11.5 dB, and Doppler frequency spread of 28 Hz. Moreover, the last three parameters were shown to be highly correlated with each other. This is easy to understand since a large K-factor occurs with concentrated power delay profile and power Doppler frequency profile. The delay spread and Doppler frequency spread followed a similar trend in most cases. This is consistent with the common belief that a larger delay spread is caused by a large number of scatterers at either the Tx, or Rx, or both, which naturally results in larger Doppler frequency spread when either end of the link moves.

3.4.4 New paradigm for realizing realistic HST channels in the 5G mmWave band

In the era of 5G mmWave HST communications, it is not feasible to conduct extensive channel measurements on HST, due to the lack of a wideband channel sounding solution for mmWave with high mobility (without cable). Although some vehicular measurements have been performed in the mmWave band; the collected data are still neither with sufficient resolution nor with enough degrees of freedom for channel characterization. Thus, traditional channel modeling with measurements at the core needs essential evolution to realize realistic 5G mmWave HST channels. Thus, rather than conducting mmWave HST channel sounding directly with high mobility, in Guan et al. [GAP+18a] the 5G mmWave HST scenario is defined in terms of links, applications, bandwidth requirement, Key Performance Indicators (KPIs), and environment. Barrier, tunnel, building, train, traction, track, and ground are considered as the main objects, which are mainly made of metal, concrete, resin, tempered glass, and brick. Then, all the six main materials in the HST scenarios are measured in the lab to estimate their EM and scattering parameters. The significance of the objects and materials in the 5G mmWave HST channels is then determined as follows.

1. Train and ground are the main deterministic objects for both outdoor and tunnel environments. Barriers are significant when they appear in outdoor environments, and tunnel walls are always significant in the tunnel environment.
2. For the outdoor HST environment with barriers when both the Tx and Rx are lower than the barriers, random objects outside the barriers are insignificant. Therefore, they can be excluded.
3. For the outdoor HST environment without barriers random objects made of concrete, tempered glass, or metal are significant. This conclusion also holds for the case when either the Tx/Rx is higher than the barriers.
4. For the tunnel environment random objects such as furniture made of resin or metal are significant if its area is larger than 7 m^2.

Based on these conclusions, two outdoor HST scenarios with/without barriers and one tunnel 3-D environment models were reconstructed. Then through RT simulations, the main propagation mechanisms were determined:

1. For the lower 5G mmWave band (30 GHz) in a straight tunnel: the LOS ray and reflected rays up to the 10th order.
2. For the higher 5G mmWave band (90 GHz) in an outdoor HST environment: the LOS ray, the reflected rays up to 3rd order, and the scattered rays whose differences to the strongest path are smaller than 45 dB.

The new paradigm was validated by examining two case studies; one in the outdoor HST at 90 GHz and one in the tunnel at 30 GHz [GAP$^+$18b]. The validation included RT simulations, stochastic modeling and realization, verification with RT simulations, and with a reduced set of measurements. Here, the high-performance RT platform is the key engine. The advantages of high computing power, cheap cost of services, high performance, scalability, accessibility as well as availability, makes cloud computing a popular service. Developing RT over a High Performance Computing server (HPC) platform can significantly improve its efficiency without sacrificing the accuracy. An open access high-performance RT called CloudRT was developed by Beijing Jiaotong University and detailed in [HAG$^+$18] as well as its website http://raytracer.cloud. The hardware components of CloudRT include a HPC cluster which contains 96 computing nodes (1600 CPUs and 10 GPUs), 1 managing node and 1 web server, a data storage server and UEs. The RT engine is installed on the computing nodes to realize parallel processing via Message Passing Interface (MPI). The architecture design of the CloudRT simulator is shown in Fig. 3.41.

Based on the RT simulations in both the outdoor HST and tunnel scenarios, two stochastic channel models were validated by a reduced set of measurements at 90 GHz and 30 GHz, respectively, in terms of path loss, shadow fading, small-scale fading, and power delay profile. The good agreement demonstrated that the new paradigm provides a viable way to generate realistic CIRs for 5G mmWave communications in high-speed railways, even when extensive channel measurements are not available.

The CIRs generated by these models can be input to link-level simulators to evaluate the physical layer technologies. In addition, with the aid of these RT-based

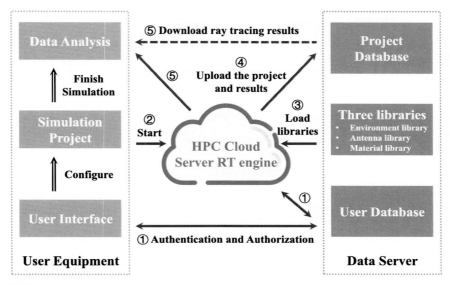

FIGURE 3.41

Workflow of the high-performance RT platform – CloudRT.

channel models, the time-varying parameters of 5G mmWave HST channels can be characterized. Table 3.14 summarizes the minimum, mean, and maximum values of the mean Doppler shift μ_v, the RMS Doppler spread σ_v, and the coherence time T_{coh} of the 5G mmWave HST channels in both cases at the speed of 500 km/h. These parameters are crucial to evaluate the non-stationarity and the correlation of the channel. If OFDM is applied, the sub-carrier spacing should be greater than the maximum Doppler spread σ_v, which is 0.14 kHz inside a tunnel at 30 GHz and 2.62 kHz for the outdoor environment at 93.2 GHz. Moreover, the packet length or the equalization procedure should be shorter than the minimum T_{coh}, which is 0.47 ms inside a tunnel at 30 GHz and 0.13 ms for the outdoor HST environment at 93.2 GHz.

Table 3.14 Doppler characteristics and coherence time in the 5G mmWave HST scenarios.

Scenario	Tunnel at 30 GHz			HST outdoor at 93.2 GHz		
Parameter	μ_v [kHz]	σ_v [kHz]	T_{coh} [ms]	μ_v [kHz]	σ_v [kHz]	T_{coh} [ms]
Maximum	13.89	0.14	4.00	43.21	2.62	3.70
Mean	13.48	0.02	3.00	42.68	0.38	1.90
Minimum	1.04	$3e^{-5}$	0.47	7.12	0.06	0.13

3.5 Electrical properties of materials

As most of the cellular mobile users are in indoor environments and the transmitters can be located in outdoor or in indoor premises, knowledge of penetration loss, as well as electromagnetic characteristics (permittivity and conductivity) of different materials have to be accurately known to be embedded into wireless channel simulators. In this subsection, specific measurements are described to identify penetration loss as well as different methods to derive electromagnetic characteristics of materials.

3.5.1 Transmission loss above 6 GHz

Material penetration loss measurements can be achieved through on-site [KLHNK+18] or laboratory isolated material measurements [DM18], including anechoic chamber measurements. On-site measurements are important, as real building materials are often inhomogeneous and it is not always simple to detach a sample for laboratory measurements. Whereas laboratory measurements are effective when specific penetration loss models need to be validated against measurements or analysis for several different materials have to be carried out. Laboratory measurements were carried out in [DM18] to validate the penetration loss model in ITU recommendation [ITU15] for transmission through a single pane of glass, in the 5-60 GHz range. Results show good agreement between the measurements and the model, making the model suitable to be embedded into simulation tools. In [KLHNK+18] three measurement methods were considered and compared for on-site penetration loss evaluation, namely: i) outdoor-to-indoor channel measurements, based on the identification of the direct path amplitude and subtracting the free space path loss with the same propagation delay; ii) far-field penetration loss measurements, based on the measured gain difference between LOS and an obstacle shadowing the LOS; iii) near-field penetration loss measurements, where probe antennas are very close to the material under test and subtracting the free space connection loss. Measurements conducted in an office building through a window and brick wall show good agreement of penetration loss estimates from the three methods. A band-stop characteristic in the window penetration loss was observed which could be due to the frequency dependence of loss where at low frequencies, the wall dimension is comparable to or smaller than the wavelength, whereas the penetration loss increases as the frequency increases.

3.5.2 Material properties

Several methods for electromagnetic parameter characterization have been developed over the years for specific frequency ranges, materials, and applications. Amongst these, the relative permittivity which can be evaluated from on-site measurements of the Fresnel reflection coefficient as in [BGLF+16], for the 60 GHz band which exhibited for the window the highest relative permittivity value of (6.01), whereas rougher materials tend to have a smaller permittivity (e.g. 1.29 for the bath stone as sum-

Table 3.15 Estimated relative permittivity of building materials at 60 GHz, as reported in [BGLF+16].

Measured material	Relative permittivity
Internal block wall with plaster finish	4.71
Block wall, plaster and write-on sheet	2.40
Dressed stone block	1.75
Bath stone	1.29
Penzance sand stone and red oxide	2.53
Window (small meeting room)	6.01

Table 3.16 Estimated relative permittivity of building materials at 60 GHz, as reported in [VNHW18].

Measured permittivity	ITU-R values	Classified materials
2.6	2.9	Plasterboard
2.5	2.4	Playwood
6.7	6.27	Glass
6.1	5.31	Concrete
12.7	-	Floor (NI)
18.3	-	Ceiling (NI)
>20	-	Metal

marized in Table 3.15). Another on-site permittivity evaluation method is reported in [VNHW18], where the inverse reflection problem is solved. Multipath estimates from on-site measurements at 60 GHz and the identification of scatterers leading to the multipaths, allows estimation of the relative permittivity. The method is based on a limited number of channel sounding measurements to extract the reflected multipaths and compare them with the outcome of a ray tracing tool, using accurate point cloud description. To this aim an accurate geometrical database of the environment for identifying flat and smooth surfaces producing reflections is required. The permittivity database obtained from the isolated measurements of composite is consistent with the ITU recommendation [ITU15], as reported in Table 3.16. Finally, a method for the determination of the complex permittivity (both real and imaginary part) of construction materials based on the Fabry-Pérot resonance of the material's slab is proposed in [DEZV+17]. The method is suitable for slab measurement not only in an anechoic chamber, but also in a room or on-site, provided that the measurement setup avoids multipath due to the surrounding environment or diffraction over the slab's edge. As the method is applicable to wideband measurements, it is required that the material's characteristics are constant over the measurement bandwidth. Moreover, the thickness of the slab should be enough for the Fabry-Pérot resonance to take place.

Over-the-Air testing

4

Wim Kotterman[a,b]**, Moray Rumney**[c]**, and Pekka Kyösti**[d]

[a]*Technische Universität Ilmenau, Ilmenau, Thuringia, Germany*
[c]*Rumney Telecom Limited, Edinburgh, United Kingdom*
[d]*Keysight Technologies, Oulu, Finland*

4.1 Introduction

Over-the-Air (OTA) testing has the objective to test the radio communication performance of Objects-under-Test, including the influence of antennas. Antennas have directional radiation patterns that shape transmitted fields and filter received fields. This spatial filtering by antennas through their patterns, influences both the transmission loss and the fading on the received signal, and with multiple elements, the correlation between received signals. The latter is of importance for MIMO operation, be it for multiplexing or diversity purposes. Essential is that incorporating antennas in the testing automatically requires emulating EM-fields that resemble realistic radio scenarios.

The last years have seen the finishing of the OTA testing of MIMO equipment (MIMO OTA) standardization and the onset of the discussion on extension of OTA technology for a multitude of other applications and for the upcoming mobile systems of the 5th generation. The projected extensions regarding applications, system concepts, and frequency bands more or less share a common trait; the OTA technology developed for MIMO OTA will only partly suffice. Cars and especially trains are physically too large for many anechoic rooms, indispensable facilities in the present methods, and are electrically too large for the current field emulation practice. The latter is true for massive MIMO (base station) antennas too. With respect to 5G, multiple problems occur: the millimeter-wave bands at which virtually any device becomes electrically large, and the beam-steering mode of operation for which yet neither a propagation model nor a model for beam dynamics is available. Proposals for new methods are put forward, be it rather dedicated to test situations, not as a general solution. That means, solutions are tailored to the type of Device under Test (DuT), the functionality to test, and the frequency band(s), as shown for the case of a small multi-antenna device in Fig. 4.1. The following sections will highlight some of these newly proposed methods.

---

[b] Chapter editor.

Inclusive Radio Communications for 5G and Beyond. https://doi.org/10.1016/B978-0-12-820581-5.00010-9
Copyright © 2021 Elsevier Ltd. All rights reserved.

FIGURE 4.1

Over-the-Air testing of a small multi-antenna device by 3D WFS (*courtesy: Fraunhofer IIS, Ilmenau*).

4.2 Field emulation for electrically large test objects

Electrical size denotes the size of an antenna array in wavelengths at the considered frequency band. OTA testing for Long Term Evolution (LTE) devices typically targeted field emulations for electrical size of a fraction of one wavelength, e.g., 0.7. Now, in 5G the electrical size of DuTs is increased both because of the shorter wavelengths at higher frequencies and the increased physical size of arrays. The latter has at least two reasons:

- at mmWave bands also BSs are to be OTA tested due to the lack of RF antenna connectors [KFPL16],
- at mmWave bands also UE will apply antenna arrays.

The field emulation refers to reconstruction of fading channel conditions that contain an antenna model of the other link end and statistically or exactly specified propagation conditions in the time, frequency, angle, and polarization domain. The target channel model can be either WSS [FZW19] or dynamic [KHK+18].

Different OTA test methods for large devices are analyzed in [FZW19]. The radiated multi-stage method is used in LTE with small devices. However, inverting the calibration matrix for the radiated part is found difficult for large MIMO configurations and even impossible for time variant antennas of analogue/hybrid beamformers.

Reverberation Chamber (RC) based systems are incapable of emulating angular and polarimetric controlled fields. These characteristics, on the other hand, are envisioned to be essential in testing of, e.g., massive MIMO BSs and other beamforming devices. The third alternative is the Multi-Probe Anechoic Chamber (MPAC). It has benefits, but field synthesis over a large test zone (in terms of wavelengths) might require prohibitive high numbers of probes [KHF+18] and fading emulator resources.

Angular spread of the target channel has an impact on the probe placement and the number of probes in the MPAC method. The angular dispersion of a radio channel depends both on the type of the observing device and the operating frequency band [FZW19]. Firstly, BS antennas are directive and arrays are typically designed for a sectored configuration, while UE antennas cannot afford high directivity as they must be capable of capturing signals from any direction. Secondly, angular profiles from a BS point of view are often substantially less dispersed as compared to UE profiles. This is a consequence of typical BS placement on higher elevation and farther from interacting objects than UEs that are on the street level or indoor with many objects in their vicinity. BSs perform vertical beamforming, thus making 3-D field emulation essential, which is different from the standardized Two-Dimensional (2-D) emulation for LTE UE. Thirdly, the propagation channel at mmWave bands is found to be more sparse and specular compared to lower frequencies [ABK+17]. At mmWave bands both BS and UE are expected to steer beams towards dominant propagation paths, in order to compensate the high propagation loss. As a result, the effective angular spread of the channel will be further limited.

4.2.1 **Sectored MPAC**

From the previous, we can summarize that:

- field emulation for electrically large DuTs requires sampling by a high number of probes,
- a 3-D probe configuration is needed, and
- the overall angular spread as observed by the considered devices is limited.

These three aspects make a sectored MPAC attractive, where only a confined azimuth and elevation sector is sampled by probes within an anechoic chamber [KHF+18]. An illustration of such a test setup is given in Fig. 4.2.

Relevant research and design questions on MPAC are [FZW19]:

1. what is the minimum acceptable range length,
2. what is the minimum number of probe and fading emulator resources,
3. how to design a cost-effective probe configurations for representative propagation scenarios?

For answering these questions and assessing the performance and accuracy of MPAC setups, numerous evaluation metrics have been specified in [KFPL16,FZW19, FZJ+16,KFK17,KHF+18]. In the range length evaluations, metrics like the fixed beam power loss, the direction of arrival estimation accuracy, and the multiuser-

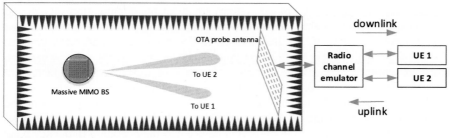

FIGURE 4.2

Sectored MPAC set-up: restricted solid angle for field emulation.

MIMO sum rate capacity error were proposed. In a field performance evaluation, the magnitude, angle, and phase deviation from the ideal far field condition were determined as well as the spatial correlation error, beam peak distance, and total variation distance of beam allocation distributions. The spatial correlation between simplified phased arrays is studied in [Kyo19]. This is an interesting metric for hybrid beamforming devices that may use several sub-arrays simultaneously for spatial multiplexing. Correlation functions for both ideal and MPAC OTA cases were derived and correlation errors with certain range length and other parameter ranges are shown.

The number of metrics can be minimized by choosing the fixed beam power loss as metric for range length evaluations and the total variation distance of the Power Angular Spectrum (PAS) for both range length and probe configurations. The first metric captures the amount of power lost due to curved wave fronts, assuming DuT beamformers tuned to planar waves. The latter metric contains the same angular profile information as the spatial correlation error, but with more detailed assumptions about the DuT arrays. These metrics have been used in 3GPP contributions as exemplified in [Tec19] for choosing minimum range lengths and number of probes with specific 3GPP channel models and assumptions.

It was found that the Fraunhofer distance need not be applied as the range length can be within the radiative near field of the DuT [KFK17]. Overall, the far field condition is a requirement contradicting the practical link budgets available for MPAC setups [KFPL16]. Basic channel gain and Error Vector Magnitude (EVM) measurements at various distances with 28 GHz BS antennas are reported in [MIHK19]. The purpose was to measure how an 8×8 phased array receives signals with decreasing range lengths. The measurement confirmed that on approaching the lower edge of radiative near-field distance, the channel gain starts fluctuating and ceases following free-space path loss. Initially, the EVM improves with decreasing range length, due to lower transmission loss, but saturates at a certain minimum distance and starts fluctuating with progressively decreasing range length.

The requirements for field synthesis must be relieved for electrically large devices [KHF⁺18]. The Nyquist sampling by probes may lead to tens and hundreds of probes, which is generally not feasible. At mmWave frequencies probes may be cheap, at least if manufactured on PCB. However, the other RF components and fading emulators can be costly and their numbers scale with the number of probes. As a consequence, an MPAC setup with a uniformly spaced probe configuration and having all probes equipped with dedicated RF components and fading emulator resources aiming for field emulation of any channel model, is not affordable. That is, not by directly scaling up the existing setups. Thus, some trade-offs must be made. One trade-off is abandoning omni-directionality and equip just a sector of space angles with probes [FZW19]. This offers savings not only in the number of Hardware (HW), but also in the floor space of anechoic chambers. Another is having many probes on PCB and using a switching unit between the probes and the fading emulator for selecting only a sub-set of probes for the field emulation [KHF+18]. Due to channel sparseness and directivity in mmWave bands, it might be sufficient to select only a few active OTA antennas. The selection would be based on the target Power Angular Spectrum (PAS) and would enable flexible and effective use of fading emulators. However, development of such switches poses practical design and cost challenges.

Validation of physical MPAC setups has been standardized for LTE technology in [RAN18], with procedures for power delay profile, Doppler power spectrum, cross polarization power ratio, and spatial correlation measurements and analysis. An important measure in 5G, especially at Frequency Range 2 (in 5G context: mmWaves 24.25–52.6 GHz), is the PAS. A new validation method of the joint delay-angular power profile is proposed in [FZW19]. A virtual uniform circular array is used to measure temporal snapshots of frequency responses, see Fig. 4.3. Then, delay and angle profiles are estimated by a sequential one-dimensional search method. The proposed method is a good candidate for 3GPP NR to substitute the spatial correlation validation at FR2.

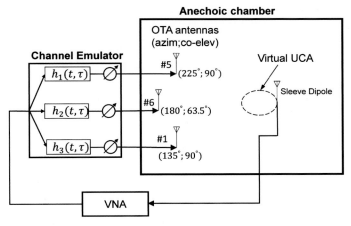

FIGURE 4.3

Schematic of test set-up with virtual circular array for measuring angular-temporal dispersion of emulated fields.

4.2.2 Other methods

Other methods of field emulation may become attractive with sub-6 GHz massive MIMO BSs. A method is described in [KH19b] that is not OTA, but has similar research questions and algorithms as MPAC systems. Virtual probes, their number and placement in angular-polarimetric domains are the optimization problems in a conductive phase matrix setup aiming to emulate BS fading channel conditions. There the excessive need of fading emulator resources, due to numerous BS antennas and UEs, is reduced by using a phase shifter and combiner unit in between the DuT and the fading emulator. The limited degrees of freedom are analogous to probes of physical MPAC setups. One conclusion is that a fully faded (conductive) emulation approach remains to be the most accurate method, but the reduced capability phase matrix method may provide lower cost or practical scalability in some use cases and with suitable channel models.

4.3 Emulation of mmWave channels

The OTA testing challenges presented by the migration from 4th generation to 5th generation radio access technology are considerable and were initially covered in [Rum16a] with further details in [Rum18c]. The primary reason driving the need for new test methods is the increased reliance on advanced antenna technology, in particular the use of large antenna arrays at both ends of the radio link. Large antenna array technology is being increasingly used by the base station in sub-6 GHz systems while at mmWave frequencies, arrays are required at both ends of the link, although the arrays in the UE are typically much smaller than in the BS. It is useful to review that basic SISO OTA performance metrics and test methods were introduced during the 3rd generation. These focused on transmitter and receiver isotropic efficiency. The move from cabled to radiated testing proved challenging, however the SISO OTA test ecosystem did stabilize after a few years with the publication of the CTIA Test Plan [CTI01] in 2001, followed in 2008 by 3GPP TS 25.144 [3GP08b] and 3GPP TS 34.144 [3GP08a]. The 4th generation (Evolved UMTS Terrestrial Radio Access (E-UTRA), also known as LTE) saw the introduction of MIMO as a baseline UE capability and this led to a new initiative in 2009 to develop MIMO OTA test methods. This work proved challenging and it took until 2016 for CTIA to publish a test plan [CTI16] and 3GPP to publish TS 37.544 [3GP17c]. Two test methods were approved by 3GPP: the reference MPAC method described earlier and the Radiated Two-Stage (RTS) method which is considered harmonized with the MPAC method. These methods cover the basic 2×2 MIMO in 3GPP Release 8 but did not address many of the features in Release 8 nor the new MIMO features introduced through 3GPP Release 14. As such, there are no UE performance requirements defined beyond basic Release 8 for MIMO, and in addition, 3GPP did not specify any LTE SISO requirements. COST Action IC1004 covered the period when most of the research on LTE MIMO OTA test methods took place, however, there was one paper early in COST IRACON that dealt with a still unresolved issue regarding the channel model validation of the Spatial Channel Model Extended (SCME) fields emulated by the MPAC and RTS test methods [RJS16]. This paper focused on the consequences of sub-sampled Laplacian distributions of the desired SCME field, and how these, particularly the high correlation Urban Macro (UMa), are vulnerable to the choice of starting phase for each subpath. Mathematical models, simulations, and measurements all show a residual error exists for sub-sampled fields as used in the MPAC method, which is a function of the number of sinusoids and the antenna pattern. Simple antenna patterns e.g. dipoles, show minimal error but known problematic antenna patterns from real phones show considerable error. Channel model validation procedures assume a dipole and are therefore insensitive to the problem. CTIA who use UMa for MIMO OTA are continuing to study unresolved differences between MPAC systems, and one of the explanations for these is this unresolved issue of subsampled Laplacian distributions and sensitivity to starting phase.

By the time of the 5th generation New Radio (NR) in 3GPP Release 15, the gap between capabilities enabled through advanced antenna technology and the lack of radiated performance requirements was large. Without radiated performance requirements beyond 2×2 MIMO there was no mandate to develop more advanced test methods. In 3GPP Release 15 the demands on antenna systems grew considerably with the introduction of massive MIMO and the extension to mmWave frequencies. Due to the loss of antenna connectors, the move to mmWave frequencies resulted in a Release 15 study item [Cor16] tasked with developing basic OTA connectivity. The study outcome is in 3GPP TR 38.810 [3GP18] that identified basic test methods for LOS radiated connectivity for the testing of RF, demodulation, and non-spatial Radio Resource Management (RRM) requirements. This solved the essential need for OTA connectivity to a mmWave device in the same way that SISO OTA test methods did for the 3rd generation, but provided no equivalent of the MIMO OTA capability of the 4th generation where the test signal has spatial properties. In the 4th generation the spatial properties were modeled as static 2D, but for the 5th generation (and in particular at mmWave frequencies) the spatial properties are 3D and highly dynamic. The need to address this higher level of spatial emulation resulted in a Release 16 study item [COS18]. This study item extends the static Line-of-Sight connectivity from Release 15 towards the more challenging and realistic operating conditions that are known to exist at both sub-6 GHz and mmWave frequencies as discussed in [FPK$^+$18].

The Release 16 study item covers both sub-6 GHz and mmWave frequencies. The former can be considered as the first extension in scope to the original MIMO OTA study item of 2009, and considered massive MIMO BS antennas, higher MIMO rank as well as the challenges of moving to the 6 GHz band. For the mmWave frequency band the main focus of the study is on demodulation and RRM testing in spatial fields, and it is here where the widest gap exists between 3GPP radiated requirements and test methods, and the actual propagation conditions that will be encountered during normal operation. This gap should be considered as a major risk to the successful operation of mobile devices at mmWave frequencies and was the motivation behind [Rum16a], [FPK$^+$18], [Rum16b], [KPB$^+$17], and [RKH17]. The spatial emulation challenges for 5G NR at sub-6 GHz and mmWave are quite different. For the low frequencies, the primary challenge is to expand the size of the MPAC test zone which is a function of the distance between probes. An eight-probe (cross-polarized) 2D system can support a test zone of 0.7λ which at 2.6 GHz is only 10 cm. Extending such a system to 7 GHz requires more closely spaced probes. The alternate RTS method has no test zone limit but is restricted to non-active antenna systems. The expectation is that the test signal will need to model a specific angular spread although this is expected to be static and narrower than for LTE. The 3GPP working assumption for 5G NR channel models is based on TR 38.901 [3GP17b]. There has been significant debate about the veracity and completeness of this study which is covered in the channel modeling sections of this book, however it is still necessary to turn the templates provided in TR 38.901 into specific channel models. This topic was discussed in [Rum18a] with further developments in [Rum19]. Despite the fact that the TR 38.901 [3GP17b] channel models are based on 23 clusters independent of frequency, it turns out that the factor most affecting the channel models being chosen to develop performance requirements is the base station antenna assumptions. These assume an 8×8 Uniform Rectangular Array (URA) with 12.9 degree half power beamwidth and 22.75 dBi directivity. This is significantly narrower than the assumptions used for 4G MIMO OTA in TR 38.977 based on a 2-element cross polarized array with 120 degree beamwidth. Using the 8×8 URA model to spatially filter the TR 38.901 23-cluster models results in significant channel hardening such that only one dominant direction of arrival occurs. This has a big impact on the test system since at least for the static geometry case, the test system no longer needs to emulate wide angles of arrival using widely spaced probes. The Release 16 study did address demodulation aspects for static spatial fields as defined in TS 38.901 [3GP17b], but did not have time to consider the dynamic aspect of spatial fields required to test spatial mobility (RRM).

The complexity of the many different combinations of test environments and test methods was discussed in [Rum18b]. By far the most challenging test environment to emulate is where the spatial properties of the channel are dynamic. The simpler static geometry cases can easily be emulated using a discrete number of fixed probes but the challenges in emulating dynamic environments as identified in [Rum18a] for the cases of UE mobility or cell acquisition and beam management introduce the possibility of rapidly changing Angle Of Arrival (AOA) (64 BS transmit directions in 5 ms during uplink beamsweeping for cell acquisition) which rule out any mechanical system of realigning probes. The obvious choice is to use some form of MPAC system such as the proposal in [KHF+18] using a sectored approach to limit the complexity and cost. However, a novel approach to emulating a 3D dynamic AOA test environment was first discussed in [RBM+18]. This OTA test method is based on the reflective properties of an ellipsoid chamber utilizing a probe antenna at one focal point and the DUT at the other focal point, Fig. 4.4.

[RBM+18] includes theoretical analysis of the properties of the ellipsoid. The method was further developed in [RBR+19] with the use of elliptical cylinders as a means of increasing the useful test zone volume. Measurements were made on a

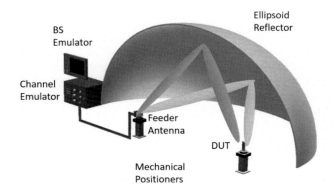

FIGURE 4.4

5G OTA test method setup with an ellipsoidal reflector.

prototype shown in Fig. 4.5. Based on a path length of 4 m, this prototype was able to create a usable test zone of $6 \times 20 \ \text{cm}^2$ where the signal level was within ± 1.5 dB. The method has been further developed to improve the test zone size, however, this work was not available to be included.

FIGURE 4.5

Schematic and prototype of the proposed mmWave OTA test method.

The issue of UE antenna design for use in mmWave spatially dynamic fields was discussed in [HHJ17]. This TD evaluated two types of 60 GHz UE antenna design, an eight-element uniform rectangular array (URA) and an eight-element Distributed Array (DA). The evaluation environment was an airport check-in hall. The study showed that although the URA provided the best performance in LOS conditions, the existence of body shadowing and NLOS conditions meant that the less directive DA provided around 3 dB better performance on average. The 3GPP assumption now expects two arrays per UE, so this may change the balance of performance towards the URA. In addition, 3GPP has now specified spherical coverage requirements in TS 38.101-2 [3GP19] that may be difficult to meet using a DA.

4.4 Extending the present framework

Given the increasing number of Over-the-Air tests to be performed in the near future, beside research into radically new solutions for 5G, the present technology is still being expanded and extended.

4.4.1 Complexity reduction for field emulations

On complexity reduction, either for electrically large test objects or a general lowering of costs, two alternatives to Wave Field Synthesis were offered. To be fair, no orders of magnitude of increase in sweet spot diameter or volume will be achievable. One is an ab initio approach, describing the test procedure as a multidimensional operation modeled by tensor equations [LKDG17] and subsequently reducing complexity, searching for sparse dimensions. Progressively reducing complexity along these lines would also mean that OTA testing shifts away from emulating radio environment to emulating the aggregate response of the radio environment and receiver functionality, the latter being non-linear. Another alternative is stochastic emulation that does not aim at the spatial correlation for very large test objects, but aims at providing correct temporal fading [KLDG16]. Still, the number of OTA antennas in the setup needed to keep the errors small is large. Possibly, the setup must be tuned to the radiation pattern of the vehicle, which would imply non-transparent testing. Also, for SCME type of channel models, some care is needed to ensure sufficient randomness.

4.4.2 Testing specific performance parameters

The stirred-mode reverberation chamber was ruled out for the standardization of MIMO OTA as the method is not able to sufficiently shape the spatial distribution of incoming waves [Car16]. However, a recognized capability is precisely determining total output power or total isotropic sensitivity, and therefore well-suited for measuring power emissions of 5G base stations like Total Radiated Power (TRP), Adjacent Channel Leakage power Ratio (ACLR), and spurious emissions [PAHG16], conform to 3GPP TS 37.104. The size of the reverberation chamber required for optimal testing is an issue because of the generally narrower beamwidths of base station antennas, when compared to a typical UE. This requires both a larger chamber and attention to be paid to optimally placing the chamber (sensing) antennas. In that case, measurement uncertainty is only minimally increased by the directional antennas of the BS. Testing massive antenna arrays for mmWave systems on deficient antenna elements is also an application of stirred-mode reverberation chambers [Gla19]. In a scenario of narrowband multi-user MIMO operation in a single cell, the achievable ergodic sum rate capacity of a downlink Massive MIMO system is evaluated. Zero Forcing (ZF) and Matched Filter (MF) precoders were used. A similar scenario was also evaluated in a compact range setup for creating Random LOS channels, as opposed to the channels of the reverberation chamber. The choice of these two channels is based on the conjecture that these two types form extremes and satisfactory operation in both channel types would guarantee good operation in any other channel [Kil13]. Based on simulations, different responses to array deficiencies were noted of the two precoding strategies, depending on the channel type, and deficient array performance could be detected.

An evaluation is proposed of the performance of vehicular antennas, while in operation in mobile networks, by three different metrics [DAB+19]:

- Coverage Indicator,
- Power Indicator, and
- Throughput Indicator.

The metrics are all based on recorded Reference Signal Received Power (RSRP) values, either from measurements or advanced network simulators (in this case SiMoNe). In the latter case, the antenna patterns from both Base Stations and vehicles must obviously be included in the simulations. Per route, the linear average of the received power is determined for area elements of specified size along the route, "Grid Squares", of 25 m^2. For the Coverage Indicator, the number of squares exceeding a set minimum value of average received power are counted. The Power Indicator is determined by summing the average powers of Grid Squares. A weighted sum of Signal to Noise, and Interference Ratio per Grid Square to calculate the Throughput Indicator. Given the highly dynamic and spatially inhomogeneous nature of interference, typically a number of assumptions have to be made. Rather than painstakingly measuring SAR in a phantom, a numerical method for computing SAR values can be applied using the DuT itself [HHPB17]. The method relates impedance changes at antenna ports to minute perturbations of a phantom by volume elements with slightly contrasting material constants (permittivity and/or permeability). The contrasting elements sense the local fields and the impedance changes can be converted into SAR distributions. Good accuracy is achieved, with the smaller the volume and material contrast of the elements, the smaller the intrinsic error but also the smaller the impedance change, potentially resulting in larger measurement errors. Note that this method requires access to antenna ports. On the other hand, when equipped with the right chip set, complex antenna port voltages could be measured inside the DuT and be made available to external processing, making application in practical testing feasible.

4.4.3 Emulating human influence

A serious challenge for emulating realistic radio environments is mimicking the human influence on radio transmission. One example is the proximity of antennas to human tissue that detunes antenna elements by its permittivity and conductivity. A recipe is developed for controlling the main electrical properties of phantom materials (liquid or as gel) over a huge bandwidth (up to over 20 GHz) [CGF+16]. The main ingredients of the phantoms are acetonitrile, salt, and ethanol. The phantom's impedance is determined by a patented calibration procedure with open-ended coaxial systems, using three reference substances: air, as "open circuit", short-circuit, and water [FLGPC+17]. Another known effect is shadowing and blocking by human bodies. By measurements and ray-tracing, body shadowing effects outdoors, in "Talk mode" and "Data mode", (in an industrial urban area) at 15 GHz and 28 GHz were determined [ZGL+17]. Antenna patterns of three different antenna designs were measured in an anechoic chamber in free-field and in combination with a user for both modes. These patterns, in combination with local angular power spectra of incoming fields, were then used to analyze the partly considerable losses incurred by body shadowing compared to free-field. Talk-mode was generally the worst with the largest variance, but it was noted that large personal variations in shadowing due to actual handling will occur, depending on positions of hands, proximity of the device to the body, etc. Another approach is based on extensive measurements of blocking by humans at mmWave bands performed in an anechoic chamber [VH19]. The aim is a simple but efficient 3-D blocking model. A bit surprisingly, knife-edge diffraction at (all edges of) a rectangular opaque screen suspended in air outperformed for instance cylinder models.

4.4.4 Testbeds, additional equipment

Over-the-Air system tests do not only have to emulate the radio environment but also have to provide a suitable representation of the corresponding radio system itself with network functionality and protocol stacks. Such emulators are typically not commercially available during standardization. With this in mind, a configurable wideband full-duplex reference radio system with mmWave capability was built [DGR+19]. One application is benchmarking User Equipment. Several salient extensions are planned. Less aiming at reference qualities than at scalability, low complexity, and small size is a test setup for mmWave systems, proposed in [ZMG19]. Two dampened, EM-tight boxes isolate the two communication partners on a single link and the Over-the-Air links are those between the simple antennas in the boxes and the respective DuT through a channel emulator that can only influence the temporal dispersion. As a result, the angular dependences of the fields in the boxes are invariant. For the channel emulator, paralleled Software Defined Radios (SDRs) are used to increase bandwidth, in a process that is called frequency-stitching [SST16]. Up and down convertors frequency-shift the mmWave communication signals into RF bands the SDRs can handle. More focused on PHY-layer tests is an extremely wideband arbitrary waveform generator, generating a 60 GHz wide signal [CBS17]. The bandwidth is achieved by multiplexing three 20 GHz wide sub-bands. Each of the sub-bands is generated by opto-electronic means, i.e. detecting the beat-note of two laser signals, detuned by 20 GHz, one of which is modulated by Mach-Zehnder modulators. Multiple sources of deviation from frequency linearity exist, for which suitable correction methods are sought.

4.5 Concluding remarks

The primary OTA challenge identified during the Action was the need to address the continued sophistication of the exploitation of the spatial domain through massive MIMO and the move to FR2. In particular, the problem domain from Fourth Generation (4G) is compounded by spatially dynamic operating environments resulting from mobility in the context of active antenna systems. The scientific progress in this field during the Action was somewhat sobering, for a number of reasons. Due to the complexity of this subject and the pressure on 3GPP to complete the 5G specifications, the 3GPP Release 15, the first full set of standards for 5G New Radio, did not develop specifications to address aspects related to spatial dynamics. The complexity of the system combined with the lack of knowledge of the spatial dynamics of the actual propagation posed such problems that 3GPP was left no other choice than to postpone the handling of mobility to future releases. Effectively, standardization is dealing with the very basics of connectivity and has not had time to focus on the spatial dynamic consequences of how the system will behave or be specified and validated other than in LOS scenarios.

The core of the lack of knowledge of spatial dynamics of the radio propagation channel is the lack of supporting data from extensive measurement campaigns. Earlier measurement campaigns for Second Generation (2G), Third Generation (3G), and 4G system development were performed in the context of large projects of the European Framework Programs 5, 6, and 7 that supplied resources for these campaigns. During these programs, the frequency band of interest did not change significantly, merely the measurement bandwidth was gradually increased as was the complexity of the scenarios. Very few of such projects were granted under Horizon 2020 due to the changed scope of this program. In the meantime, completely new measurement equipment is required, that is, test systems of completely new designs in very expensive technology, with many RF problems to solve, for instance coherence, phase noise, receiver noise, and output power. It is difficult to commit much money and implementation effort to a test system that is not guaranteed to be suited to future tasks, given the uncertainty in standardization of potential systems-under-test, apart from being able to gather the funds. The latter will remain a general problem, in view of the costs of hi-tech equipment, in case research institutes can only dispose of local funds instead of supra-national funding like the past European Framework Programs.

Coding and processing for advanced wireless networks

Jan Sykora[a,b]**, Andreas Czylwik**[c]**, Laurent Clavier**[d]**, and Alister Burr**[e]

[a]*Czech Technical University in Prague, Prague, Czech Republic*
[c]*University Duisburg-Essen, Duisburg and Essen, Germany*
[d]*Université de Lille, Lille, France*
[e]*The University of York, York, United Kingdom*

Introduction

Coding and processing for advanced wireless networks presents a wide area of PHY related aspects of radio communication systems. This includes theoretical design and analysis of coding and signal processing schemes, their practical implementation and simulation performance evaluation, and also their practical connection the channel modeling and network layer. The scope is very wide, so the actual research contributions form a large number of building blocks generally applicable in many diverse scenarios. Despite the wide range of topic, there are clearly emerging two paradigms present in the majority of works — massive MIMO and cooperative algorithms (signal processing, coding, decoding).

5.1 Advanced waveforms, coding, and signal processing
5.1.1 Models and bounds

The complexity and diversity of environments encountered in 5G and beyond make fundamental limits in wireless communications an important topic. Some measurements campaign allows to address these limits in specific cases like mmWave communications or using relays. Interference is also important and is significantly impacted by the densification of networks and the heterogeneity of systems. Secrecy of communications, viewed from the information theory point of view; is also discussed in this section.

---

[b] Chapter editor.

Inclusive Radio Communications for 5G and Beyond. https://doi.org/10.1016/B978-0-12-820581-5.00011-0
Copyright © 2021 Elsevier Ltd. All rights reserved.

Capacity from measurements

The channel capacity and the Spatial Degrees of Freedom (SDoF) in a street canyon are analyzed in [HNK16]. Measurements have been carried out at carrier frequencies of 15, 28, and 61 GHz channels have been performed. The analysis showed that the inherent multipath richness of the channel (given by SDoF) decreases as frequency increases. However, line-of-sight (LOS) channels at the different frequencies show an almost comparable capacity if the physical aperture of the receiving antenna and the transmit-receive antenna distance are fixed.

The channel capacity of a MIMO indoor channel at a frequency range from 55 GHz to 65 GHz is analyzed in [BMG+17]. The dependence of mutual information on antenna separation is investigated. For spatial modulation applications the antenna separation of a uniform linear array (ULA) should be $\zeta = \sqrt{D\lambda/N}$ where D denotes the distance between the antenna arrays, λ is the wavelength, and N is the number of antenna elements [LDS]. The mutual information in a measurement scenario with LOS and a low amount of multipath components is shown in Fig. 5.1 as a function of the SNR for three different antenna separations $\zeta/2$, $\zeta/\sqrt{2}$, and ζ and a 4x4 MIMO system. The result is compared with the mutual information of four completely orthogonal LOS channels. It is found that the mutual information in the measured real scenario for antenna separation ζ is only by 8.9% smaller than for the ideal orthogonal LOS channels.

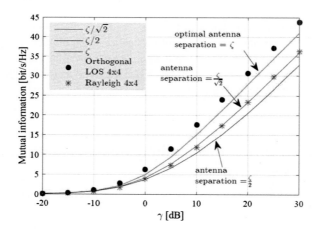

FIGURE 5.1

Mutual information for a 4x4 MIMO LOS channel (ULA) for different antenna spacings.

Furthermore, it is found that the higher the bandwidth is, the lower is the mutual information due to orthogonality disturbance caused by increasing suboptimality of antenna element separation. The mutual information can be estimated from the mean singular value of the LOS component with good precision.

Capacity using a relay

Relay stations (RSs) may help to increase the channel capacity of individual links between a base station (BS) and a mobile station (MS) in a cellular network. In [CMC16] the capacity of the direct link between BS and MS is compared with a relay-assisted transmission. Whether or not an improvement can be achieved mainly depends on the used relaying protocol. In a downlink scenario it is shown that direct transmission without relay outperforms relaying when separate time slots for BS transmission and relay transmission are needed. Even an optimization of the transmit periods according the capacity does not help. Only in a full duplex system when the relay is able to transmit and receive simultaneously significant gain can be achieved, especially for the case when the MS is close to the RS. The analysis is based on simple path loss models as well as on measurements in an urban area. The capacity gain provided by the relay reaches more than 30% depending on the antenna relay height and location. The analysis assumed omnidirectional antenna at RS and was performed at 2.2 GHz.

Interference

For a fixed active set of transmitting devices, the interference is well-modeled as Gaussian. This is the basis of the standard interference modeling methodology in wireless cellular networking. For M2M networks, the set of active transmitter can be rapidly changing and the interference model has to be modified [ECdF$^+$17].

    We consider a worst-case scenario where the network is interference-limited, uncoordinated, and the locations of interferers are governed by a homogeneous Poisson point process. This setup is relevant for networks supporting the internet of things and in large-scale sensor networks, where transmitting devices are very simple and have limited ability to coordinate. We also assume that the active set of transmitters varies symbol-by-symbol, which contrasts with the Gaussian model where the active transmitter set is fixed.

    For a reference receiver at the origin, the interference at time t from the other devices is given by

$$I_t = \sum_{k \in \Phi} r_{k,t}^{-\eta/2} h_{k,t} x_{k,t}, \tag{5.1}$$

where η is the path loss exponent of the interfering links. The circularly symmetric complex normal distributed random variable $h_{k,t}$ is the Rayleigh fading coefficient for the link from device k to the reference receiver. The baseband emission of each interferer k is denoted by $x_{k,t}$. We assume that the real and imaginary parts of $h_{k,t} x_{k,t}$ are symmetric and $e^{j\phi} h_{k,t} x_{k,t} \overset{(d)}{=} h_{k,t} x_{k,t}$ for all $\phi \in [0, 2\pi)$, which means that $h_{k,t} x_{k,t}$ is isotropic. This is not a strong assumption and is satisfied, for instance, in the case of Rayleigh fading with circularly symmetric complex Gaussian baseband emissions. The distance from the device to the reference receiver is denoted by r_d. Φ is the set of active transmitters apart from the useful one.

It can be shown in this context that the interference is an isotropic α-stable random variable, with $\alpha = \frac{4}{\eta}$ and parameter [ECdF$^+$17]

$$\sigma_{\mathbf{N}} = \left(\pi \lambda C_{\frac{4}{\eta}}^{-1} \mathbb{E}[|\mathrm{Re}(h_k x_k)|^{\frac{4}{\eta}}] \right)^{\frac{\eta}{4}}, \tag{5.2}$$

where $\Gamma(\cdot)$ is the Gamma function and

$$C_\alpha = \begin{cases} \frac{1-\alpha}{\Gamma(2-\alpha)\cos(\pi\alpha/2)}, & \text{if } \alpha \neq 1 \\ 2/\pi, & \text{if } \alpha = 1. \end{cases} \tag{5.3}$$

The proof is based on the Lepage series representation of α-stable random variables. To characterize the capacity of the channel with an additive isotropic α-stable noise is not a trivial question. To obtain a lower bound we modify the constraints and consider the capacity optimization problem given by

$$\underset{\mu \in \mathcal{P}}{\text{maximize}} \; I(X; y)$$

$$\text{subject to} \qquad \mathbb{E}[|\mathrm{Re}(X)|^r] \leq c, \tag{5.4}$$
$$\mathbb{E}[|\mathrm{Im}(X)|^r] \leq c,$$

where $\mathcal{P}$ is the set of probability measures on $\mathbb{C}$ and $0 < r < \alpha$. The unique (see [EdFC$^+$16,dFEC$^+$17]) solution to (5.4) can be lower bounded: for fixed r_d and h_d, the capacity of the additive isotropic $\frac{4}{\eta}$-stable noise channel subject to the fractional moment constraints in (5.4) is lower bounded by:

$$C_L = \frac{\eta}{4} \log \left(1 + \frac{\left(\sqrt{2|r_d^{-\frac{\eta}{2}} h_d|^2} \left(\frac{c}{C(r,\frac{4}{\eta})} \right)^{\frac{1}{r}} \right)^{\frac{4}{\eta}}}{\sigma_{\mathbf{N}}^{\frac{4}{\eta}}} \right), \tag{5.5}$$

where $\Gamma(\cdot)$ is the Gamma function and

$$C\left(r, \frac{4}{\eta}\right) = \frac{2^{r+1}\Gamma\left(\frac{r+1}{2}\right)\Gamma(-\eta r/4)}{\frac{4}{\eta}\sqrt{\pi}\Gamma(-r/2)}. \tag{5.6}$$

The next step is to analyze the interference structure in between the two extreme cases: static set of interferer or independent symbol by symbol. Dealing with dependence structure and impulsive noises, especially models with infinite variance, is not however a trivial question. It is proposed in [ECZ$^+$18,ZEC$^+$19] to rely on Copulas to take this dependence into account. Consider that a packet is transmitted over several resource blocks that can be defined in time or frequency. In each block an interferer will transmit with probability p. The higher p, the more probable a transmitter will use several times the same resource blocks as the desired user, creating dependency in the interference vector. If $p = 1$, we will have the fully dependent case (the set of interferer is the same on each resource block) which results in a sub-Gaussian random vector for interference. This means that the Gaussian assumption on each packet is adapted. On the other hand, when p tends towards 0, the set of interferers will with high probability be different on each resource block leading to the α-stable independent case studied previously.

To adapt to others value of p, we propose to use a t-copula model. Let us assume we consider K resource blocks and consider the random vector $\mathbf{I}_{2K}$, the $2K$ dimensional interference vector formed by the real and imaginary parts of the interference on each of the K resource blocks. Let $F_{\mathbf{I}_{2K}}$ be its joint distribution. Each element of $\mathbf{I}_{2k}$ is an α-stable random variable and follows the continuous Cumulative Distribution Function (CDF) F. In that case, there exists a unique copula $C : [0, 1]^{2K} \rightarrow [0, 1]$, such that the joint distribution of $\mathbf{I}_{2k}$ can be written as

$$F_{\mathbf{I}_{2K}}(z_{1,r}, z_{1,i}, \ldots, z_{K,r}, z_{K,i}) = C(F(z_{1,r}), F(z_{1,i}) \ldots, F(z_{K,r}), F(z_{K,i})) \quad (5.7)$$

by Sklar's theorem [Nel99]. This divides the density into two components, the copula component and the independent component, and provides a basis for optimizing receiver structures and other system components, for instance if we can select a copula C that allows an analytical tractability. The t-copula offers such a tractability but also a good flexibility in the dependence structures it can model, especially for the tail dependencies. Fig. 5.2 shows the Kullback-Leibler (KL) divergence between a simulated set of interference and different models: the sub-Gaussian (fully dependent) approach, the independent on each subband case and the t-copula with α-stable marginals. As expected, the copula approach offers a good adaptability to a wider range of p and is then appropriate to model the dependent interference situation.

Secrecy

Security issues are also a crucial question in wireless communication and a trade-of between transmission capacity and secrecy has to be made. This is studied through the Wire Tap Channel (WTC) [Wyn75]. On another hand, the State-Dependent Channels (SDC) [Sha58] allows to study the impact of imperfect channel knowledge on the link capacity. In [KPVC16], it is shown that given an SDC, there always exists a WTC (called dual) whose secrecy capacity achieving scheme is identical to the capacity achieving scheme of the SDC.

This results shows that compensating the lack of Channel State Information (CSI) at the receiver has an equivalent cost as combating an eavesdropper in terms of channel capacity, meaning that efficiently confusing the eavesdropper in the WTC requires the same amount of redundancy as ensuring reliability over the SDC.

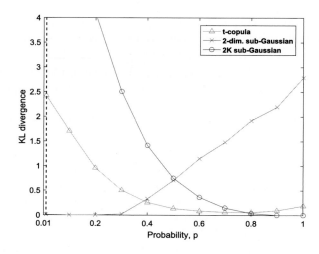

FIGURE 5.2

KL divergence between simulated interference and proposed models ($K = 8$).

5.1.2 Pre-coding and beam forming

Focusing the signal power towards only the target terminal allows to improve the link quality but also to reduce interference impact. This raises a complexity/accuracy trade-off. In [AHM19b] a beam selection is proposed based on a hybrid beamforming setup comprising an Analog Beamforming (ABF) network followed by a ZF baseband processing block. Machine learning and deep learning schemes are used to select the optimal ABF. The inputs of these schemes are given by the AOA and received powers estimated using the Capon or MUltiple SIgnal Classification (MUSIC) methods. This last method is significantly more accurate, which impacts the quality of the classification. This work compares different selection schemes: Support Vector Classifiers (SVC), k-Nearest Neighbors (kNN), and Multi-Layer Perceptron (MLP). Two references are also given, a random selection of the beam and the optimal choice. Performance is given as the accuracy of the method in comparison to the optimal scheme. This can however not directly be translated into a loss in sum-rate capacity because classification errors often fall into adjacent regions, associated to beamformers with similar spatial response due to the codebook construction. Fig. 5.3 shows tat the deep learning (MLP) approach is significantly better than the machine learning ones when the training set gets larger than 3000 samples, at the price of complexity.

Besides, the 80% accuracy in fact results in a very limited sum-rate capacity loss.

5.1.3 Channel estimation and synchronization

Carrier frequency offset between transmitter and receiver is an important source for degradation in an OFDM based transmission system. Because of the dense spacing between subcarriers in OFDM, even low frequency offsets lead to a significant degra-

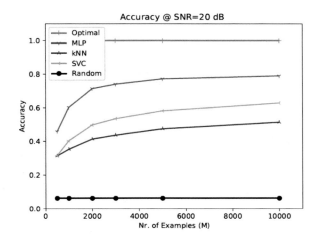

FIGURE 5.3

Classification accuracy vs. number of examples in the training set, 16 receive antennas, 6 RF chains, 4 active users. A set of 28 different configurations for the ABF is possible MUSIC estimator.

dation which depends on the carrier frequency offset normalized to the frequency spacing between subcarriers. Due to the frequency offset, the received symbols are multiplied with a subcarrier-independent complex gain factor U. The frequency offset furthermore creates Inter Carrier Interference (ICI) between subcarriers at the receiver. In [OV18] this ICI is reviewed and it is proposed to reduce the ICI by a precoding method. In order to carry out the precoding at the transmitter, it is assumed that the frequency offset is known at the transmitter.

The precoding method consists of two steps: First, the complex transmit symbols of a Quadrature Amplitude Modulation (QAM) alphabet are multiplied with the inverse gain factor $1/U$ in order to adjust the amplitude and phase of the received signal. Second, ICI is compensated by simply subtracting it. The second step can be formulated by multiplication of the transmit symbol vector by a precoding matrix. Numerical analysis shows that the direct neighbor subcarriers create the largest contribution of ICI. Therefore, it is sufficient to take for the precoding matrix only neighbor subcarriers into account. For a normalized carrier frequency offset of 10%, precoding that considers only six adjacent subcarriers gives already a SINR gain of 3 dB.

The advantage of carrying out the frequency synchronization at the transmitter by precoding is that computational load is moved from the receiver to the transmitter.

5.1.4 New waveforms

Conventional OFDM partitions the communication channel in orthogonal bins in the time and frequency domain, defining a two-dimensional grid. OFDM transmits each data symbol on a single grid location in the Time-Frequency (TF) plane. Orthogonal Time and Frequency Signaling (OTFS) is a transmission method based on OFDM which promises a full utilization of time and frequency diversity for a time-variant and frequency-selective channel. OTFS uses a 2D Discrete Symplectic Fourier Transform (DSFT) to precode the data symbols at the transmitter side [HRT+17]. After this precoding operation the data is transmitted with conventional OFDM modulation. Hence, the information of each data symbol is linearly spread on all available grid points of a data frame in the TF domain. By appropriate equalization and decoding, the full time- and frequency-diversity of the wireless communication channel can be utilized on the receiver side.

OTFS encodes data symbols using orthogonal 2D basis functions. Their orthogonality is destroyed at the receiver side, due to the effect of the time and frequency selective channel as well as due to a possible windowing at the transmitter and receiver side. This lost orthogonality leads to Inter-Symbol Interference (ISI) of the data symbols contained in a frame. Hence, an appropriate equalization technique to remove it is required. A low-complexity equalization for OTFS is proposed in [ZHL17]. It is an iterative maximum likelihood approach which cancels out inter-symbol interference recursively. In the first step Minimum Mean Squared Error (MMSE) equalization is carried out. The following iterations show a significant improvement with respect to MMSE equalization.

5.2 Distributed and cooperative PHY processing in wireless networks

Traditionally the physical layer of wireless communications, as for any other communication system, was supposed to apply to signaling on a single link between one transmitter and one receiver. Any signals received by other receivers had to be treated as interference, which was minimized by using orthogonal signals for different receiving nodes. However as wireless networks become denser it is increasingly difficult to separate these signals, and such orthogonal multiplexing becomes increasingly inefficient. Hence the physical layer is of necessity becoming cooperative, operating over the whole network rather than over discrete links within the network. This section discusses cooperative techniques within wireless networks, including relaying, wireless physical layer network coding (WPNC), distributed cooperative access networks, and distributed sensing.

5.2.1 Cooperative relaying

The relay-assisted Hybrid ARQ is regarded as a mean of further improving the coverage and reliability performance of future wireless networks. Most of the algorithms implementing this approach use the source-destination link, besides the source-relay and relay-destination links. They employ either Chase combining or space-time coding. In [AB17b], we study the reliability, spectral efficiency and delay performance of a Repetition Redundancy Hybrid ARQ algorithm that does not use direct source-destination link. The analysis is made over block-faded Rayleigh channels using decode and forward relaying and practical modulation and channel code. The results show that the algorithm provides smaller block-error rate and ensures the target performance goal and smaller delay at low SNR ratios. The overall Block Error Rate (BLER) is much smaller than in the non-cooperative case. These results are further extended in [AB17a] by exploiting some preliminary considerations regarding the adaptive use of the three algorithms under maximum delay and BLER constraints, so that a maximum possible spectral efficiency would be obtained.

In [BV18a], we aim at establishing the maximum spectral efficiency and the corresponding average delay provided by the relay-assisted cooperative algorithm (CHARQ), which operates under imposed target Block-error Rate (BLER) and maximum delay. The analysis is performed using adaptively a set of QAM constellations and convolutional coding rates, used in a generic OFDMA transmission scheme, over Rayleigh block-faded channels. To this end it presents an algorithm (Minq-Max Beta) that configures the CHARQ by selecting the modulation and coding rate (MC) pair which would require the smallest number of retransmissions and provide the greatest spectral efficiency, while keeping the probability of a lost block under an imposed target value BLER and observing a maximum number of retransmissions allowed. The algorithm selects the MC pair adaptively, out of a predefined set, according to the SNRs of the three links involved, i.e. Source-Destination (SD), Source-Relay (SR), and Relay-Destination (RD) that are affected by path loss, Rayleigh fading and Gaussian noise. The performances of CHARQ using the proposed configuration algorithm are compared to the performances of an HARQ algorithm, operating only on the S-D link that uses the same configuration algorithm. The results obtained show that, when configured with Minq-Max Beta, the CHARQ algorithm ensures BLER greater at smaller SNR values than the non-cooperative HARQ, and requires a smaller average number of rounds at low SNR values. The SNR gain of CHARQ over HARQ increases with the decrease of the target BLER. These improvements come at the expense of smaller spectral efficiency.

[JJHK18] presents a novel approach to cooperative Non-Orthogonal Multiple Access (NOMA) using distributed space time block coding (STBC) known as STBC-NOMA. In conventional NOMA, the strong users detect the messages of weak users through successive interference cancellation (SIC). In cooperative NOMA, these copies are then forwarded by strong users to weak users at the expense of extra time slots. However, the proposed scheme exploits this feature of cooperation using STBCs to enable cooperation among the users. The STBC-NOMA renders less complexity as lesser number of SICs are performed at each user. To this end, we derive the outage probability of the STBC-NOMA scheme which involves finding the distribution of signal-to-interference ratio at the receiving terminals. The numerical results show that STBC-NOMA outperforms conventional NOMA and conventional cooperative NOMA in terms of outage probability and average sum rate.

5.2.2 Wireless physical-layer network coding
Network code maps and error rate performance

In [Bur16], we discuss the requirements of a general (not necessarily linear) Wireless Physical layer Network Coding (WPNC) mapping function for application in a network containing multiple sources and multiple relays. We consider the definition of linearity in such a mapping, then the requirements for unambiguous decoding of the output of the network, and finally the effect of singular fading in the links to a given relay. We show how a mapping scheme can be designed for a specific topology and a specific set of channels. We note also however that such schemes are not likely to be readily compatible with a layered network coded modulation scheme, in which error control coding is separated from the network coding function.

The impacts of a specific bit-wise form of the network coding map in WPNC are addressed in [Syk17]. We analyze achievable hierarchical rates in one stage hierarchical MAC channel for higher order component constellations with bitwise Hierarchical Network Code (HNC) maps. This is motivated by a possible application of state-of-the-art binary (e.g. LDPC) codes over higher order constellations. We show that the bitmapped binary codes do not have the same achievable rates as direct higher-order codes. On top of this, the individual bits in the HNC map might provide very uneven performance and it strongly depends on the combination of component alphabets.

The implication of using bit-wise network code maps is further exploited in [Syk18b]. We focus on using binary component codes in bit-wise mapped higher-order hierarchical constellations. This creates a complicated nonlinear block-like inter-bit structure binding and the bit-wise observation model cannot be any longer described by single-letter entities. The nonlinear inter-bit structural binding strongly depends on the source constellation bit mapping, actual bit-wise a priori probability mass functions, and also on the actual relative channel parametrization. We derive bit-wise soft-aided Hierarchical Soft-Output Demodulator (H-SODEM) and analyze the bit error rate properties of the dual loop soft-aided bit-wise hierarchical decoding in 2-source H-MAC HDF scenario. This scheme is shown to significantly improve the performance w.r.t. classical uniform a priori case.

System-level simulation has been widely used recently to evaluate the performance of wireless networks. In [CB16] we consider the simulation of a dense, multihop wireless network which applies WPNC. The system-level simulator calculates the packet error probability (PER) at the relay against the signal and noise power. The above PER could be deduced from the relay's symbol error rate (SER). Here we analyze the SER of QPSK modulated Physical Layer Network Coding for the case where two source signals are received at one relay, since this forms a building block of many typical networks.

In [BC17], we consider a hierarchical wireless network in which two convolutionally coded BPSK sources transmit to a single relay which performs hierarchical decode and forward (HDF), decoding the exclusive OR (XOR) combination of the source data. The sources use the same convolutional code, so that the XOR combination of the coded data is a code sequence of this code, which decodes to the

XOR combination of the uncoded data. We derive a tight upper bound on the BER of the decoded hierarchical data, as a function of the channel fade parameters for the source-relay channels.

Channel and network topology exploitation

The channel and network topology estimation is an important aspect of all cooperative and WPNC based systems. Work [Syk18a] focuses on the channel state estimation (CSE) problem in parametrized Hierarchical MAC (H-MAC) stage in WPNC networks with HDF relay strategy. We derive a non-pilot based H-MAC channel phase estimator for 2 BPSK alphabet sources. The CSE is aided only by the knowledge of H-data decisions. At HDF relay, there is no information on individual source symbols available. The estimator is obtained by a marginalization over the hierarchical dispersion. The estimator uses a gradient additive update solver and the indicator function (gradient) is derived in exact closed form and in approximations for low and high SNR. We analyze the properties of the equivalent solver model, particularly the equivalent gradient detector characteristics and its main stable domain properties under a variety of channel parameterization scenarios.

WPNC network nodes determine their front-end, back-end, and node processing operation depending on the knowledge of the global Hierarchical Network Transfer Function (H-NTF). The H-NTF [Syk16] can be derived from the network connectivity state and it is essential in determining how to perform hierarchical decoding, what hierarchical network coding function to use, and how to encode the Network Coded Modulation at each node. The data payload in WPNC network uses a common signal space of mutually interacting signals that superpose on the receive nodes. There are no orthogonalized communication links that would allow traditional control-data signaling based approaches to determine the network state. Also, the procedure where each node establishes a complete picture of the network state must be performed before the payload phase and with superposing nonorthogonal pilot signals that fully respect the WPNC paradigm. This paper proposes a distributed consensus-based estimator of the network state which operates on a common signal space shared by all nodes in the network. The algorithm allows each node to find the global network state based only on local neighbor superposed pilot observation.

The impact of the side-information for a broadcast coding is analyzed in [Ska17]. In a broadcast channel with one transmitter and several receivers, where receivers possess some partial data, the information rate can be increased by using coding techniques. This problem is known in the literature as index coding with side information. If some data packets are lost in the course of transmission, then properly designed coding scheme could be useful in increasing reliability of the system and decreasing the latency. In this work, the efficiency of coding in such systems is studied.

5.2.3 Distributed cooperative access networks

Distributed cooperative processing can substantially improve the performance in Massive-MIMO and Cloud-RAN (C-RAN) networks. Work [BBM18] focuses on cooperative radio access network architectures, especially distributed Massive MIMO and Cloud RAN, considering their similarities and differences. We address in particular the major challenge posed to both by the implementation of a high capacity fronthaul network to link the distributed access points to the central processing unit, and consider the effect on uplink performance of quantization of received signals in order to limit fronthaul load. We use the Bussgang decomposition along with a new approach to MMSE estimation of both channel and data to provide the basis of our analysis.

The combination of compute and forward strategy in Massive MIMO scenario is investigated in [HB17]. We consider the uplink of cell-free massive MIMO systems, where a large number of distributed single antenna access points (APs) serve a much smaller number of users simultaneously via limited backhaul. For the first time, we investigate the performance of compute-and-forward (CF) in such an ultra dense network with a realistic channel model (including fading, path loss and shadowing). By utilizing the characteristic of path loss, a low complexity coefficient selection algorithm for CF is proposed. We also give a greedy AP selection method for message recovery. Additionally, we compare the performance of CF to some other promising linear strategies for distributed massive MIMO, such as small cells and maximum ratio combining. Numerical results reveal that C&F not only reduces the backhaul load, but also significantly increases the system throughput for the symmetric scenario.

Cloud radio access networks (C-RAN) are a promising technology to enable the ambitious vision of the fifth-generation (5G) communication networks. In spite of the potential benefits of C-RAN, the operational costs are still a challenging issue, mainly due to the centralized processing scheme and the large number of operating remote radio head (RRH) connecting to the cloud. In [AS17] we consider a setup in which a C-RAN is powered partially with a set of renewable energy sources (RESs), our aim is to minimize the processing/backhauling costs at the cloud center as well as the transmission power at the RRHs, while satisfying some user quality of service (QoS). This problem is first formulated as a mixed integer non linear program (MINLP) with a large number of optimization variables. The underlying NLP is non-convex, though we address this issue through reformulating the problem using the mean squared error (MSE)-rate relation. To account to the large-scale of the problem, we introduce slack variables to decompose the reformulated (MINLP) and enable the application of a distributed optimization framework by using the alternating direction method of multipliers (ADMM) algorithm.

5.2.4 **Distributed sensing**

Distributed sensing is quite specific form distributed algorithms. One of the most fundamental tasks in sensor networks is the computation of a (compressible) aggregation function of the input measurements. What rate of computation can be maintained, by properly choosing the aggregation tree, the TDMA schedule of the tree edges, and the transmission powers? This can be viewed as the convergecast capacity of a wireless network. We show [HT17] that the optimal rate is effectively a constant.

This holds even in arbitrary networks, under the physical model of interference. This compares with previous bounds that are logarithmic. Namely, we show that a rate of $\Omega(1/\log^* \Delta)$ is possible, where Δ is the length diversity (ratio between the furthest to the shortest distance between nodes). This is achieved using the natural minimum spanning tree. Surprisingly, this barely non-constant bound may be best possible, since we can show that there are MSTs that cannot be scheduled in fewer than $\Omega(1/\log^* \Delta)$ slots. Our method crucially depends on choosing the appropriate power assignment for the instance at hand, since without power control, only a trivial linear rate can be guaranteed. We also show that there is a fixed power assignment that allows for a rate of $\Omega(1/\log\log \Delta)$.

[CKB18] presents an analysis of correlation based clustering procedure in cooperative spectrum sensing. The motivation is to assess if it is more beneficial in terms of energy efficiency to group nodes according to the received SNR on the link between Primary User and node or according to the distance between nodes. To this end, a merged clustering measure is introduced. Then a number of network topologies are compared via computer simulations. Moreover, the temporal analysis to the correlation-based procedure is introduced.

An underdetermined multi-measurement vector linear regression problem is considered in [GBM19] where the parameter matrix is row-sparse and where an additional constraint fixes the number of nonzero elements in the active rows. Even if this additional constraint offers side structure information that could be exploited to improve the estimation accuracy, it is highly nonconvex and must be dealt with caution. A detection algorithm is proposed that capitalizes on compressed sensing results and on the generalized distributive law (message passing on factor graphs).

5.3 Massive MIMO

The Massive MIMO, where many transmit antennas are used relative to the number of users, thus providing a greater opportunity to use the spatial characteristics of the channel for spatial diversity and multiplexing, is clearly one of the core technologies for 5G networks. The Massive MIMO based practical systems need to address a number of the issues going beyond the classical MIMO. From the perspective of this section, we divide them into two large groups. The first is obviously related to the signal processing and coding for Massive MIMO. The second large area of interest is related to advanced channel modeling and associated actual performance evaluation.

5.3.1 Processing and coding for Massive MIMO

[HHDB16] investigates a novel uplink power control scheme for Massive MIMO. The proposed algorithm achieves the highest signal to interference-plus-noise ratio (SINR) for each user by considering the mutual interference arising from other users. It also minimizes the required uplink transmit power to achieve the highest possible uplink SINR per user without increasing the complexity at the user equipment. The base station controls the user transmit power using only two bits in a Transmission Power Control command and the newly introduced algorithm is expected to increase both the energy efficiency and spectral efficiency whilst simultaneously decreasing the transmission overhead, latency and complexity of the receiver.

In [HHDB17], a novel user grouping scheme is developed for massive MIMO in a single cell. The base station divides the users into multiple groups that are served in different time slots. The proposed algorithm increases the number of users and the transmission reliability in each group by considering the number of users and the mutual spatial correlation arising from all users. It provides the required signal to SINR for the desired modulation scheme based on a simple assigning process. The proposed algorithm can enhance the performance of several popular linear decoders and precoders such as Minimum Mean Square Error, Zero-Forcing and Matched Filtering. The effect of the inaccurate channel estimation is also illustrated. Potential solutions are proposed which reduces the effect of the inaccurate channel estimation.

In [KO17], a low computation complexity based direction of arrival estimation with unknown mutual coupling for massive MIMO in 5G is proposed. Our work extends the use of joint iterative optimization algorithm proposed by earlier studies with the unique difference being that it assumes that the mutual coupling is also present and requires to be calibrated. It is shown that the computation complexity of the proposed method is $\mathcal{O}(M^2 + (\frac{180}{\Delta})r^2 + r^3)$ or $\mathcal{O}(M^2)$ as compared to subspace based methods like MUSIC, which has computation complexity of the order of $\mathcal{O}(M^3)$, where M is number of antennas and r is steering matrix reduced rank. This gives a big advantage of using joint iterative methods with rank reduction optimization in massive MIMO rather than classical or subspace based methods.

Hybrid analog and digital beamforming structure is a very attractive solution to build low cost massive MIMO systems. One major challenge in releasing its potential is the acquisition of accurate channel state information at the transmitter (CSIT). In [JK17], we propose a CSIT acquisition method based on the channel reciprocity property under time division duplex mode. Especially, we show how to calibrate hardware non-symmetry in the hybrid beamforming structure. Different with existing CSIT acquisition methods, our approach does not require any assumption on the channel model and can obtain near perfect CSIT.

Orthogonal Frequency Division Multiplexing and massive MIMO are key techniques in the deployment of 5G, due to the fact that they can significantly increase the overall capacity of a multiuser system with low-complexity equalization techniques. In [CHA18], we compare in terms of signal-to-interference and noise ratio (SINR) the single-carrier and multi-carrier schemes combined with the use of massive MIMO. We provide some analytical expressions of the signal model for downlink and uplink scenarios and the SINR expressions for each case. Moreover, some simulation results are shown to provide a better understanding and offer some insights on the best combinations.

The hybrid beamforming architecture wherein the number of radio-frequency chains is much smaller than the total number of antennas has been reported as an appealing solution but efficient antenna selection strategies are yet to be defined.

A strategy relying on the receiver spatial correlation is investigated [CMG$^+$18] to select the best antenna subset. It is evaluated from ray-traced radio channels at the channel level using propagation metrics and also at the system level with precoding and sum-rate capacity computations. The results demonstrate that a distributed array configurations with wisely selected antennas outperform collocated ones for all metrics with performance close to the full array. Moreover, the best antenna subset presenting the lowest Rx correlation values maximizes the sum-rate capacity with only a third of the initial radio-frequency chain hardware complexity.

In [BB18], we consider a cell-free Massive MIMO system and investigate the system performance for the case when the quantized version of the estimated channel and the quantized received signal are available at the central processing unit, and the case when only the quantized version of the combined signal with maximum ratio combining detector is available at the central processing unit. Next, we study the max-min optimization problem, where the minimum user uplink rate is maximized with backhaul capacity constraints. To deal with the max-min non-convex problem, we propose to decompose the original problem into two sub-problems. Based on these sub-problems, we develop an iterative scheme which solves the original max-min user uplink rate. Moreover, we present a user assignment algorithm to further improve the performance of cell-free Massive MIMO with limited backhaul links.

In the not too distant future, the onset of a wide range of new services is expected. Between their requirements greater data transmission capacity or strict real-time operation can be highlighted. In [BAF17], we investigate the uplink achievable rates of massive MIMO systems by means of using a wearable hub which brings the benefits of a very large number of antennas directly to the end user. Specifically we compare the achievable rates applying several precoding schemes such as matched filter, zero-forcing or optimum precoding based on perfect channel state information when using a large textile antenna array in different simulation scenarios. As a baseline for analysis we also compare with the results achieved in the absence of channel knowledge. Some examples of new services enabled by the high achieved rates are discussed.

5.3.2 Performance evaluation and modeling for Massive MIMO

Massive MIMO techniques have mainly been investigated with the independent and identically distributed Rayleigh channel. However, Massive MIMO techniques with realistic channel models are required. Hence, a realistic COST 2100 channel model is considered in [BBC17]. Most MIMO techniques such as user scheduling, precoding design and optimization algorithms require channel state information at the transmitter (CSIT) which in the Massive MIMO case is very difficult to obtain in perfect form. The paper considers the time division duplexing mode. The base station estimates the channel of all users in the uplink and exploits the estimated channel for downlink transmission. However, the complexity of the minimum mean square error estimator is cubic with the size of the covariance matrix of the channel. Hence, updating CSIT in every time-slot of data transmission is one of the main challenges in large MIMO systems. In this paper, we suppose that the base station does not need to estimate the channels of the users, and selects users and designs the beamforming matrix based only on the channel correlation matrix. We propose a new correlation-based user scheduling and precoding design and investigate the degrading effect of no CSI estimation at the base station on the average throughput of the system. Analysis and simulation results show that while the system throughput slightly decreases due to the absence of CSIT, the complexity of the system is reduced significantly. Moreover, the paper considers large MIMO simplifications and provides quite tight approximations for the system throughput. Analysis and simulation results show throughput the superiority of the proposed scheme over the graph-based scheme.

In [HHB+16b], we provide an overview of the first real-time massive MIMO mobility measurements conducted at Lund University, and present results for a spatial domain power control algorithm applied in one of these scenarios. The 100-antenna Lund University testbed was used on the roof of an engineering building in Lund with several user clients served in line-of-sight within a car park below.

The OpenAirInterface Massive MIMO testbed [JK16] is built on the open source 5G platform OpenAirInterface. It is one of the world's first LTE full protocol stack compatible base stations equipped with large antenna array, which can directly provide services to commercial user equipments. It shows the feasibility of using Massive MIMO in LTE standard, indicating the possibility of smoothly evolving the wireless network from 4G to 5G. It provides an innovation platform in solving 5G challenges, by giving the possibility of advanced algorithm testing, concept validation, channel measurements.

The first measured results for massive MIMO performance in a line-of-sight scenario with moderate mobility are presented in [HMJV+17], with 8 users served by a 100 antenna base Station at 3.7 GHz. When such a large number of channels dynamically change, the inherent propagation and processing delay has a critical relationship with the rate of change, as the use of outdated channel information can result in severe detection and precoding inaccuracies. For the downlink in particular, a time division duplex configuration synonymous with massive MIMO deployments could mean only the uplink is usable in extreme cases. Therefore, it is of great interest to investigate the impact of mobility on massive MIMO performance and consider ways to combat the potential limitations. In a mobile scenario with moving cars and pedestrians, the correlation of the MIMO channel vector over time is inspected for vehicles moving up to 29 km/h.

The performance and modeling of Massive MIMO scenarios with mobile users is of great importance and work in [BMB17] presents results from both simulated and practical mobility campaigns with multiple-users and large base station antenna arrays. The evolution of the Condition Number in time for the simulated and practical scenarios is examined as a way of quantifying the multipath richness of the channel and the rank deficiency of the channel correlation matrix, an important feature as a full rank correlation matrix represents the ideal scenario for Massive MIMO based spatial multiplexing, within a mobility scenario. Features of the time series for both measured and simulated data are compared and the correlation functions of the series investigated in order to consider the possibility of modeling the time evolution of the channel stochastically using Auto-Regressive Integrated Moving Average Processes. Some initial suggestions for the processes are presented and certain features of the time evolution of the channel are shown to be present in both the practical and simulated data.

Downlink beamforming in Massive MIMO either relies on uplink pilot measurements—exploiting reciprocity and TDD operation, or on the use of a predetermined grid of beams with user equipments reporting their preferred beams, mostly in FDD operation. Massive MIMO in its originally conceived form uses the first strategy, with uplink pilots, whereas there is currently significant commercial interest in the second, grid-of-beams. It has been analytically shown that in isotropic scattering (independent Rayleigh fading) the first approach outperforms the second. Nevertheless there remains controversy regarding their relative performance in practice. In [FRT+17], the performances of these two strategies are compared using measured channel data at 2.6 GHz.

5.4 Full-duplex communications and HW implementation driven solutions

This section is dedicated to a wide range of very diverse topics having in common their relationship to HW implementation. A significant part of these topics is related to full-duplex communications while the focus of the remaining ones ranges from on-the-chip design, over mixed analog-digital design, to HW specific models and implementations.

5.4.1 Full-duplex communications

Electrical Balance Duplexers (EBDs) can achieve high transmit-to-receive isolation, but can be affected by interaction between the antenna and environment. Circuit simulations incorporating measured time-variant antenna impedance data have been used in [LZB+16a] to quantify performance variation and determine circuit adaptation requirements for EBDs operating in vehicular scenarios at 875 MHz and 1900 MHz. Results show that the interaction between the antennas and the external environment is limited and vehicle motion does not necessitate high speed EBD adaptation, however the impedance of dashboard mounted antennas can vary due to interaction with the windscreen wipers, causing substantial variation in the Tx-Rx isolation and requiring EBD re-balancing intervals of 5 ms or less to maintain performance.

Similar experiments in high speed rail application are conducted in [LZB+16b]. Results show that electromagnetic interaction between the antenna and the environment outside the train is limited, and thus that high speed circuit adaptation is not required in this environment. However the results may have been affected by the metalized tinted window on board the train, and therefore the investigation should be repeated on older rolling stock without metalized windows to determine what effect this may have had. In [LZB+17], this work is extended, presenting a comparison of simulation results from two train types: a British Rail High Speed Train (HST) (presented in previous TD), and a British Rail Class 158 train. Simulation results show that passing trains can influence the antenna reflection coefficient and cause variation in the duplex isolation provided by the EBD. Variation was more substantial on board the class 158 trains, due to the narrower separation between passing trains compared to the HST. However in all scenarios, it was noted that re-balancing the EBD at intervals of 5 ms is sufficient to maintain performance.

In-band full-duplex (IBFD) radio transceivers, which transmit and receive on the same frequency at the same time, may potentially double the spectral efficiency of future radio networks. However, to achieve this requires high isolation between the transmitter and receiver in order to suppress the self-interference (SI) which results from the co-located co-channel transmitter and receiver. Work [ZLB+17] analyzes the phase noise (PN) influence on digital non-linear self-interference cancellation when using a shared local oscillator (LO) between the transmit and receive radios in IBFD applications. The SI channels nonlinearity, memory effect, delay and the LOs performance are considered. Simulation and experiment results show that the LO PN cannot be completely canceled, even when using a shared LO, however the performance of digital baseband SI cancellation is dominated by the LOs thermal noise floor, rather than the LO PN 3 dB bandwidth, and with practical low noise oscillators the residual SI may be well below the receiver noise floor.

5.4.2 HW specific models and implementations

The potential of applying different strategies to reduce the energy consumption of a wireless access network, powered by a photovoltaic panel system, during energy shortages is investigated in [DRM+]. The goal is to reduce the amount of energy that should be bought from the traditional energy grid during a renewable energy shortage. Three different strategies are compared and the results are looking very promising. Depending on the strategy, up to 72% less energy should be bought compared to the fully operational network for a worst case scenario and a time period of 1 week. However, applying such a strategy has also its influence on the network performance. The influence on the user coverage is limited, with a reduction of 3% at maximum, but the capacity offered by the network decreases significantly with 51% up to even 71%.

Despite the progress made in digital signal processing during the last decades, the constraints imposed by high data rate communications are becoming ever more stringent. Moreover mobile communications raised the importance of power consumption for sophisticated algorithms, such as channel equalization or decoding. The strong link existing between computational speed and power consumption suggests an investigation of signal processing with energy efficiency as a prominent design choice. In [OMT+17], we revisit the topic of signal processing with analog circuits and its potential to increase the energy efficiency. Channel equalization is chosen as an application of nonlinear signal processing, and a vector equalizer based on a recurrent neural network structure is taken as an example to demonstrate what can be achieved with state of the art in VLSI design. We provide an analysis of the equalizer, including the analog circuit design, system-level simulations, and comparisons with the theoretical algorithm. First measurements of our analog VLSI circuit confirm the possibility to achieve an energy requirement of a few pJ/bit, which is an improvement factor of three to four orders of magnitude compared with today's most energy efficient digital circuits.

The analog VLSI processing concept is further developed in [Tei19], where we consider the problem of decoding. Specifically, we designed and realized an iterative threshold decoder for self-orthogonal convolutional codes. We report on a student project where we used standard off-the-shelf electronic components to realize the analog decoder circuit. Starting point is the iterative threshold decoder, the structure of which corresponds to the structure of a high-order recurrent neural network (HORNN). Thus, the structure as well as the weights of this HORNN are defined by the problem. A training phase is not required. The dynamics of the HORNN can be implemented in discrete-time, this corresponds to the iterative threshold decoder, or in continuous-time. Both implementations lead to the same asymptotic state, which corresponds to the desired decoder output. The dynamical evolution of the continuous time HORNN is governed by a set of first-order differential equations. Based on that we design an analog electronic circuit which shows a similar dynamical evolution as the continuous-time HORNN, and especially also the same asymptotic state.

The realization of digital radio link for high-speed data transmission was presented in [RCK$^+$17]. Its concept and practical realization, using USRP devices from National Instruments, were described. Developed software for generation and reception of digital signals in baseband, including description of modulation types, and time and frequency synchronization mechanisms, was presented. Moreover, an operation of designed radio link in laboratory conditions was also subscribed.

In [Zog17], we analyze energy-efficiency vs. performance trade-off in low power wireless communications at component level and check hypothesis that the trade-off can be improved by introducing power supply voltage as design variable compared to the case when power supply voltage is fixed. We set two testing formulations and analysis based on SI4455 transceiver show that the hypothesis holds true for both. Moreover, we show that almost 130% less power can be consumed when optimal setting is used in SI4455 transceiver, compared to the case when very low output power is produced under the high power supply voltage.

Inspired by our brain, neuromorphic systems are designed to have a very low energy consumption: a recently developed CMOS based neuron was proven to consume only few femto-Joule per operation which is several orders of magnitude lower than what standard processors would need to execute similar operations. In [SCL$^+$18], we propose to study the theoretical performance of a neuro-based digital communication, where the signal detection is performed using an Integrate and Fire neuron. The main contribution is to derive the error probability of such a detector, either with one single neuron or with multiple neurons in parallel.

A simplified version of massive MIMO with few Universal Software Radio Peripherals (USRPs), where each of them can implement a 2x2 MIMO is exploited in [dAVCA18]. By synchronizing them and with appropriate signal processing techniques and algorithms we propose to have a low complexity version of massive MIMO, while implementing the hybrid analogue-digital solutions we are proposing in hardware and software within our research. We have configured the UC3M MIMO testbed, with the longer-term goal to conduct joint experimental research on massive MIMO and hybrid analogue-digital precoding schemes. These USRPs have been synchronized, starting from a SISO communications system, by considering 16 Quadrature Amplitude Modulation (QAM). The implementation of the reception code is based on interpreting the signal sent by the transmitter, in a certain time, upon receiving the data in wireless communication, by means of either antennas connected to the USRP or by means of a cable connection.

The use of photonic phase shift beamforming is highly attractive in terahertz transmitters due to its seamless integration with terahertz generation by photomixing. By feeding several photodiodes with coherent infrared signals with well defined phase relationships, highly directed and steerable terahertz radiation can be generated. At the same time, the total radiated power is increased by free-space power combining. Since optical phase modulators are readily available in planar lightwave circuit technology and do not need to scale with RF frequency, planar optic integration is simplified compared to photonic true time delay beamforming. However, phase fluctuations due to thermal and mechanical influences become a major issue if fiberoptic components are used and necessitate the implementation of a phase control circuit. Since direct detection of the terahertz phases is not feasible, an alternative approach is needed. In [KLHC18], a concept for controlling the terahertz phases using a lower-frequency reference tone generated by a second set of infrared lasers is proposed and a simplified system-theoretical model is developed and analyzed.

In [DTC+19], we consider two transmitter designs for symbol-level-precoding (SLP), a technique that mitigates multiuser interference (MUI) in multiuser systems by designing the transmitted signals using the Channel State Information and the information-bearing symbols. The considered systems tackle the high hardware complexity and power consumption of existing SLP techniques by reducing or completely eliminating fully digital Radio Frequency (RF) chains. In these architectures, which we refer to as RF domain SLP, the processing happens entirely in the RF domain, thus eliminating the need for multiple fully digital RF chains altogether. Instead, analog phase shifters directly modulate the signals on the transmit antennas. The precoding design for all the considered cases is formulated as a constrained least squares problem and efficient algorithmic solutions are developed via the Coordinate Descent method. Simulations provide insights on the power efficiency of the proposed schemes and the improvements over the fully digital counterparts.

5G and beyond networks

Silvia Ruiz[a,b]**, Hamed Ahmadi**[c,b]**, Gordana Gardašević**[d]**, Yoram Haddad**[e]**,
Konstantinos Katzis[f]**, Paolo Grazioso**[g]**, Valeria Petrini**[g]**, Arie Reichman**[h]**,
M. Kemal Ozdemir[i]**, Fernando Velez**[j]**, Rui Paulo**[j]**, Sergio Fortes**[k]**, Luis M. Correia**[l]**,
Behnam Rouzbehani[m]**, Mojgan Barahman**[m]**, Margot Deruyck**[n]**, Silvia Mignardi**[o]**,
Karim Nasr[p]**, and Haibin Zhang**[q]

[a]*Universitat Politècnica de Catalunya, Barcelona, Catalonia, Spain*
[c]*University College Dublin, Dublin, Ireland*
[d]*University of Banja Luka, Banja Luka, Bosnia and Herzegovina*
[e]*Jerusalem College of Technology, Jerusalem, Israel*
[f]*European University Cyprus, Nicosia, Cyprus*
[g]*Fondazione Ugo Bordoni, Rome, Italy*
[h]*Ruppin Academic Center, Hadera, Israel*
[i]*Istanbul Medipol University, Istanbul, Turkey*
[j]*Universidade Beira Interior, Covilhã, Portugal*
[k]*Universidad de Malaga, Malaga, Spain*
[l]*INESC-ID / IST, University of Lisbon, Lisbon, Portugal*
[m]*Universidade de Lisboa, Lisbon, Portugal*
[n]*Ghent University, Ghent, Belgium*
[o]*University of Bologna, Bologna, Italy*
[p]*University of Greenwich, London, United Kingdom*
[q]*TNO, The Hague, Netherlands*

6.1 Introduction

The aim of this chapter is to investigate the Network Layer aspects that will characterize the merger of the cellular paradigm and the IoT architectures, in the context of the evolution towards 5G-and-beyond.

In recent years, researchers have been working to identify and assess the network architecture of 5G-and-beyond systems; studying the impact of the "fog" networking/computing approach foreseen for 5G on the evolution of the radio access technologies; evaluating RRM approaches compatible to the new requirements; proposing new concepts and paradigms to take into account the plethora of new applications arising from the IoT context. The different chapter sections structure and summarize the main contributions on these topics.

---

[b] Chapter editors.

Inclusive Radio Communications for 5G and Beyond. https://doi.org/10.1016/B978-0-12-820581-5.00012-2
Copyright © 2021 Elsevier Ltd. All rights reserved.

In this chapter we first will have a look at the issues from previous generations carried to 5G and then introduce more recent challenges mainly concerning 5G and beyond. We first review recent contributions on ad-hoc networks and then continue with spectrum management, resource management, and scheduling.

Small cell networks, Heterogeneous Networks, and network densification are key issues of 5G. We look at different aspect concerning them and summarize the theoretical and hands-on contributions in these areas. Cloud-based densification is a 5G technology which has attracted a lot of attention in recent years. We have a section on Cloud Radio Access Network (C-RAN) which will be followed by a section on Software Defined Network (SDN) and Network Function Virtualization (NFV). In the later section we discuss network virtualization and softwarization, its challenges and recent contributions.

Finally the chapter includes two sections devoted to promising technologies, and emerging services and applications that will become more and more important in next years as: the use of Unmanned Aerial Vehicles (UAVs) as Unmanned Aerial Base Stations (UABSs), real-time smart grid state estimation, vehicle to vehicle and V2X communications to improve road safety and efficiency, or new public protection and disaster relief wireless networks standards.

6.2 Ad-hoc and V2V networks

One of the critical questions, when it comes to V2X communication, is which communication technology to use. With the rise of autonomous vehicle this is a critical issue which requires the best possible communication means both in term of throughput and latency. There has been many standards proposed for V2X, but among the most famous we can mention IEEE 802.11p which is a dedicated standard for wireless access in vehicular environments (WAVE) in the licensed 5.9 GHz band. Another possibility could be the use of the legacy overlaying LTE cellular technology. However, due to possible overload in classical cellular networks this should be done in a wise manner. One possible implementation is to combine both the flexibility and low latency of the ad-hoc direct vehicular communication of IEEE 802.11p with the wide range coverage and high data rate which can be potentially realized with LTE. In [DMM+16], a dynamic data traffic algorithm to enable hybrid vehicular communication is proposed. To steer the traffic between the cellular LTE network and the ad-hoc 802.11p communication a novel steering algorithm was developed that is based on an optimized utilization of the dedicated spectrum of each system. The goal of this algorithm is to react on an impending 802.11p channel congestion as well as on an overloaded LTE cell to which the vehicle is connected to. The proposed solution was implemented in a system-level simulator and evaluated using a realistic simulation scenario set in the city of Doha, Qatar where a road congestion at an urban intersection is emulated.

When considering V2V, one frequent question arises about the way it can be simulated. In [BBG+18] an accurate simulator for VANET is proposed. Classical platforms rely on many assumptions that fail to catch the right mobility of the vehicles and some basic physical layers parameters. For instance one of the most popular approaches is to treat the vehicles as a Poisson Point Process (PPP) along a straight line, resulting in uniformly distributed vehicles along the line whereas we need to account for inherently inhomogeneous vehicle distributions. Thus, in [BBG+18] the mobility simulator SUMO is used in combination with the vehicular network extension VEINS of the network simulator OMNET++. In addition, Karedal's pathloss model is used to calculate communication ranges.

Within the framework of V2V and ad-hoc networks, the topic of Delay/Disruption-Tolerant Network (DTN) is frequently mentioned and considered as a challenge since they have dynamic topology with frequent disconnections. Due to this sparse and intermittent connectivity, inference and learning over DTNs is much more complicated than in traditional networks. In [LBC+16], the authors consider the problem of distributed defective node detection in DTNs. A node is considered as defective when one of its sensors frequently reports erroneous measurements. The identification of such defective nodes is very important to save communication resources and to prevent erroneous measurements polluting estimates provided by the DTN. It is assumed that nodes are not aware of the status (good or defective) of their sensors. Distributed Fault Detection (DFD) for Wireless Sensor Networks (WSNs) has been well-investigated; however, the networks considered in most of the literature are dense and have a static topology whereas DFD in DTNs has been under-investigated. In [LBC+16], the authors present a fully distributed and easily implementable algorithm that allows each node of a DTN to determine whether its own sensors are defective. The theoretical results provide guidelines to properly choose the parameters of the algorithm. In the simulations, a jump motion model and a Brownian motion model are considered. The results show a good match with theory.

6.2.1 Prediction and reliability

We are all interested in ubiquitous connectivity but without interruption and disconnection. To avoid these kinds of situations the cellular operators can perform some important tasks. First, the network has to be designed to be reliable. For instance, a single point of failure should be avoided. Then prediction should be used to know ahead of time that overload and lost of coverage can occur in order to trigger in advance a handover and/or strengthen the deployment.

Network topologies are sometimes assessed according to their resilience to failure. For instance, a star topology with a central node is considered as very vulnerable whereas a tree topology is better while some aggregation links should be doubled (also known as 1+1 links) in order to ensure connectivity if one link fails. One of the most resilient topologies is the ring topology. The reason is that even if a link fails the node is not completely disconnected since there is always a second link to which the node is connected. However, one important and complex issue is how to split the nodes in the network into several rings to be connected to the aggregation node. This NP-hard problem is considered in [EHD17]. The authors considered two different cases. First, when the probability of failure is the same for each link, the authors propose an optimal solution where the main condition is that the empirical variance of the ring size distribution should be as small as possible. Then, when it is assumed that links may have various failure probabilities, the problem is further split into two different cases. The first one is when fixed 3-nodes ring sizes are assumed for each ring. An optimal solution is proposed based on a search of a perfect matching in a weighted graph. While in the more general cases with ring of different sizes, which is proved to be NP-complete, some approximation methods are proposed.

In [MHRA18,RHMA18], it is proposed to use Machine Learning theory to predict the wireless coverage in any given point in space. Up to now, cellular coverage is predicted by cellular operators thanks to propagation model and radio planning tool based on ray-tracing. These tools are however limited when networks are ultra dense. Moreover if we consider the M2M scenario to be incorporated with future 5G, this kind of planning is almost impossible. In these works, the authors proposed to gather sample SINR measurements from smartphones and/or dedicated low cost IoT devices. Then, thanks to Artificial Neural Networks based on radial basis function activation, the SINR can be predicted in other places. This kind of methods proved to be very accurate and can be run inside a smartphone, since machine learning processing is now incorporated in the chip of smartphones. Another kind of prediction is considered in [Bit19] with a mmWave mesh networks. Cognitive Radio capabilities are assumed where the goal is to predict SINR at the primary users where interference is generated by secondary users. Thanks to SINR prediction, some pre-processing over the interference (such as interference cancellation) can be performed. The SINR can be predicted using a Piecewise Cubic Hermite Interpolating Polynomial (PCHIP) method.

Given the availability of sources that can provide detailed information about social events, and the interest of using this in the management of cellular networks, in [FPSB18b] a novel general framework for the automatic acquisition of events data, its association with cellular network information, and its application in Self Organizing Networks (SON) is proposed.

Recently, the direct communication between mobile terminals, also called D2D, has attracted a lot of attention as an extension of the conventional connectivity. The aim is to increase the range of services, making it possible to lower the load offered to networks and to provide coverage extension (via relay or other means). [IOS+19] addresses a time-dependent analysis of the Signal-to-Interference Ratio (SIR) in D2D communications scenarios in order to build a uniform theory of mobility-dependent characterization of wireless communications systems. Analytical expressions for the variance and Coefficient of Variation (CoV) of SIR for a link between two communicating devices in D2D scenarios have been derived which use the kinetic-based mobility model capable of capturing a wide variety of users' movement pattern. Moreover, the effect of the variance and CoV on link stability in terms of the number of outages for non-stationary mobility patterns of users has been investigated.

6.2.2 **Network simulation/emulation platforms**

The use of tactical radios to communicate in case of emergency and rescue situation is a typical real use case of Mobile Ad-Hoc Networks (MANETs). Due to the operating environment, these radios experience specific characteristics such as limited bandwidth, in addition to high delay and packet losses. In order to assess the value of the optimizations and the performance of the new generation radios, many open-source and commercial network simulators/emulators exist to evaluate the Quality of Experience (QoE) of the different services. Authors in [NPJ15] provide a survey on best known emulation testbeds for MANETs. They outline differences between simulators and emulators and the complexity of deploying and managing emulators for research purposes. A comprehensive list of frameworks can also be found in [Ope]. However, most of these frameworks do not offer out-of-the-box features that allows for the use of customize proprietary applications in the system or means for quickly reproducing emulated experiments in the field. Authors in [RGW$^+$18] present the work achieved to develop, evaluate and test a novel application-layer routing algorithm specifically designed for tactical MANETs. In order to test and evaluate the proposed system, they developed two different platforms. The simTAKE platform has been developed to facilitate the development and analysis of the results in laboratory conditions, without the need to deploy the system on the field, which is very time-consuming and might produce almost no reproducible results. The platform provides the possibility to set the number of communication systems (called nodes) deployed in the scenario, their topology and the parameters of the network such as data bandwidth, packets delay, and packet drop rate as well as the distribution of messages emitted by each node during a simulation. The emulTAKE platform offers a system intended for the next level of tests, namely under field conditions. It allows to control the emulation of users for each node. With this platform, the field test manager can define scenarios, deploy them to each node, in order to generate messages according to specific distributions. Finally, results can be collected once a simulation is completed from a single web page. In order to evaluate the proposed algorithm in real conditions, two platforms and specific QoE metrics have been developed. This contribution might be of interest to those involved in V2V communications, especially the emulation of such networks. In order to emulate the MANET's radio performance in a more realistic way, authors in [NWB$^+$18] used the Extendable Mobile Ad hoc Network Emulator (EMANE) framework to reproduce the behavior of real tactical radio benchmarked in a laboratory. This was done by measuring the throughput and round-trip time of real tactical radios using wideband or narrowband Time Division Multiple Access (TDMA)-based waveforms. The wideband and narrowband measured performances were reproduced very closely using the same 2 second multiframe (M1D7) TDMA structure as the radios under test. The rate adaptation featured by the real radios could also be reproduced by changing the slot data rate using a Python agent external to EMANE. All contributions to EMANE and OSLRd1 required to obtain the presented results are available in a gitlab repository [Pre]. Authors in [NBB19] extend their previous research in order to investigate the performance results (Round Trip Time and message Completion Rate) of a 24-nodes MANET using TDMA radios and OLSR in a real time emulator based on the open-source EMANE framework. This contribution presents EMANE based emulation results using realistic TDMA radios including the use of fixed data rate or a rate adaptation scheme. The open source OLSRd1 or OLSRd2 implementations have been used.

Detection of social relations ties among nodes is important because of mobile devices in the MANET-DTN are carried by humans which are social creatures that live and move in social groups. For those reasons, mobility models should follow human behavior. Authors in [HMDP18] propose tools for evaluation of social relations from movement in mobility models which are described as new evaluation methods. Based on the proposed evaluation methods, it is possible to decide when the used mobility model shows signs of social behavior. The proposed evaluation method was created based on the Louvain method for community detection and other network graph parameters (average weighted degree). Simulations of evaluation methods were made as a comparison between two random mobility models and one social based mobility model. All models were simulated with a different number of nodes and radio ranges and evaluated by proposed method and other existing protocol-dependent and independent methods. Results of simulations prove that social based mobility models should have modularity quality values above 0.3, which is a good indicator of significant community structure. On the other hand, small values of modularity quality were obtained when random mobility models were used.

In the last years WiFi has seen many improvements which gave rise to new IEEE 802.11 standards such IEEE 802.11ah specification (also called "Wi-Fi HaLow") [IEE17] which plays a key role in the integration of a Wi-Fi oriented technology to operate in TV white space spectrum (TVWS) in the UHF band. Most of existing simulators have been developed with specific networks protocols in mind. In [MPSK17] a novel and universal link-level simulator for well known IEEE 802.11g/n/ac and also for upcoming 802.11 ah/af/ax has been developed. Basic functions of the simulator for the IEEE 802.11n standard were verified by simulations. The functionalities of the proposed simulator were demonstrated by several simulation outputs. The advantage of the proposed WLAN simulator, besides supporting wide range of IEEE 802.11 technologies, is also that is compatible with Vienna LTE/LTE-A simulator [Vie] which allows to investigate possible coexistence scenarios which can occur between these systems in shared ISM bands (2.4 GHz and 5 GHz), which will likely happen in the forthcoming 5G networks.

6.2.3 Energy harvesting

Typically, a Wireless Sensor Network (WSN) comprises a number of sensor nodes deployed in a given geographical area which collect and send measurements to a Fusion Center (FC). Current technological advances make it feasible to deploy inexpensive sensors in large numbers. In this context, the problem of optimally selecting a subset of sensors to perform a given task naturally arises. This often stems from resource (e.g., bandwidth), interference level or energy consumption constraints, which make massive sensor-to-FC communications barely recommended (or simply not possible).

In the literature, this is referred to as the sensor selection problem. Besides, Energy Harvesting (EH) is becoming a promising technology capable of extending the operational lifetime (or even allowing self-sustainable operation) of WSNs. Authors in [CMA16] have proposed a novel EH-aware sensor selection policy that outperforms a well-known EH-agnostic policy. A selection is needed due to the reduced number of available sensor-to-FC channels. The goal is to minimize the distortion in the reconstruction of the underlying source at the FC, subject to the causality constraints imposed by the EH process. Performance in terms of reconstruction distortion, impact of initialization, actual subsets of selected sensors, and computed power allocation policies is assessed by means of computer simulations. To that aim, an EH-agnostic sensor selection strategy, a lower bound on distortion, and an online version of the SS-EH and JSS-EH schemes are derived and used for benchmarking. The proposed sensor selection policy attains a lower bound on the reconstruction distortion of the optimal (joint) solution, as soon as the percentage of selected sensors is above 40%.

6.2.4 Measurements for specific applications

Operational activities of public entities, like border patrols, force development of communication solutions. The reinforcement of radio-communications or IT (Information Technology) systems increases the security of the officers on duty and allows to carry out all activities more effectively. Authors in [KRCJS17] provide results of research conducted to test, design, and build digital radio link for high-speed multimedia data transmission in maritime environment. The measurement campaign was carried out in the Vistula River Lagoon area near the Baltic Sea coastal waters in Poland. In addition, to verify the behavior of the developed digital radio link in real maritime environment, the selected measurement location is characterized by the possibility to ensure the LOS (Line of Sight) propagation conditions over a distance of several kilometers. It must be noticed, that entire radio link is realized in SDR (Software Defined Radio) technology and each change must be deeply verified. The key change was the hardware modification in USRP (Universal Software Radio Peripheral) devices. The positive results of the measurements verify and confirm the rightness of some minor changes in the software and hardware layer of the realized technological demonstrator.

6.3 Spectrum management and sharing

6.3.1 IoT/machine type communications

IoT has been widely used as a new technology that offers a revolutionary way to interact with the environment connecting objects and devices to improve the everyday lives of citizens around the world. ITU estimates that 25 billion devices will be connected to the Internet by 2020. Nowadays, new research is being carried out to define the technology and frequency bands for IoT. The requirements expected to be fulfilled are the support a massive number of devices, at low transmission rate and reduced power consumption while maintaining long battery life. In addition, it is expected that there will be reduced complexity with extended coverage and limited latency and prices that minimize the cost of the equipment and maintenance [Eva15]. For the operation of such a large number of devices, [MPLC16] proposed the use the TV White Spaces (TVWS) of the Digital Terrestrial Television (DTT) band through spectrum sharing between DTT and IoT networks. The aim of the work presented in [MPLC16] was to evaluate whether the coexistence of both technologies in co-channel or adjacent-channel is possible by determining the margins of protection and EIRP that could be transmitted by base stations and IoT devices. Authors considered the Digital Video Broadcasting Terrestrial Access 2 (DVB-T2) standard as the most widely used in the world. The proposed scenario considered a DVB-T2 network offering fixed rooftop reception as a primary service and Narrow Band LTE (NB-LTE) network IoT as a secondary service. Five representative scenarios to deploy IoT were evaluated while both uplink and downlink were considered. The first two scenarios considered urban or suburban environments characterized by buildings of 2 or 3 floors with wide and narrow streets looking into smart parking and traffic congestion. The third and fourth scenarios considered smart agriculture looking into smart farming and animal tracking. Finally, the fifth scenario was about eHealth – Indoor, looking into patients' surveillance inside a hospital. Results demonstrated that the coexistence between DVB-T2 and NB-LTE-IoT is feasible in adjacent channel. Measurements set the maximum allowable EIRP for a NB-LTE Small Cell between 10 and 15 dBm, considering a 1 MHz band guard. Hence, NB-LTE could be efficiently used as a secondary service when having a 6 or 7 MHz white-space between two DTT channels. The maximum EIRP that could be transmitted by a NB-LTE device varies in a range between 9 and 14 dBm for the best scenario (Smart Parking), whereas this power is reduced to a range of 3 and 8 dBm considering the worst one (Traffic congestion), from a 2 MHz guard band. Considering high duty cycles and a 2 MHz guard band, the maximum EIRP could be reduced up to 4 dBm if the waiting time between transmissions is lower than the duration of the DVB-T2 frame (250 ms). For a 6 MHz guard band, this impact is almost negligible. Given that there is an inversely proportional dependence between throughput and battery life, for a specific waiting time between transmissions and long transmission time high throughputs and short battery life are achieved. Authors in [US17] describe how 5G networks should be able to support Machine Type Communication (MTC) traffic generated by the IoT as well as traffic generated from the traditional Massive Broadband (MBB). They also state that both types of traffic can be achieved by utilizing media access schemes based on Random Access (RA) while resource partitioning is necessary to avoid collisions between both traffic types. The paper is proposing a dynamic partitioning scheme that utilizes a control loop to manage the amount of reserved radio resources for MTC traffic. Authors employed Slotted ALOHA to exploit its characteristics for adjusting the amount of available Small Packet Blocks (SPB) for the one-stage access. This introduced discrete time-slots where IoT devices could only try to access the medium at the beginning of a time-slot. This reduced collisions and therefore increased the maximum throughput. The control loop was then used to measure the average collision probability of the RA based MTC traffic and estimated the MTC traffic load. More specifically, the control loop observed the SPB collision probability and adjusted the amount of SPB such that a certain target SPB collision probability with a hysteresis of 2.5 percent was achieved. This estimate was then used to control the amount of reserved radio resources. The performance of the control loop was evaluated using constant as well as variable MTC traffic load. Results suggest that the proposed control loop can be applicable for low load situations with small load variations. As the medium access scheme was Slotted ALOHA, the target SPB collision probability of the control loop determined the operation point of the system. The paper concluded that there are three possible target Packet Block (PB) collision probability regions which result in different operation points:

   $< 44\%$: more successful transmissions, bad resource utilization
   $= 44\%$: less successful transmissions, best resource utilization
   $> 44\%$: increasing SPB collisions, should be avoided.

Furthermore, the control loop achieves a better control quality for short optimization interval duration, but too short duration leads to many SPB reconfiguration and thus to decreased control quality and throughput.

Another contribution [MHD$^+$17] compares non-intelligent and intelligent methods of channel assignment and their influence on optimal path selection in the process of message delivery from the source to the destination. The work focuses on finding the path between the source and the destination node in a Cognitive Radio Mobile Ad-Hoc Network (CR-MANET) environment, where the single channel and multi-channel assigning were implemented with and without primary user (PU) activity. The paper describes how Cognitive Radio (CR) technology is implemented to Mobile ad-hoc network and how spectrum sensing is an important process before channel assignment. Authors present mobile nodes in CR-MANET as self-organized, decentralized, and autonomous routers, which can select different frequency channel between a couple of nodes

6.3.2 Coexistence and sharing

The advent of 5G will cause a massive increase in required spectrum, that will be necessary to accommodate new, bandwidth-hungry applications. For this reason, alongside bands already assigned to mobile services, new spectrum chunks have been identified for 5G at both European and global level. In this complex scenario, coexistence must be ensured both with other services in adjacent bands as well as with legacy mobile communications systems using the same bands in neighboring countries, given that the time schedule for 5G roll-out might differ significantly among different countries. The latter issue is particularly significant in Europe, owing to the high number of nations in a relatively small area, often sharing terrestrial and maritime borders with several neighbors.

The issue of coexistence has, naturally, attracted a lot of interest within the COST-IRACON community, which is reflected in the remarkable number of documents devoted to the subject, that in some cases evolved in papers published in scientific journals.

An analysis of various 5G bands may be found in [CGMP18], which examines the three pioneer bands identified to offer 5G services in Europe by 2020 on a large scale, namely the 700 MHz, the 3.5 GHz, and the 26 GHz band, using for each band appropriate propagation models.

In regards to the 700 MHz band, in [CGMP18] the focus is on the study of the technical conditions that allow the coexistence between the victim LTE M2M base station in the 733-736 MHz and the interfering LTE Supplemental Downlink (SDL) base station in the 738-758 MHz. Results are derived in terms of separation distances, that guarantee the Interference to Noise Ratio (INR) requirement, as a function of SDL transmitted EIRP. Separation distances ranges between 0-50 km for [IR16], 0-2.5 km for [DC99], 0-18 km for [IR09], and 0-5 km for [IR13].

The 26 GHz analysis in [CGMP18] studied the impact of an International Mobile Telecommunication 2020 (IMT-2020) BS interferer on a victim incumbent Point to Point (P-P) link for the propagation model defined in ITU-R P.452-16 [IR15] in combination with ITU-R P.2108 [IR17]. It considered an IMT-2020 outdoor urban hotspot base station and three Fixed Service (FS) receivers at different heights (15 m, 30 m, and 60 m). The minimum separation distances that allows to meet the constraints on INR are about 27-44 km for the different FS antenna heights. However, the FS antennas at 26 GHz are strongly directive, so that even a small angular separation between the FS and the P-P radiation pattern will considerably reduce the protection distance.

It is well known that the Radio Spectrum Policy Group (RSPG) of the European Commission considers the 3400-3800 MHz band to be the primary band suitable for the introduction of 5G-based services in Europe even before 2020, noting that this band is already harmonized for mobile networks, and consists of up to 400 MHz of continuous spectrum enabling wide channel bandwidth [EC16]. However, there are some issues related to this band, owing to the fact that it can be used also by other services, and in particular FS and Fixed Satellite Service (FSS). In [CGMP18] tests for the 3500 MHz band analyzed the coexistence between an interferer LTE eNodeB and a victim incumbent FS for both co-channel and adjacent channel interference scenarios and various cell types considering the transmitted parameters suggested by [ECC13]. In [CGMP18] results are provided for different propagation models and for different combinations of horizontal and vertical angular discrimination. Angular discrimination considerably reduces the separation distances. As an example, in a Macrocell scenario the separation distance reduces from about 120 km for 0° of horizontal angular discrimination to about 23 km for 30° of horizontal angular discrimination.

Concerning the FSS in the 3400-3800 MHz band, a remarkable example is the reception of broadcast television by means of Very Small Aperture Terminal (VSAT), widely used in Africa, Asia, and Latin America. The issue of coexistence between land mobile services (and particularly LTE Time Division Duplex (LTE-TDD) and VSAT-based television reception) is analyzed in [CGP+17], which presents the results of an extensive measurement campaign and laboratory tests. Proper separation distances for both co-channel and adjacent channel were identified in different real propagation conditions, LOS and NLOS. The characteristics of VSAT equipment, studied by laboratory measurements, show a relatively small variation in selectivity between different receivers.

In addition to the three pioneer bands identified to offer 5G services in Europe other frequency bands have been analyzed for a possible introduction of future mobile systems. In [SSTV18] the viability of 5G New Radio (5G NR) spectrum sharing in Ultra High Frequency (UHF), Super High Frequency (SHF), and mmWave in outdoor environment is investigated. Performance evaluation includes the study of the behavior of Physical Layer (PHY) throughput as a function of the coverage distance for 2.6, 3.5, 28, 38, 60, and 73 GHz. The preliminary analysis shows that the highest system capacity and the highest modulation and coding schemes are achievable for the shortest coverage distances at mmWave whereas the supported throughput for long coverage distances is more favorable for UHF and SHF bands.

Even the 2.3-2.4 GHz band attracts a lot of interest for 5G deployment. However, this band is used in various European countries for incumbent services such as FS, Programme Making and Special Events (PMSE), Governmental use and WiFi above 2.4 GHz. Protection shall be guaranteed to all the incumbent users. One promising technique towards this aim is Licensed Spectrum Access (LSA). To investigate its potential, Italy hosted a world-first LSA pilot, promoted by the Italian Public Administration and the Joint Research Centre of the European Commission under the technical coordination of Fondazione Ugo Bordoni. It counted on the participation of research and industrial partners from Italy, France, and Finland. The achievements of the pilot were reported in [GCP+16] and in [GCP+17]. Seven LTE-TDD radio base stations (BS) at 2.3-2.4 GHz have been installed both indoor (5 Femto BSs) and outdoor (2 Macro BSs) in the building of the Italian Ministry of Economic Development (MiSE). Fig. 6.1 shows the LSA Pilot end-to-end architecture. The various elements of the pilot were located around Europe. The indoor and outdoor LTE nodes located in Rome were connected to the Evolved Packet Core (EPC), which allowed the communication toward user equipment (i.e. commercial smartphones equipped with authenticated test SIMs). An Operation Administration & Management (OAM) communicated with the LSA controller (both elements were located in Finland) and was capable of managing the mobile network in order to cope with the requirements imposed by the sharing rules stored in the LSA repository, located in France. The Sharing Tool used to determine the sharing rules was in the administration domain.

Tests were made about the coexistence between LTE-TDD base stations and two types of incumbent services, namely FS and PMSE. Coexistence is achieved thanks to sharing rules properly identified based on the incumbent services characteristics and tested in a real scenario. Compliance with the sharing rules for the FS has been verified on the field through measurements, both for the protection zone and restriction/exclusion zone approach.

Subsequently, tests on channel evacuation were realized considering a possible incumbent PMSE user, requesting a channel for its operations in a given location.

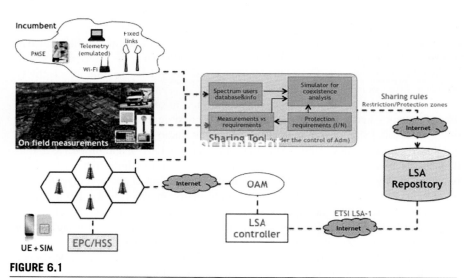

FIGURE 6.1

Italian LSA Pilot end-to-end architecture.

Results show that it was always below 40 seconds. Results from the pilot indicate that LSA is a straightforward approach that provides high predictability and certainty for both the incumbent and the licensee while preserving existing ecosystem and business models. Space limitations don't allow to provide a detailed account of tests results here, and the interested reader is referred to the quoted references for details.

Also related to the 2.3-2.4 GHz band was the experimental campaign carried out in Italy in response to the invitation of the European Commission to the Member States to perform specific coexistence analysis on developing conditions for the introduction of Wireless Broad Band (WBB) in this band. Both on-air measurements in an indoor environment and laboratory tests were performed aiming at investigating the possible interference towards WiFi due to out-of-band emissions from LTE equipment in the adjacent bands. Results of this activity were reported in [GCC+17]. The main trends in terms of interfering effects of the LTE signals are:

- The effect of LTE interference decreases as frequency separation between the interfering and victim signals increases.
- The interfering effect of LTE Down-Link (DL) onto WiFi victim receivers is stronger than LTE Up-Link (UL).
- Although different WiFi equipment may have different performance (e.g. in terms of throughput), similar effects of the LTE interference can be highlighted due to LTE signals (e.g. effect on different WiFi channels, effect of LTE DL vs LTE UL, etc.).

The 2.4 GHz band is also of interest for the future introduction of Low Power Wide Area Network (LPWAN) in the ISM band. The coexistence between LTE and

LPWANs has to be analyzed. In [PPMK19], experimental laboratory measurements were realized in order to test the robustness of the LTE-DL full data load to interference, caused by a Long Range (LoRa) signal, in the 2.4 GHz band. Experiments were performed in order to determine C/I ratios needed to provide LTE-DL service with good quality. Guard bands were varied between 1 MHz and 5 MHz. The obtained results show that 64QAM modulation has the lowest robustness against interfering LoRa signal. It was also observed that a LTE-DL signal with wider bandwidth is more sensitive on the interfering LoRa signal.

Another dynamic spectrum sharing approach in a quite different context is presented in [Lys17], which contains an overview of the achievements and ongoing work in the domains of TVWS, Dynamic Spectrum Access (DSA), compatibility and protection against interference, and smart antennas in South Africa. The rationale behind the work is the need to provide wireless Internet access, especially in rural areas characterized by a sparse population distributed over a large territory. This situation is common also to several other African countries.

The South African government set the target of universal broadband coverage, including broadband access for every school in the country. The identified solution was to exploit the TVWS in a cognitive way, by means of DSA. To this aim, the Council for Scientific and Industrial Research (CSIR) built a Geolocation Spectrum Database (GLSD). This approach was validated since 2013 in a number of large scale trials of TVWS technology, together providing Internet connectivity to over 20 schools with over 20,000 pupils and teachers. They were also followed by trials in other African countries, such as Ghana and Botswana. The trials included compatibility testing and demonstrated the viability of the approach based on DSA and using the GLSD to exploit TVWS.

Another key issue to be considered is the coexistence between mobile radio and other services such as Public Protection and Disaster Relief (PPDR). In Colombia, as well in other and other American countries, the 806 to 824 MHz and the 851 to 859 MHz bands are allocated for PPDR services (UL and DL respectively), with a combination of technologies like P25, TETRA, and LTE-PPDR, creating a scenario where technologies with different bandwidths and technical characteristics must coexist. At the same time, the lower PPDR band is adjacent to the 700 MHz LTE band, while the upper PPDR band is adjacent to the 850 MHz LTE band. The issue is further complicated by the fact that both narrowband and wideband PPDR systems are in use and it is necessary for the regulator to guarantee their coexistence. An experimental study, based on both laboratory testing and simulations, is reported in [NCM17].

The authors show the results related to three different scenarios, considering both interference between PPDR and LTE bands, and internal interference between PPDR systems using the same band. Key findings are that it is possible to reduce the guard band between LTE 700 MHz and PPDR systems by up to 2 MHz, while more critical cases are located in the upper part of the 850 MHz band where the LTE-DL is adjacent to the PPDR-DL.

6.3.3 Field monitoring

One practical problem to be addressed in planning and rolling out a new network lies in the field exposure limits that may prevent adding new radiating elements, e.g. for 5G, in sites where already other radio technologies are present.

This is further complicated by the fact that there is no harmonization between exposure limits throughout Europe (let alone worldwide). Therefore, operators with a wide international footprint, which is the case for various European operators, must face the additional challenge of adapting their roll-out strategies to regulations changing from country to country and, in some cases, also at regional level within the same country.

Finally, the advent of new technical features, such as Massive MIMO and Carrier Aggregation, will also require an adaptation of field measurement criteria and equipment, as well as of computation algorithms, currently in use.

Therefore, having reliable and accurate measurement and simulation procedures is necessary in order to correctly plan and roll-out new systems and networks. This issue was dealt with by various contributions presented at IRACON meetings ([GCV+16], [CGV+19], [CPB+19]).

The authors of [GCV+16] presented both Narrow-Band (NB) and Wide Band (WB) measurements taken during the operation of the above mentioned LSA pilot. Measurement were performed both indoors and outdoors and over the frequency band from 87.5 MHz to 3 GHz, in order to include FM radio and terrestrial TV broadcasting along with mobile and wireless systems. Both the NB and the WB measurements showed that the field levels were always well below the limit of 6 V/m even when the LTE-TDD femtocell network was deployed at 2.3 GHz. In particular, the highest measured values were slightly below 1.2 V/m and results were corroborated by simulations. As an example, Fig. 6.2 shows indoor measurement results for three bands in several locations within the premises of the Italian Ministry of Economic Development. Additional results can be found in the paper.

The introduction of Massive MIMO causes a high variability of exposure levels, owing to the usage of smart antennas. Therefore, the actual exposure levels are lower than those experienced with traditional antennas, and the methodologies to assess them (both measurements and simulations) must be adapted accordingly [CPB+19]. A method adopted in Italy defines a reduction factor named "Alpha24" that evaluates the average level of electric field during 24 hours of a base station under test. This methodology allows a more realistic evaluation of actual exposure levels and, therefore, allows the introduction of base stations using Massive MIMO even under very restrictive limits.

A related contribution [CGV+19] presents a methodology for carrying out large-scale measurements of radiofrequency electromagnetic field and low frequency electric and magnetic field. The system consists in performing measurements using a control unit equipped with the appropriate probes positioned on the roof of a car moving along the territory (dynamic measurements). The possibility to perform massive measurements in a short time could become extremely useful in the future, when the implementation of the new 5G services will lead to an increased use of the

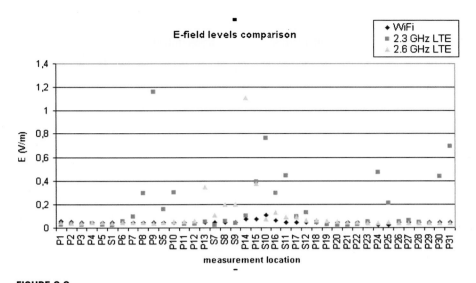

FIGURE 6.2

Measured E-field comparison.

electromagnetic spectrum, with consequent increase of electromagnetic field levels. Measurement activities carried out in different contexts and scenarios showed an excellent agreement between dynamically measured radio frequency electric field levels and the Electro-Magnetic Field (EMF) levels obtained by carrying out measurements with static probes. This agreement, makes it theoretically possible to perform large-scale radiofrequency field monitoring using the dynamic procedure, thus making the task much quicker and straightforward.

6.3.4 Virtualized networks

Wireless network virtualization is a key technology for 5G, both as an enabler for operational sustainability and business agility [AMG+16]. Wireless network virtualization requires the abstraction of physical resources such as spectrum, infrastructure, power, etc., into virtual resources. These resources can then be partitioned and allocated to different Virtual Network Operators (VNOs). Authors in [AMG+16] propose a virtualized network architecture for an infrastructure provider that shares the physical resources of a Massive MIMO cell among several VNOs using spatial multiplexing as well as orchestrate the interaction between the VNOs and the infrastructure provider through an auction-based mechanism that allocates spatial streams to the VNOs.

Furthermore, authors in [AAC+17] present a customizable resource virtualization algorithm for multi-user data scheduling in a LTE C-RAN deployment. The algorithm is based on the hypervisor specific dynamic assignment of air resources to the VNOs, based on either joint scheduling or per-cell schemes. The objective is to improve the resource allocation mechanism for two scenarios linked with night-time and daytime.

Another contribution [CB18] presented the detection methods in the energy-efficient context in 5G networks. It was shown that the considered context-awareness information has enormous volume. Simulations performed showed that a power consumption reduction is possible if intelligent network organization is implemented. Authors suggest that the correlation between nodes used for finding clusters in the network allows to reduce the information exchanged in the network. However, mobility of nodes limits the clustering gain since the correlation in the network is limited in high-velocity scenario.

Finally, contribution [PKM18] presented the importance and the benefits of employing a geo-location spectrum database for deploying a TVWS network as a vital tool to limit harmful interference to primary TV spectrum users. This work outlines how a fully operational web-based TVWS system has been designed and developed for Cyprus using the web API framework and adhering to the methods defined by CSIR.

6.4 Radio resource management and scheduling

RRM refers to the efficient use of radio resources so that the system capacity is kept at maximum and the Quality of Service (QoS) of the users are met. Any mobile communication system should be able to offer services with different QoS requirements depending on each user subscription level. To fulfill different request service types such as voice, video streaming, web-browsing, etc., a mechanism to classify those types of bearers into different QoS Class Identifier (QCI) is needed. 3GPP specifies that each QCI is characterized by priority, packet delay budget, and acceptable packet loss [GCS17].

Due to the huge increase in traffic and services in mobile networks, network management has defined an alternative quality indicator known as QoE, which is defined as the overall satisfaction of the user of a service as it is subjectively perceived. QoE is widely measured using the Mean Opinion Score (MOS) scale, in which a particular value is assigned to the experience perceived according to the user's opinion raging from 1 (bad) to 5 (excellent) [GLGP13]. Therefore, when RRM approaches are developed, it is important to perform the allocations considering QoE.

When a cell becomes overloaded, the QoS is degraded, because incoming users or services suffer high blocking rate and call dropping probabilities due to the lack of available resources. Therefore, load balancing is required in order to transfer traffic from heavy loaded cells to the lightly loaded neighboring cells [GCS17]. It involves exchanging information periodically between evolved Node B (eNB)s and users. The load of the cells is compared, and, if needed, a Hand-over (HO) procedure is initiated. In frequency domain, load is typically defined as the percentage of used Resource

Blocks (RBs) over the total available in the system, as it is the smallest combination of data that can be transmitted or received to/from a terminal.

In addition, SON techniques have been developed to automate network management so that Load Balancing (LB) is achieved. The aim of load balancing techniques is to alleviate congestion problems caused by unevenly distributed traffic. This is achieved by sharing traffic among neighbor cells through the modification of HO margins. Different network/cell parameters can be modified in order to implement LB, with HO margins being one of most used options [RlBMB11]. This balance is expected to decrease the overall blocking ratio, thus increasing the total carried traffic in the network [ARG17].

When performing RRM, the user location knowledge is important from many points of view, not least in understanding and optimizing network performance. Mobile operators plan their network coverage dividing the area into cells and sectors. Cell edge users consume on average much more radio resources than users near the cell center, and even though the cell edge problem is well known and different techniques are addressed to mitigate this, very little work has been found regarding the actual user distribution within the cell, while some work has been found discussing different effects of it. To find the distance and the direction from the base station to the phone, the location of the serving cell is retrieved from the cellular network. The Cell ID from the smartphone and the cell list from the network are then correlated in order to calculate the distance and direction (see Fig. 6.3). Data collected from a live network clearly indicates that the angular direction follows a symmetrical distribution with the mean and median close to the center of the cell sector. The von Mises angular distribution is shown to be a good model for the angular distribution of the users [LG17].

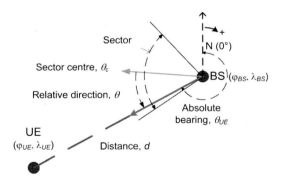

FIGURE 6.3

Definition of directional data within a sector [LG17]. The symbols φ and λ refer to the latitude and longitude.

In order to take the maximum profit of radio spectrum, higher frequency bands are exploited for low coverage areas, while lower frequency bands are used to provide great coverage areas but with a smaller capacity. This difference between coverage and capacity of different frequency bands brings some challenges in network planning, as it needs to cope with the resulting interference [GCS17]. Hence, the use of frequency bands together with the power control plays an important role in RRM and scheduling. In the sections below, we present different approaches that target to utilize the radio resources effectively, scheduling the users according to a resource utilization plan.

6.4.1 Resource allocation in wireless mesh networks

This section focuses on the frequency and time resource allocation problem of the Orthogonal Frequency Division Multiple Access (OFDMA) Wireless Mesh Networks (WMN) system which differs from cellular network allocation due to the spacial architecture and because the OFDMA channel can be split into sub-bands of subcarriers. Using graph theory models and algorithms an optimization scheme was proposed to achieve the goal of minimum resource utilization over a network and allow frequency reuse [PR16], [RWM18]. The design of WMN considered is based on the following assumptions [RWM18], [RWP18]:

1. The resource allocation is initially performed, and routing will be done on a network which has resource allocations.
2. The resource allocation is performed by a centralized algorithm which has knowledge of all the connections in the network.
3. Connection between two nodes is possible if the signal of noise and interference ratio, SNIR is above a minimal level. A signal below this level does not contribute to the interference.

The following shared resources are considered:

- Time slots: The system operates with time slots division, having time synchronization among nodes.
- Frequency: The frequency division can be among different OFDMA bands, or among different sub-bands in an OFDMA band.
- Separation Rules

The non-collision rules assumed are:

- A node may not transmit and receive in the same time slot in the same frequency band
- All transmissions from one node to other nodes in the same time slot should have different frequency bands, or alternatively in a specific band should have different frequency sub-bands.
- All receptions at a node from other nodes in the same time slot should have different frequency bands, or alternatively in a specific band should have different frequency sub-bands. This includes intentional transmissions to this node as well as interference, such as transmissions from another nodes in the area of reception not intended for this node.

Each combination of time slots frequency band, and frequency sub-band is called a Resource Element (RE). The algorithms have the ability to obtain an assessment of the resources required globally in the system. An example of allocations produced with the simulation is given in Fig. 6.4. The links on the graph are marked with the allocated REs in both direction. We denote $|V|$ as the number of vertices, and $|E|$ as the number of edges. D_{max} is the maximum degree of the graph and $|C|$ (for "colors") is the number of resource elements allocated by the algorithm.

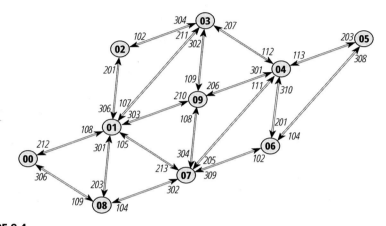

FIGURE 6.4

Example of resource elements allocation. $|V| = 8$, $|E| = 36$, $D_{max} = 6 \longrightarrow |C| = 24$.

6.4.2 **RRM for D2D scenario**

RRM also plays an important role in D2D communication. D2D communication in cellular systems is defined as direct communication between two mobile users without traversing the BS or core network [AWM14]. D2D in this case, typically re-uses UL resources since they are under-utilized. When allocating resources to D2D users, the interference and the user power levels play an important role. Most of the existing studies focus on a simple case of single user pair and hence are not practical. It is therefore important to develop approaches that can consider the intercell interference and enables the resource sharing between more than two users which promises great potential.

By adapting and expanding the interference structure of [ZCYJ15], an interference-graph based resource allocation algorithm for the LTE UL that respects the Single Carrier Frequency Division Multiple Access (SC-FDMA) constraint of continuous resource assignment to individual users, resource sharing between multiple users while controlling the interference is achieved [BK16]. Here, the interference situation in each cell is modeled as an interference graph $G = \{V, E\}$ consisting of vertices V and edges E. Each communication pair forms a node, represented by a vertex in G and the edges between them are used to model the potential interference. Simulation results have shown that such an approach increases the system capacity greatly. It should be noted that such capacity increase comes at the cost of a lot of information (pathloss and interference) required at the scheduler. Besides, the additional overhead to communicate the determined schedule depends on the dynamic of the scenario and may prohibit the straightforward application in an actual network.

6.4.3 RRM via frequency reuse

By reusing the existing frequency bands, either the same or different bands, the mobile users can be allocated to different bands in order to meet their QoS requirements. This is feasible as the coverage is frequency dependent as shown in Fig. 6.5. With Signal-to-Noise Ratio (SNR) thresholding and QoS requirement, an algorithm that minimizes the HOs is developed in [GCS17] calculating the required user data rate, expressing its SNR as [3GP10]

$$R_{b \text{ [bits/s]}} = \frac{A}{B + e^{-C\rho_{[dB]}}} \tag{6.1}$$

where A, B, and C depend on the modulation type. By calculating the SNR for different frequency bands visible to the mobile users, a user is then allocated to an available BS. Users' load distribution to different bands provides high QoS with gain in throughput exceeding 8% [GCS17].

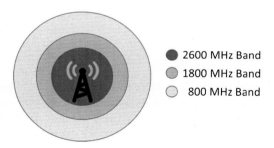

FIGURE 6.5

Coverage of different frequency bands.

Traditional load balancing techniques, which do maximize QoS statistics, may show certain limitations when optimization algorithms alternatively focus on QoE as the main figure of merit. In [ARG17], the cell load and QoE balancing techniques are investigated in an LTE network. A simulation tool is developed that offers performance statistics of every user connection, from which QoE values are estimated by using the so-called utility functions using MOS scale. These utility functions allow to quantitatively model the relationship between objective performance QoS indicators taken directly from the network and QoE, highlighting the QoS impact on the user subjective perception of the service. Moreover, HO thresholds are also modified so that LB is better achieved, as shown in Fig. 6.6. In the tool, the QoE is measured through different equations defined for voice, video, and Web services. Results show that load balance algorithms have limitations and may even worsen the experience perceived by the users when comparing QoE performance indicators [ARG17]. Hence, the future systems shall also use QoE as the network performance.

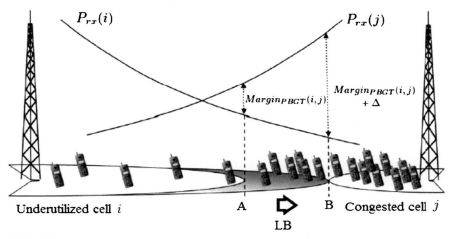

FIGURE 6.6

Handover margin modifications by Load Balancing [ARG17].

The effect of HO is further analyzed in [GLRT17], where a sensitivity analysis of throughput according to HO margins is presented. An alternative criterion for tuning HO margins is then introduced, focusing on end-user throughput. A mobile LB approach that modifies the service area of highly congested cells by decreasing HO margins, while simultaneously low congested cells increase their HO margins is introduced [GLRT17]. The assessment of this approach is carried out in a trial LTE network. Results show that the proposed indicator improves network performance in terms of end-user throughput from that obtained with classical mobile LB algorithms. The proposed algorithm changes cell service areas aiming to improve end-user throughput from that of classical mobile LB techniques. Unlike most research in this field [DTRM15], method assessment is not carried out in a simulator but in a trial LTE network [GLRT17].

RRM for DUDe scenario

RRM coordination can also happen between different BSs, where one BS is typically a macro BS while the other can be a pico BS. This is depicted in Fig. 6.7 and is termed as DL and UL Decoupling (DUDe) [SA18]. Some works claim that the gains of DUDe could reach up to 200%-300% in 5th percentile UL throughput when compared to a baseline case without Cell Range Expansion (CRE) + enhanced Inter Cell Interference Coordination (eICIC) [EBDI14].

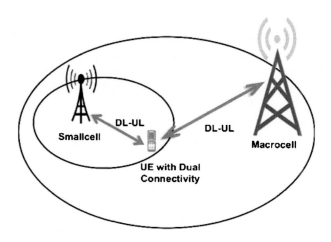

FIGURE 6.7

The diagram for the sample case of DUDe. Reprinted from [SA18].

In a DUDe scenario, a given user chooses the macro BS for its DL traffic, while for the UL the pico BS is chosen. Not only does DUDe split the UL and DL, but it uses also different cell selection criteria such as pathloss information or interference levels [WGLM+17]. When simulations for different user cases are run, it can be concluded that the CRE + eICIC slightly outperforms DUDe in improvement of UL user throughput when a dense user environment is deployed [WGLM+17].

6.4.4 PCA for higher capacity

Frequency reuse in cellular systems requires utilization of different methods for minimizing the interference [MH17]. One of these methods is to place a Power Control Algorithm (PCA) in charge of monitoring the link gains and managing the transmission power levels accordingly [Zan92]. There are two types of PCA: centralized and decentralized (distributed). Centralized PCAs have a single control center and calculate the best transmission power levels for an entire network. Decentralized PCAs have independent distributed control centers which are each in charge of a small region. Each of these control centers calculates the best transmission power levels for their region based on the live status of their local network links and sometimes their neighbors' as well. While decentralized PCAs generally provide less accurate results [Zan92], they are preferred today over centralized PCAs due to the fewer control links required and their real-time management speeds.

The decentralized PCA algorithm offered in the technical document [MH17], exploits the natural propagation of radio waves in a populated environment. It basically operates in a merge sort fashion; taking a large problem and splitting it into smaller overlapping ones, solving them with brute-force and then merging them back together. This Merging Brute Force Algorithm (MBFA) provides extremely accurate minimum acceptable threshold. Simulations have shown that the MBFA, in certain scenarios, can provide surprisingly high accuracy even though it is an approximation of the optimum algorithm, the brute-force (a centralized PCA). Accuracy of 99.34% is achievable when the algorithm is configured correctly for each particular scenario.

6.4.5 **Resource allocation and sharing for HetNets**

RRM across different networks has been studied extensively as the future 5G-grade NB-IoT systems will be utilized for the sensory data collection. In this heterogeneous network, it is expected that 5G networks will carry heavy traffic like video streaming, while the NB-IoT systems will carry low-load data traffic. In [SG18], an analytic performance evaluation framework that captures both service and network centric metrics of interest as well as the peculiarities of resource sharing between LTE and NB-IoT transmissions is developed. Using this tool, three feasible strategies for RA and sharing between LTE and NB-IoT sub-systems, namely: static, dynamic and dynamic with reservations are developed. The RA for each of these cases is represented in Fig. 6.8. A unique property of the considered model is that the basic channels are assumed to be scheduled sequentially by taking into account the features of in-band NB-IoT technology [WLA+17].

The analysis shows that static resource sharing between LTE and NB-IoT is highly sensitive to the offered traffic load. On the other hand, dynamic resource sharing between LTE and NB-IoT drastically improves the overall system performance. However, it does not automatically provide any guaranteed performance to the over-the-top services and may as well, unpredictably degrade performance, especially when the offered traffic load across sub-systems is non-uniform.

The dynamic RA strategy with resource reservation allows to satisfy the feasible reliability requirements needed by big data analytics. The proposed dynamic strategy with reservation is therefore strongly recommended for the forthcoming IoT deployments over 5G mobile networks.

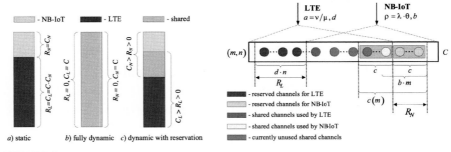

FIGURE 6.8

RA strategies and the process of sequential resource allocation in an LTE cell. Here, C represents the channels and R represent the minimum number of channels, with subscript L indicating LTE, and subscript N indicating NB-IoT.

6.4.6 A RRM tool

The previous sections presented different approaches and cases for RRM testing, where the researchers develop their own simulation platform, because the existing simulators are unlikely to find a one-fits-all solution. The Vienna LTE Simulator [MIŠ⁺11] is implemented in MATLAB® and it is rather focused on the lower layers, so it cannot appropriately model different service types. Besides, the simulation is rather time-consuming [DA17]. The NS3 [NS3] tool provides network level simulations but it also suffers from long simulation time [DA17].

To provide a better answer for such question, in Generic Wireless Network System Modeler (GWNSyM) a flexible platform that allows easy-configuration and easy-analysis of rather large and complex system deployments is developed. The tool has been designed with the main goal of being easily extended with either new functionalities or strategies. Unlike other tools that follow event driven simulation, the simulation with GWNSyM is based on snapshots, where the previous snapshot is used to feed the following one, which is required to analyze service evolution and dynamics (e.g. varying load) [DA17].

6.5 Heterogeneous networks and ultra dense networks

The ever-increasing demand for bandwidth hungry broadband services, and enhanced QoE, increased spectrum efficiency, and reduced energy consumption, have resulted in several challenges in designing next generation 5G wireless networks [HH15]. The use of network densification through the deployment of low power small cells, whether by a mobile network operator or an end user, is recognized as one of the key strategies towards achieving the 5G vision and targets. By densely deploying additional small cell nodes within the local area range and bringing the network closer to end users, the performance and capacity are significantly improved [SZHA15,FVV⁺14]. The energy levels decrease, while retaining seamless connectivity and mobility, resulting in improved QoE and user satisfaction.

6.5.1 Scenarios and capacity evaluation for small cell heterogeneous networks

Scenarios and Architectures for RRM and Optimization of Heterogeneous Networks have been discussed in [SVHR17], where the state of the art information related to scenarios and architectures from Energy efficient High-speed Cost Effective Cooperative Backhaul for LTE/LTE-A Small-cells (E-COOP) technologies is discussed. One of the proposed architectures exploits infrastructure based on small cell deployment using RRU technology which is connected to the core network using backhaul technology based on fiber optic links. Fig. 6.9 presents the general architecture for Cloud RAN.

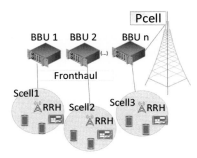

FIGURE 6.9

Cloud RAN architecture.

In Fig. 6.9, the Macro base station (PCell), operating in a low frequency band, e.g., 800 MHz, provides full coverage and Small Cells (SCells), operating in a upper frequency (e.g., 2.6, 3.5, or 5 GHz), provides local high data rate.

6.5.2 System level evaluation of dynamic base station clustering for coordinated multi-point

The authors from [Sch17a], [Sch17b] introduce promising techniques to achieve Coordinated Multi-Point (CoMP), which are partly standardized for LTE networks. The authors focus on Joint Transmission (JT) in the DL direction of the C-RAN, where the same data is transmitted from multiple eNBs, organized into CoMP clusters, so that the signals interfere constructively at the receivers. Under these clustering groups, cooperation is only allowed between eNBs belonging to the same cluster.

The realistic scenario under study accounts for user mobility patterns, Web-traffic modeling, and dynamic reconfiguration of clusters. Although LTE has been considered by the authors, as long as the concept of OFDM and the segmentation of time and frequency resources into resource blocks is utilized, the considered approaches are also applicable to the upcoming 5G.

Authors adopted the dynamic clustering algorithm (Algorithm 1) from [BBB14]. The simulation results of a realistic network including a vehicular mobility model indicate that the algorithm significantly improves the total performance in terms of sum data rate but also the individual data rates perceived by the UEs, even in the presence of mobility or highly dense scenarios.

6.5.3 Comparison of the system capacity between the UHF/SHF bands and millimeter wavebands

The viability of spectrum sharing in UHF, SHF, and mmWave in outdoor environments have been discussed in [BSV18]. In the mmWave the linear cellular topology is considered, while in the UHF/SHF bands cells with hexagonal shape are assumed. Performance evaluation includes the study of the behavior of PHY and supported throughput for 2.6, 3.5, 28, 38, 60, and 73 GHz. While the two-slope model is considered for the 2.6 and 3.5 GHz frequency bands, the modified Friis propagation model, with shadow fading and different values for the standard deviation, is considered in the millimeter wavebands.

The variation of the supported throughput with the coverage and reuse distances is analyzed in [TV18], for different values of the Channel Quality Indicator (CQI) and reference Carrier to Noise-plus-Interference Ratio (CNIR) requirements recommended by 3GPP, and given International Telecommunication Union-Radiocommunication Sector (ITU-R) propagation models. The PHY throughput is computed through the implicit function analytical formulation that was already applied in [SVP17], [TV18]. One example of the variation of the PHY throughput that corresponds to the curves of CNIR for the UHF/SHF and mmWave bands is presented in Fig. 6.10.

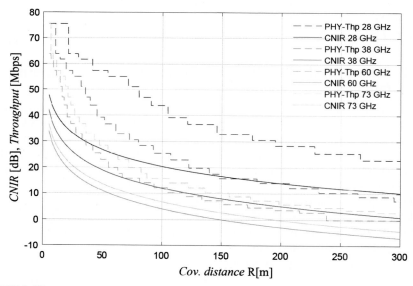

FIGURE 6.10

Variation of the CNIR/SINR and PHY throughput with d for 28 GHz, 38 GHz, 60 GHz, 73 GHz, for $R = 300$ m.

As there are limitations in the availability of dedicated spectrum for each Mobile Operator (MO), this work considers sharing without coordination, which means that each operator adopts the same frequency reuse strategy. Spectrum sharing access in heterogeneous networks assumes that two or more MOs have dedicated spectrum for macro cellular layer while SCells will share the access to spectrum in an opportunist manner.

In fact, scenarios where the frequency bands are shared are very relevant and an opportunity for new entrants, e.g., service providers, to own their light infrastructure.

These SCell networks will likely be deployed in a near future. The intermittent nature of traffic access will facilitate that usage of resources from different networks can temporarily access the infrastructure with the same resources. The authors from [BSV18] considered, as a case study, that one cell from MO #2 overlapped with cells from MO #1, as shown in Fig. 6.11(a), both using the shared band, as shown in Fig. 6.11(b).

In the sharing scenario from [BSV18], the supported throughput is highly reduced by the interference caused by the cell from a different MO. However, the ubiquitous coverage of the mobile communication system remains, as the PHY throughput does not reach zero inside the cell. To understand the practicability of considering spectrum sharing in UHF/SHF and mmWave, Fig. 6.11(b) puts together all the curves from the analysis of these frequency bands with and without sharing.

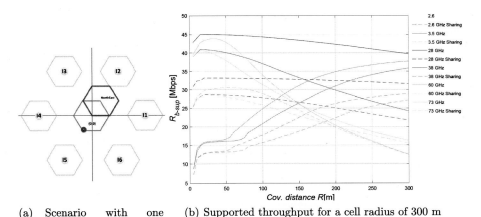

(a) Scenario with one UHF/SHF SCell interferer (Northeast)

(b) Supported throughput for a cell radius of 300 m

FIGURE 6.11

Comparison of the system capacity with/without sharing between the mmWave and UHF/SHF bands, with $K = 3$ and 20 MHz of bandwidth.

The supported throughput is higher for the 28 GHz frequency band compared to the remaining mmWave frequency bands (38, 60, and 73 GHz). One can also conclude that the system capacity decreases for increasing coverage distances. For shorter coverage distances up to circa 100 m, the supported throughput is higher at the millimeter wavebands, mainly due to the reduction that characterizes the application of the two-slope propagation model at the UHF/SHF bands.

It is noticeable that the supported throughput performance for the 28 GHz frequency band is the best compared to the other considered frequency bands, with values varying between the 45 Mbps and 39 Mbps for the scenario without sharing, and values varying between 30 Mbps and 33 Mbps for the scenario with shared spectrum. For all the other frequency bands, due to the behavior arising from the two-slope model (Umi LOS) applied to 2.6 and 3.5 GHz the supported throughput at the mmWave is higher than for the UHF/SHF bands, for the shortest coverage radii.

As the coverage distance increases the upper MCSs are supported up to shorter distances inside the cells. These results allow for interpreting why at the mmWave bands the system capacity is clearly higher for the shortest coverage distances, ceasing to be better for longer distances (except for the 28 GHz band, whose supported throughput is always the highest one).

For all the other frequencies considered, we have learned that the highest system capacity and the highest MCSs are achievable for the shortest coverage distances at mmWave bands, which have a higher throughput up to 156 m, after this coverage radius, the 2.6 and 3.5 frequency bands have higher supported throughput values, reaching a maximum of 38 Mbps. In fact, due to the behavior arising from the two-slope propagation model (Umi LOS) applied to 2.6 and 3.5 GHz, for shorter R the mmWave have the highest values for the supported throughput. One concludes that the mmWave band a higher supported throughput is achievable for shorter radii ($R \leq$ 156 m) while at the UHF/SHF highest values of the supported throughput only occurs for longer radii ($R > 156$ m). For cell radii up to 300 m, higher supported throughput always occurs for the 28 GHz frequency band.

6.5.4 Cost/revenue trade-off in small cell networks in the millimeter wavebands

Mobile cellular communications in the millimeter wavebands can support very high bit/data rates within small cells with short-range coverage. Based on the results for the equivalent supported throughput presented above, cost and revenues have been studied, [TV18]. One considers the cost/revenue model described in [VC03]. The revenues per cell, $R_v/cell$ [Euro], can be attained as a function of the throughput per Base Station, $thr BS$ [kbps], and the revenue of a channel with a data rate Rb [kbps], Rrb [Euro/min], and Tbh corresponding the equivalent duration of busy hours per day.

Revenues are considered on an annual basis. Six busy hours per day, 240 busy days per year [VCMV12], and the price of a 144 kbps "channel" per minute (corresponding to the price of approximately 1 MB) are also considered. In Fig. 6.12(a) presents the results for the overall cost per unit length per year, C_0, and the revenue per unit length per year, for only one carrier through the variation of R for $R_{max} = 300$. The revenues are noticeably higher than the costs of all frequencies. For 28 GHz band better results in terms of incomes are expressed. Fig. 6.12(b) presents the results for the Profit.

The results clearly indicate that short coverage distances should be used to optimize the system by having into consideration the maximization of the profit in percentage (mainly in 28 GHz), as the goal of operators and service providers is to enhance the profit.

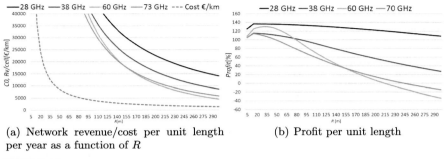

(a) Network revenue/cost per unit length per year as a function of R

(b) Profit per unit length

FIGURE 6.12

Cost/revenue trade-off with $R_{max} = 300$ m.

6.5.5 Effects of hyper-dense small-cell network deployments on a realistic urban environment

In [RK16] a realistic urban scenario for a hyper-dense small-cell network deployment is presented, using full 3D building models and 3D ray-optical pathloss predictions. In addition to the macro-cell layer, different expansion stages of the small-cell network deployment have been evaluated and compared to a 3GPP-like environment with equal site density. For this purpose, the complementary cumulative distribution functions (CCDF) of RSRP and SINR in case of a macro- and small-cell deployment on the same carrier as well as a standalone deployment of the small-cell layer, have been discussed. It has been shown that in both cases, the 3GPP-like environment provides far better coverage than the considered realistic scenario. With respect to SINR, the macro- and small-cell deployment behave similarly for the two scenarios. For the standalone case, the 3GPP-like environment provides rather steep CCDF curves, which are good to optimize, but lack certain realistic aspects. Particularly the missing degree of complexity in the building information leads to too simple predictions. Furthermore, different expansion stages of the small-cell deployments in both environments are discussed with respect to the coverage and interference situations.

6.5.6 Advanced management and service provision for ultra dense networks

Given the increasing complexity and service requirements of cellular networks, their OAM tasks have become increasingly not addressable solely by human operators' analysis. New approaches are needed for a full-stack service provision that is both optimized and reliable. In this line, the SON paradigm [NGN08] aims to automate the OAM activities and the general behavior of the network in order to optimize and automate failure management.

Location-aware self-organizing networks

Until now, classic SON approaches are based solely on the analysis of network-based metrics [AGBFM15]. However, Ultra-Dense Network (UDN) environments, with the very variable nature of their demand and high base station density imply deep limitations to those approaches [FBAGM15]. To correct this, the increasing availability of context sources [FPSB18a] and, specially, indoor localization systems [AGFCB15] provide an additional source of information to support self-optimization [AGFDB16][AGFG$^+$16] and self-healing [FGFL$^+$16][FAGB$^+$15][FBAG16] in those environments.

In this line, Fortes et al. [FAGB$^+$17][AGFMG$^+$15] proposed a location-aware detection mechanism for femtocell scenarios based on the comparison between the *expected* Received Signal Strength (RSS) values given the UE positions. Diagnosis of the root cause behind the issue is performed simultaneously including also the outcome of femtocell and router backhaul accessibility checks. The performance of the proposed algorithms was tested in a real four femtocell testbed. In this, different UE walktests were performed as shown in Fig. 6.13. These tests shown the capabilities of the proposed approach to support quick and reliable failure management in UDNs [FAGB$^+$17][AGFMG$^+$15].

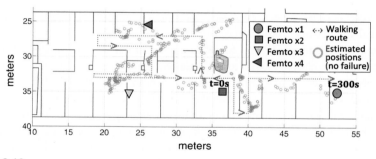

FIGURE 6.13

Testbed scenario and comparison between real and calculated positions.

Performance evaluation and packet scheduling in Home eNodeB (HeNB) deployments

Packet scheduling for UDNs is another key field of research, as studied in [PVP18b, PVP18a,PGB$^+$11] following the simulation assumptions of [3GP09]. The study presented in [PVP18b] shows that the variation of transmitter power of the HeNBs did not influence the average SINR, only the variation of the area of the apartments influences it. Further details and study in an scenario of 24 apartments is presented in [3GP09] and another scenario with a different approach to cover the same area is presented in [PVP18a]. For the analysis performed in [PVP18b], the geometry of the scenario is presented in Fig. 6.14, where a frequency reuse of two is considered.

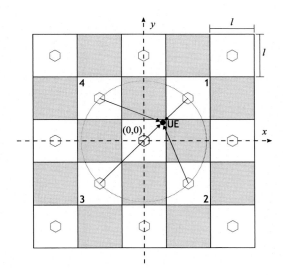

FIGURE 6.14

Simulation scenario with 25 apartments.

The type of performance evaluation defined in the standard by [3GP09] was performed by applying the LTE-Sim simulator [PGB+11,PVP16]. Results presented in Fig. 6.15 were obtained for the Proportional Fair scheduler and follow the expected theoretical behavior. For lower values of the transmitter and higher apartment side, the average throughput tends to decay more rapidly. The results for different schedulers highlighted that the best performances are not obtained with the HeNBs transmitter power to its maximum values, allowing for optimized and greener systems.

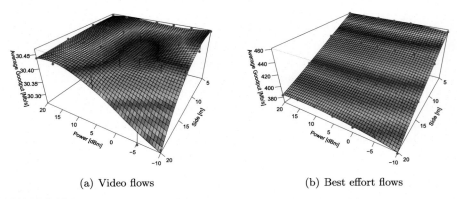

(a) Video flows   (b) Best effort flows

FIGURE 6.15

Variation of the average goodput, with PF scheduler and different values of transmitter power and room side. The fitting considers a polynomial surface with 95% confidence interval.

Optimization techniques and Knapsack optimization for MLB in small cells

Another main optimization challenge in UDN is the user association problem: the fact of determining the cell to be associated with each UE. The main modeling approaches for this are based on a "utility" cost function maximization which quantifies the satisfaction that a certain metric is met. Several techniques relying on combinatorial optimization were previously investigated [LWC+16,MAAV14,MBLM13]. Knapsack Optimization (KO) [MT90] is a new combinatorial optimization technique that is proposed for Mobility Load Balancing (MLB) for dense small cell deployments [NVM18,NM18].

An example use case based on a seven small cell scenario and 120 users unevenly distributed is used to compare the proposed MLB-KO in comparison with eICIC as defined in the 3GPP release 11 standard and with different CRE values and Almost Blank Subframes (ABS) of 20%. The Blocking Ratio (BR) is used to assess the effectiveness of the optimization as reported in Table 6.1. It is concluded that the MLB-KO technique outperforms CRE/ABS by at least 1.5 times for the 12 dB case increasing to 2.3 times for the 6 dB case and achieves more than four times improvement compared to the case with no MLB scheme.

Table 6.1 Comparison of BR for the different techniques.

Technique	Average BR (%)	Improvement factor of MLB-KO
No MLB	26.06	4.22 x
CRE = 6 dB	14.78	2.39 x
CRE = 9 dB	12.31	1.99 x
CRE = 12 dB	9.55	1.54 x
MLB-KO	6.17	-

6.5.7 IP mobility and SDN

Another important area to be addressed in UDNs is the mobility at higher layers of the protocol stack as well as the challenges open by the adoption of SDN. As studied in [BBB17,BBB17] traditional IP networking cannot solve the Layer 3 (L3) mobility problem. However SDN approaches can solve it efficiently: these works proposed to enable L3 mobility translating the client address at L3 level. This is done by using SDN for dynamic translations between the real temporal Internet Protocol (IP) addresses and virtual permanent IP address, resulting in seamless connection on both server and client sides and preventing the interrupts and subsequent reconnections.

This allows continuous data flow, which is of uttermost importance for Smart Things. Such a functionality has modest equipment requirements, the changes of the existing network are at the core layer by implementing SDN functionality. The rest of the network is implemented using the traditional approach. It does not use tunneling, but original packets with the least changes at the central router. As clients packets are not changed inside the local domain, optimal local routing and maximal link efficiency are preserved.

Compared to Proxy Mobile IPv6 (PMIPv6) via simulations in Mininet environment [Min], Fig. 6.16(a) shows the round trip time, while the maximal throughput is presented in Fig. 6.16(b). Switching time from a private IP address of one network to a private IP address of another network is reduced in the proposed solution, while the tunnel encapsulation causes visible decrease of throughput in the classical PMIPv6.

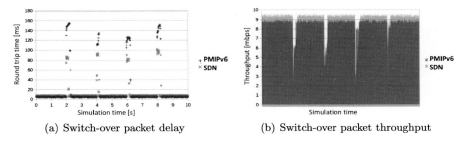

(a) Switch-over packet delay   (b) Switch-over packet throughput

FIGURE 6.16

Simulation results.

6.5.8 Digital geographical data and radio propagation models enhancement for mmWave simulation

New simulators or simulation bricks could be used by equipment manufacturers or network providers in the mmWave infrastructure dimensioning, performance assessment, automatic network design and optimization. Previous works related to mesh backhaul performance assessment and design [MCL16], adjustment and validation of propagation models [CCAL16], [ACL17] were included on the connectivity prediction for mmWave urban mesh backhaul networks.

The backhauling prediction engine must fulfill three key requirements: first, precisely estimate the visibility condition; second, predict the mmWave propagation loss for both LOS and NLOS situations; and third, support the indirect antenna alignment, i.e. when the directive antennas may be oriented not towards a dominant but towards an indirect propagation path.

Four different propagation methods on a large set of backhaul link candidates in a North American city are briefly presented in Table 6.2. The presented methods rely on distinct geographical data: Open Street Map (OSM) buildings in the first method; High-Resolution (HR) geographical data in the second one; completed with Lidar point cloud for the third and fourth methods. A direct-path propagation technique Direct-Path (DP) is employed in the three first methods. Multi-Path (MP) are enabled in the latter one.

The three methods consider geographical data models on the same one square kilometer area of the city, where most building heights are above 10 meters, and rows of trees are present along most of the streets.

Table 6.2 Tested propagation methods and LOS Percentage.

Tested Methods	DP/OSM	DP/HR	DP/Lidar	MP/Lidar
Geo data	Open building database from OSM	3D HR vectors	3D building vectors and Lidar point cloud	
Propagation model	DP only. Diffraction above buildings	DP only Diffraction above buildings. Interaction with vegetation vectors	DP only Diffraction above buildings. Interaction with vegetation point cloud	MP Interaction with vegetation point cloud
LOS percentage	59.7%	33.3%	11.7%	

The first model was extracted from an open building database, i.e. OSM. The HR database was produced by Siradel from stereoscopic aerial pictures. The last tested geographical database has been specifically elaborated for improving the accuracy of the propagation model inmmWave scenarios. Lidar point cloud was collected in all streets of the considered area, with a density and a measurement protocol that guarantee a precise 3D representation of most trees located in the street.

The propagation models under test were all obtained from different configurations of the VolcanoUrban tool [CL09]. The tool was adjusted and validated in several mmWave frequency bands, including 60 GHz. More particularly, the way to manage the vegetation obstacles has been improved [CCAL16] so both the transmission through the vegetation and the diffraction on top of the vegetation are considered.

A comparison between prediction models, Table 6.2, shows that large variations from one model to the other are mostly caused by the representation of trees. A large majority of links are found as LOS in absence of any the vegetation, while only 33.3% links remains unobstructed if main vegetation blocks and main rows of trees are added in the geographical data. Finally, the LOS percentage decreases as low as 11.7% when vegetation blockage is estimated from the rich tree representation offered by the Lidar point cloud.

Summary and conclusions

This section started by motivating the use of SCells in the context of HetNets and UDNs followed by the presentation of scenarios and general architecture for C-RAN in this framework. Combining dynamic clustering and scheduling for coordinated multi-point transmission enables to considerably enhance the total performance in terms of sum data rate but also individual data rates perceived by the UEs, not only in LTE but also in 5G networks. The comparison of system capacity between the UFH/SHF and mmWave was followed by a comprehensive study of the cost/revenue trade-off of SCell networks in the mmWave. A realistic urban scenario for a hyper-dense small-cell network deployment was considered whilst considering full

3D building models and 3D ray-optical pathloss predictions. In addition to the macro-cell layer, different expansion stages of the small-cell network deployment have been evaluated and compared to a 3GPP-like environment with equal site density, which present much better performance than the realistic one. Femtocell scenarios and advanced management and service provision for UDNs have been discussed and research on location-aware SONs has been addressed. Performance evaluation of HetNets in UDNs compared different packet schedulers, and considered Knapsack optimization and simulation methodologies. The mmWave have also been addressed by considering an approach for radio propagation modeling enhancement accounting for digital geographical data, e.g., vegetation obstacles.

6.6 C-RAN

6.6.1 Resource management in C-RAN

C-RAN as a key feature of 5G, offers a lower cost solution to the exploding data rate demands. C-RAN splits the Base-Band Units (BBUs) of BSs from their radio units, i.e. the Remote Radio Heads (RRHs), and integrates the decoupled BBUs into a central location, known as a BBU-pool. By leveraging the advantage of load fluctuation, it improves the utilization of network resources by sharing the resources of under-loaded BBUs with the over-loaded ones. An efficient resource management strategy is needed however, to distribute BBU-pool scarce resources among BBUs, especially in case of resource shortage, when not all of the BBUs demand can be served at the same time.

Brahman et al. in [BCF18a] and [BCF19a] propose an on-demand resource allocation scheme to optimize the utilization of computational resources in a BBU-pool. Resource allocation is formulated as a centralized optimization problem, based on the concept of bargaining in cooperative game theory. Having the real-time BBU demand and the weight of the active services in the BBU, a bargaining power is calculated and assigned to a BBU that prioritizes it in case of a resource shortage. Meanwhile, a minimum amount of resources is guaranteed to be allocated to prevent the BBU from crashes. Therefore, the computational resource distribution among N_B BBUs in a pool is optimized by solving the following problem:

$$\max_{\mathbf{C}^{Al}_{b,t_k[N_B \times 1]}} \prod_{b=1}^{N_B} \left(C^{Al}_{b,t_k \text{ [GOPS]}} - C^{R}_{C\,b,t_k \text{ [GOPS]}} \right)^{B_{b,t_k}} \tag{6.2a}$$

$$\text{subject to} \quad C^{Al}_{b,t_k \text{ [GOPS]}} \leq C^{R}_{b,t_k \text{ [GOPS]}}, \quad b = 1, 2, ..., N_B \tag{6.2b}$$

$$\sum_{b=1}^{N_B} C^{Al}_{b,t_k \text{ [GOPS]}} \leq C^{Av}_{BP\,t_k \text{ [GOPS]}}, \tag{6.2c}$$

where:

- C_{b,t_k}^{Al} : allocated computational capacity to BBU b at time instant t_k,
- C_{b,t_k}^{R} : required computational capacity of BBU b at time instant t_k,
- B_{b,t_k} : bargaining power of BBU b at time instant t_k,
- $C_{BP\,t_k}^{Av}$: available computational capacity in the BBU-pool.

The constraints defined in (6.2b) and (6.2c), ensure respectively that a BBU cannot be allocated with more resources than it requires and that the total allocated resources may not exceed the available ones in the BBU-pool. Simulation results for a scenario with seven BBUs in a BBU-pool and a heterogeneous service environment show a minimum 83% improvement of allocation efficiency compared to a fixed resource allocation policy based on peak requirements.

Later, in [BCF18b] and [BCF19b] the authors show that the above model for computational resource management is also applicable to a time varying network where the BBU load and computational resource demand are dynamic. Based on changes in the instantaneous demand of the BBU, the model is re-calculated and the BBU-pool computational resources are distributed among BBUs dynamically at each time instant. Simulation results for a 5-minute interval in the same scenario show that the assigned resources change in accordance with BBUs demand. Moreover, the results confirm that BBUs demand is fulfilled more than 98% without any processing delay by using only 43.5% of the BBU-pool capacity.

Gomez et al. in [GBLO19b] proposed a similar approach for computational resource management inside a BBU-pool, the cost function given in (6.2a) being replaced by the following linear function:

$$\max_{\mathbf{C}_{b,t_k[N_B \times 1]}^{\mathbf{Al}}} \sum_{i=1}^{N_B} B_{b,t_k} C_{b,t_k}^{Al} \text{ [GOPS]} \tag{6.3}$$

Later, the authors in [GBLO19c] propose the use of three different Maximum Likelihood (ML) techniques, i.e. Medium Gaussian Support Vector Machine (SVM), Time Delay Neural Network (TDNN), and Deep Learning using Long Short-Term Memory (LSTM), to provision computational resources more closely adapted to the actual needs of the BBU-pools.

Fig. 6.17 shows the error distribution of three proposed techniques for a BBU-pool resource provisioning. As the figure illustrates, SVM offers the best accuracy, with lower Root Mean Square Error (RMSE), compared to the other methods. In order to decrease the underestimation probability, negative errors in Fig. 6.17, the authors propose filtering data prior to the LSTM training process. The experimental results confirm that the underestimation probability reduces to zero however, the amount of the idle resources increases.

6.6.2 C-RAN deployment

The deployment of a reliable C-RAN capable of meeting network operators interests, is another challenging issue, since the trade-offs among conflicting objectives, e.g. delay minimization and load balancing, should be addressed.

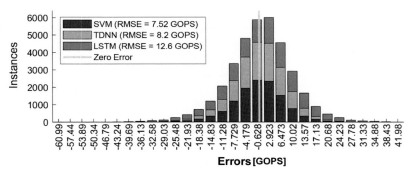

FIGURE 6.17

Histogram of the errors in GOPS for the three ML strategies.

Gomez et al. in [GBLO19a] take four different algorithms into account in the process of designing the C-RAN, each with a distinct criterion, i.e. delay minimization, BBU-pool load balancing based on the traffic, BBU-pool load balancing based on the number of the aggregated RRHs, and maximizing the multiplexing gain. Results on C-RAN deployment in a sector of Vienna city confirm that taking just the minimum delay algorithm into account minimizes round trip time, but causes lower performance and unbalanced load. On the other hand, the load balancing algorithm presents even traffic or even number of RRHs per BBU-pool. However, the round trip time is increased in both of them.

An appropriate C-RAN deployment improves both the load balancing of the network and the delay constraint satisfaction. However, the proliferation of smart IoT devices causes excessive load on the back-haul, in between massive devices and BBU-pool servers. Besides, C-RAN is an impractical solution for many delay-sensitive applications, because of the long distance between the device and the cloud center. An alternative approach is to offload some of the computing tasks from cloud servers to the network edge in the form of a new network topology known as fog computing.

Moratta et al. in [MAR+18] propose a model to determine the BBU, either in the fog node or in the cloud, that is eligible to process the workload of an RRH. The proposed model considers the round-trip delay as a composition of the propagation delay and of the processing delay. Since the propagation delay in a fiber optic link is a function of the distance between RRHs and its serving BBU, by acquiring the processing delay, the proposed model maximizes the distance of the RRH from its serving BBU so that the delay budget is not violated. Results on transferring a 6144 bits codeblock in Fig. 6.18 show that in order to be able to meet the delay constraints in case of low SNRs, either significant processing resources should be allocated to speed up signal processing, or the BBU must be located near the RRH.

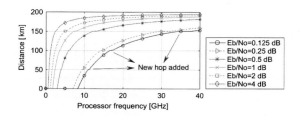

FIGURE 6.18

Maximum distance as a function of the BBU's available processing capabilities and SNR.

Besides the communication processing, the edge computing resources can also be utilized for application processing, the applications latency being reduced by offloading the computing on the edge devices, with more processing power, rather than the mobile devices with limited processing capability. Meanwhile, the trade-off between the energy efficiency and the response time, needs significant attention.

Sopin et al. in [SSC19] propose a model to offload processing tasks of a mobile device that are "heavy" in terms of the processing volume, that require more processing resources, and "light" in terms of the data size, that can be transferred fast, to a fog node. The threshold on the processing volume and the threshold on the data size filter "heavy" and "light" tasks. Fig. 6.19 depicts the impact of the offloading thresholds on the average response time in case of 20 mobile devices with random generated tasks. When the offloading probability decreases, smaller data size threshold and larger processing volume threshold, the load on mobile device becomes larger and causes an increase in average of both response time and energy consumption for locally processed tasks.

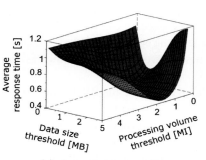
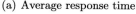
(a) Average response time

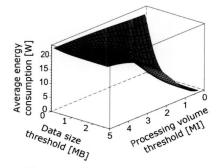
(b) Average energy consumption

FIGURE 6.19

The impact of the offloading thresholds on the average response time and energy consumption.

6.7 SDN and NFV

Network Slicing, service customization and RAN Cloudification are important enablers of 5G networks, all of which are based on the capabilities of NFV and SDN [RCC18c]. Decoupling control plane from data plane in SDN, can considerably improve programmability and customization of virtual wireless networks. Meanwhile, virtualization enhances scalability and flexibility, leading to the improvement of resource utilization in SDN.

6.7.1 Virtual radio resource management model

This sub-section presents a high-level, service-oriented model for RAN slicing and RRM. The key objective is to establish a level of performance isolation between the VNOs, which are acting as network tenants, in order to ensure that their contracted Service Level Agreements (SLAs) will not be affected by the variation of different network parameters [SGSC19]. At the same time, another main goal is to share the aggregated capacity among various service slices on-demand, and in a fair manner [RCC17a]. To achieve these goals, a centralized virtualization platform called Virtual-RRM (VRRM) is modeled as an individual management entity based on the hierarchical network architecture which was presented in [RCC17b]. VRRM does not own the infrastructure and is in charge of satisfying the demands of different service slices based on the aggregated capacity provided by multiple Radio Access Technologies, owned by Infrastructure Provider.

The VRRM is analytically modeled as a convex optimization problem, according to the criterion of weighted proportional fairness. f_{VRRM}, as a measure of efficiency, maps a portion of network bandwidth that users utilize into a real number, quantifying the expected satisfaction of users given the allocated resources. This way, VRRM makes a bridge between the functionalities of MAC and higher layers, by optimizing the allocation of radio resources for different applications. There are also some constraints associated with the model, in order to apply the customized range of serving rates considering the individual VNO's policies and the contracted SLAs [RCC18d].

$$\underset{\boldsymbol{w}^{usr}}{\text{Max}} f_{VRRM} \left(\boldsymbol{w}^{usr} \right) = \underset{\boldsymbol{w}^{usr}}{\text{Max}} \sum_{k=1}^{N^{srv}} \lambda_k \log \left(\sum_{i=1}^{N_k^{usr}} w_{k,i}^{usr} \frac{R_{k\ [\text{Mbps}]}^{srv_{max}}}{R_{[\text{Mbps}]}^{VRRM}} \right) \quad (6.4)$$

where:

- N_k^{usr}: number of users performing service k,
- N^{srv}: number of provided services,
- $R_k^{srv_{max}}$: maximum assignable data rate for service k,
- R^{VRRM}: total available capacity to VRRM,
- $w_{k,i}^{usr}$: assigned weight to user i, performing service k, ranging in [0, 1],
- λ_k: weight of service k, prioritizing data rate assignment.

6.7.2 Analysis of VRRM results

The performance of the VRRM model is presented and analyzed in this sub-section, based on the scenario which was proposed in [RCC18a], [RCC18b] for three types of SLAs: Guaranteed Bitrate (GB), Best effort with minimum Guaranteed (BG), and Best Effort (BE).

Fig. 6.20 shows the effect of the number of offered users on the total capacity share of the three VNOs. From this perspective, two parameters play decisive roles in decision making: the acceptable ranges of data rate variation for each VNO, which are indicated in the figure as one type of constraint, representing the internal policy of each VNO for service management. Another factor is the contracted capacity for each SLA type, which is fixed and does not change by the variation of traffic. The colored areas in Fig. 6.20 represent the case when all offered users in each VNO are being served. However, it is notable that when the number of offered users is very low, all users in VNO GB are served with the highest acceptable data rates, which are in fact lower than the values of SLA contract. In this special situation, the allocated capacity to VNO GB according to the VRRM algorithm, is set to the maximum acceptable data rate which can be offered to GB users.

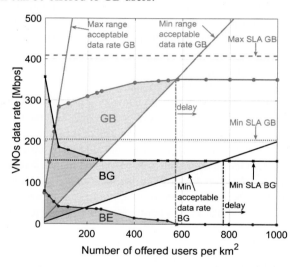

FIGURE 6.20

Capacity sharing of VRRM among the VNOs [RCC18b].

After the intersection points, when the minimum acceptable threshold of data rate surpasses the allocated capacity of each VNO, VRRM starts to delay the users from the suffered VNO based on the service priorities. As an example, when the delay in VNO GB starts, since this VNO has the highest priority among the three, all the users of VNO BE has already been delayed (as there is no minimum SLA guarantees for this VNO), and the allocated capacity of VNO BG has reached down to the minimum contracted SLA threshold.

6.8 **UAVs and flying platforms**

UAVs have gained particular interest in the literature thanks to their unique characteristic of moving almost freely in the sky. Equipping a UAV with radio devices can improve the path loss and link budget conditions. Since a UAV can travel when and where it is most needed, trajectories can be planned based on a specific task such as for example addressing user demand in a certain area or supporting the existing terrestrial infrastructure. This section discusses various opportunities of UAV-aided networks.

6.8.1 **UAV trajectory design and radio resource management**

When it comes to effectively exploit the ability of flying freely in the sky, UAVs are used for a cost-efficient deployment or real-time computed trajectories. Trajectory design is preliminary addressed by [MV17], considering one UAV as an UABS. The UABS has the purpose of supporting an underlying mobile radio network to provide service to users while accounting for the dynamic behavior of the network and user mobility. The trajectory is planned by clustering the users which remain unserved from the terrestrial network. For each cluster, a centroid point is defined such that the UABS can move from centroid to centroid by selecting the nearest to its current location. Even though, the trajectory design is straightforward, a significant improvement in the performance of the overall network can be noticed. A more advanced dynamic trajectory design is therefore introduced by [DMM+17]. Its novelty stands on the fact that it is event-driven, meaning that once a UABS arrives in a certain cluster, the algorithm is able to define its next target cluster on-the-fly. This allows the network to respond to the very challenging and quickly varying user traffic in future wireless networks. The trajectory design leverages again on the clusterization of [MV17] but chooses the next cluster centroid depending on a cost function that weights distance, user density, and direction fairness. Furthermore, an interference avoidance approach from the UABS to the terrestrial network is proposed, to allow the UABS to reuse the same frequency band in an attempt to efficiently reduce implementation costs. If efficiently implemented, RRM can both boost system performance and further decrease operational costs. [VM18] discusses the potential advantages of the integration between the terrestrial and aerial components of an UAV-aided mobile radio network, from the view point of RRM. RRM is a relevant aspect that should be addressed along with the real-time identification of the optimal route of the UAVs in a dynamic context. In fact, the UAV position, being a degree of freedom, can be seen as an additional dimension in the resource pool administered by L2 and L3 radio resource assignment algorithms. Simulation results show that a single UAV can bring an improvement in terms of network throughput of up to 1 percent. Using multiple UAVs, the performance increase can be much larger. The RRM becomes joint when the radio resources are scheduled in a coordinated manner between the terrestrial network and the UAVs, achieving a higher user throughput and an efficient network operation. The joint aerial-terrestrial resource management can be further improved by considering the decisions made in the RRM in the trajectory design as shown in [MV18]. Five factors driving the UAV to the cluster through the cost function have been identified: energy, distance, throughput estimation, density of users, and Radio Resource Units (RRUs) availability.

UABS trajectory planning can also be formulated as an optimization problem. The goal of [MBCV18] is to maximize the number of served ground users, weighted according to their priorities, while considering constraints on the battery duration, UABS speed, and data rate. Users requesting for high throughput video services in urban environments are considered as served only when the entire video is downloaded. The problem is formulated as an ILP model, that computes the optimal path for the UABS, maximizing the number of served users, while accounting for the maximum number of radio resources available. A heuristic solution method is also presented to reduce the problem complexity and time of computation.

An important aspect for UABS network design is the impact of the Air-to-Ground (ATG) channel on which the UABS operate. [MBV18] determines how the speed and height of the UABS is influenced by the radio channel parameters. The effect of different antenna systems (with different antenna gains) is tested for UAV trajectories through simulations. The path planning accounts for user distance, density, spatial fairness, and additionally the UAV energy consumption. Moreover, the UABS is assumed to know a priori extracted Radio Environmental Map (REM) data, to better estimate the capacity it can provide to ground terminals. The results bring to the general conclusions that i) different antenna gains are more effective at different UABS heights: an increase of the UAV altitude requires a decrease in the radiation angle, ii) speed and radiation angle have shown no impact nor dependencies one to the other regarding throughput outcome. Furthermore, Ray Launching simulations for the city map of Bologna (Italy) are computed to create a realistic REM and compare these output data with statistical models [SM19]. For heights between 100 and 200 m, results show a small difference on the throughput gain in percentage for an antenna aperture angle of 120 degrees, but the gap is increasing as the aperture angle decreases.

Vehicular scenarios for 5G use cases are quite challenging for a cellular network not specifically planned for it. Moreover, when the penetration of cars equipped with wireless communication devices is far from 100%, the use of UAVs, carrying mobile base stations, becomes an interesting option. It is possible to integrate aerial and terrestrial network components operating on a single carrier frequency by using an efficient joint UAV trajectory design and radio resource assignment that accounts for vehicle mobility, vehicle density, and terrestrial network operation [MBBV19]. Network performance might increase thanks to the UABS of up to 10% in the proposed scenario.

6.8.2 UAV-aided network planning and performance

The paradigm of the previous four generations of cellular technology has led to high data rate demands from mobile users envisioned in the new 5th generation technology [SJM18]. To support this, ultra-dense networks have been proposed as a promising 5G feature. However, ultra-dense networks face many challenges such as severe interference resulting in a limited capacity due to the dense deployment of small cells, site location and acquisition for the deployment of base stations, back hauling issues, energy consumption, etc. By showing that the coverage range of a UABS ranges from 50 to 200 m radius as well as the cell coverage variation with varying UABS fly height which can be as high as 45%, Sharma et al. conclude that the use of UABSs is a promising solution to overcome some of the aforementioned issues.

One of the limiting factors of UAV-aided networks is the inter-cell interference, especially at high altitudes. The study of [CVP18], which uses a realistic 3D simulator mode, conforms that UAVs at high altitudes suffer from significant interference, resulting in a worse coverage compared to ground users. By replacing the aerial base stations by mmWave cells, the ground coverage decreases to only 90%, but the UAVs just above rooftop level have a coverage probability of 100%. Unfortunately, UAVs at higher altitude still suffer from excessive interference. To reduce this interference, beamforming is a promising technology.

An interesting scenario for the applicability of UABSs is an emergency disaster scenario in which the existing terrestrial wireless infrastructure is saturated due to for example a natural disaster or a terror attack. By using UAVs, LTE femtocell base station can be brought to the disaster area to temporarily provide connectivity to the end user. Deruyck et al. have developed a deployment tool which allows not only to determine the required amount of UAVs, but also the most optimal locations of these UAVs to maximize the user coverage [DWMJ16]. The preliminary results of the study show that the envisioned scenario is a very promising application domain for the use of UABSs. Nevertheless, for the considered UAV specifications, a large amount of UABSs - approximately 1100 UABSs - is required to cover a city center of 6.85 km^2 like in Ghent, Belgium. The amount of required UABSs is highly influenced by the UAV specifications, the intervention duration, and the fly height or to limit the user coverage requirement. Since the latter is the less preferred choice, [DWP$^+$17] investigate whether it is interesting to reduce the number of required UABSs by installing femtocell base stations in the vehicles of the emergency services and public transportation. Deruyck et al. show that this is indeed feasible but the effect is rather limited since only 5% of the users in the affected area can be reconnected through the vehicles. The main reason is that the vehicle's location is not known beforehand and can not be chosen as freely as it is the case for the UABSs. One issue related to using UAV-aided communication in emergency scenarios is a proper design of the backhaul network, i.e. from the drone to the core network. One way of doing this is to provide a direct link between the UABSs and the core network by using currently unoccupied frequency bands such as 3.5 GHz and 60 GHz [CDMJ19]. For the same emergency scenario mentioned above in Ghent, Belgium, this can be a good solution, especially when using the 3.5 GHz frequency band in combination with carrier aggregation. Without carrier aggregation, the contribution is limited due to the limited amount of available resource blocks in the 3.5 GHz band. For the 60 GHz band, the main limitation is the high path loss due to the buildings in the city environment.

Mfupe et al. discuss in [MK18] the different problems that should be addressed to use UABS communication as an alternative communication for Disaster Management services, but also to provide affordable broadband connectivity to the hard-to-reach rural communities.

As mentioned in the previous section, due to the flexibility of the UABS location, a swarm of flying platforms can also be considered as an integrated part of the future cellular network. This swarm can inject additional capacity and expand the coverage for hard-to-reach areas or exceptional scenarios such as sports events or concerts. Ahmadi et al. propose a novel layered architecture where Network Flying Platforms (NFPs) of various types - such as UAVs, unmanned balloons or High Altitude Platforms (HAPs)/Medium Altitude Platforms (MAPs)/Low Altitude Platforms (LAPs) - are flying in low/medium/high layers in a swarm of flying platforms and are assumed to be an integrated part of the future cellular network as shown in Fig. 6.21 [AKS17]. The position of the LAPs in the lower layer of the architecture is defined centrally and the NFP has the ability to re-organize the lower layer to achieve its target, which can be capturing as many users, maximizing the achievable rate, and/or fairness among users. The proposed airborne SON systems that reorganizes itself as explained above outperforms an NFP with a fixed placement in the lower layer.

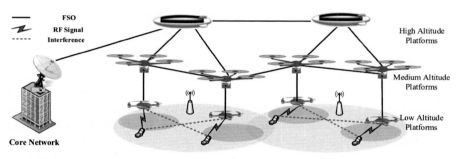

FIGURE 6.21

The hierarchical airborne self-organizing architecture where a variety of UABSs are flying at different altitudes assisting the existing wireless terrestrial infrastructure (Figure taken from [AKS17]).

When talking about UAV communication, one typically thinks about airborne scenarios, however, it can also be of major interest for maritime scenarios. In [FAZ19], three different UAV-based architectures are proposed to provide a solution for reliable, low latency cellular links for search and rescue operations over the sea by using a flying relay, a flying base station, and a flying remote radio head. Another possible maritime scenario is considered in [VNC+15] where the goal of swarms of surface autonomous vessels with multi-hop communication capabilities is to enable maritime tasks such as sea-border patrolling and environmental monitoring, while keeping the cost of each drone low enough to facilitate large-scale deployment. Experiments with their prototypes showed that LOS has critical importance at distance >100 m. The maximum obtained distance for maximum transmit power is 1400 m. Their test considering the multi-hop technology has also been successful.

6.9 **Emerging services and applications**

The introduction of 5G and beyond technologies together with the IoT will result in the proliferation of a large number of communicating devices in different smart environments targeting different services and applications. In addition to traditional voice and data services, future inclusive wireless networks will be used to fulfill the needs of a large number of verticals in a variety of smart environments and sectors. Examples include smart homes, smart health, smart manufacturing, smart grids, smart vehicles, smart agriculture, and smart cities. New technical solutions are required to fulfill the diverse requirements of these applications. Within the COST-IRACON action, a number of new and emerging applications were investigated.

6.9.1 **Smart grids**

The introduction of smart grids results in increasing needs of real-time scalable and reliable monitoring, control and protection of electric power generation, transmission and distribution systems. Monitored state information, e.g. voltages, currents, and user demands, is delivered through a number of nodes to a central unit for performance analysis and different purposes including control and optimization of the grid performance. The technology used for data exchange adopts a Wide-Area Monitoring System (WAMS). Two typical nodes deployed in a WAMS, are Phasor Measurement Units (PMUs) and Smart Meters (SMs). A PMU delivers time-synchronized values of voltage and current phasors as well as other power system related quantities. A smart meter is an electronic device that records the consumption of electric energy and communicates the information back to the electricity supplier for monitoring and billing purposes. In [LJ16] and [AT18], the feasibility of using a Long Term Evolution (LTE) cellular network for real-time smart grid state estimation is investigated. The delay performance is assessed for different cell loads and scheduling schemes with the assumption that all the nodes are connected. It is concluded that with one Physical Resource Block (PRB), 2500 nodes can be accommodated within an LTE cell if the maximum allowable delay is relaxed from 1 s to 2 s. In [AT18] the impact of LTE random access procedure on the reliability of state information delivery was studied. It is concluded that the number of contending devices and cell coverage critically affect the performance in terms of the accuracy of state estimation in smart grids.

6.9.2 **Vehicular applications**

Vehicle to Vehicle and Vehicle to infrastructure (aka V2X) communications are drivers for road safety and efficiency. Smart and autonomous vehicles pose communication challenges in terms of delay, reliability, bandwidth, and energy consumption. It is necessary to investigate penetration losses into a vehicle to evaluate network performance and to provide reliable connectivity and large capacity for vehicular use cases such as smart cars and trains. In [Ber17], the LTE coverage was analyzed and the penetration loss was evaluated for a pickup truck in a live network. Depending on the orientation of the vehicle and the environment, the penetration loss can vary from 1.88 to 10.58 dB. A V2X system level simulator was presented in [ND18] including an optimized ray tracer for path loss predictions to study IEEE 802.11p and LTE-V mode 3 and 4 systems.

6.9.3 Public protection and disaster relief systems

Characteristics of current narrowband public protection and disaster relief communication systems include the following: reliability, speech and data transmission capability, point to point, group and broadcast calls, fast call setup, coverage, long battery life, flexibility, prioritization, and security aspects. In [PS17] and [PSP18], the migration towards broadband LTE/LTE-A networks was reviewed. Proximity-based services (ProSe), group communication system enablers (GCSE_LTE) and mission-critical push to talk (MCPTT) are essential features to be implemented. Network resource prioritization, Quality of Service (QoS) provision, and spectrum management are key for coexistence with commercial services. Carrier Aggregation (CA) is presented as a potential solution for an LTE-based public safety and disaster relief network. Issues to consider include negotiation between commercial and public safety service providers, coverage overlap, prioritization, and impact on battery life.

IoT protocols, architectures, and applications

Chiara Buratti[a,b], **Erik Ström**[c,b], **Luca Feltrin**[d], **Laurent Clavier**[e], **Gordana Gardašević**[f], **Thomas Blazek**[g], **Lazar Berbakov**[h], **Titus Constantin Balan**[i], **Luis Orozco-Barbosa**[j], **Carles Anton-Haro**[k], **Piotr Rajchowski**[l], and **Haibin Zhang**[m]

[a]*University of Bologna, Bologna, Italy*
[c]*Chalmers University, Göteborg, Sweden*
[d]*Ericsson, Stockholm, Sweden*
[e]*Université de Lille, Lille, France*
[f]*University of Banja Luka, Banja Luka, Bosnia and Herzegovina*
[g]*Silicon Austria Labs, Linz, Austria*
[h]*Institute Mihailo Pupin, Belgrade, Serbia*
[i]*Universitatea Transilvania Braşov, Braşov, Romania*
[j]*Universidad de Castilla-La Mancha, Ciudad Real, Spain*
[k]*Centre Tecnològic de Telecomunicacions de Catalunya, Barcelona, Catalonia, Spain*
[l]*Gdańsk University of Technology, Gdańsk, Poland*
[m]*TNO, The Hague, Netherlands*

The proliferation of embedded systems, wireless technologies, and Internet protocols have enabled the IoT to bridge the gap between the virtual and physical world enabling the monitoring and control of the environment by data processing systems. IoT refers to the inter-networking of everyday objects that are equipped with sensing, computation, and communication capabilities. These networks can collaboratively interact and perform a variety of tasks autonomously.

A large variety of communication technologies has gradually emerged, reflecting a large diversity of application domains and requirements. This chapter describes some research activities performed with reference to such technologies and solutions. Section 7.1 reports the research in the framework of LPWANs, with emphasis on LoRa and Narrow Band IoT (NB-IoT) technologies. Studies on centralized approaches for IoT, mainly based on the IPv6 over the TSCH (6TiSCH) standard, are addressed in Section 7.2; while Section 7.3 deals with vehicular communications. Energy-efficient solutions are presented in Section 7.4, being energy consumption one of the main issues of IoT; Section 7.5 reports some architectural solutions for the

---

[b] Chapter editors.

Inclusive Radio Communications for 5G and Beyond. https://doi.org/10.1016/B978-0-12-820581-5.00013-4
Copyright © 2021 Elsevier Ltd. All rights reserved.

187

application of the SDN and NFV paradigms to IoT. The chapter ends with research on specific applications and drawing some conclusions.

Most of the research described in the chapter has been conducted via experimentation, using testbeds described in IRACON White Paper on Experimental Facilities [AB18].

7.1 Low power wide area networks

The LPWAN technology has recently emerged specifically focusing on IoT applications which require low cost device, long battery life time, small amounts of data exchanged and long distances to be covered. The key features of LPWANs can be summarized as follows: (i) wide area coverage (up to some tenths of kilometers), (ii) low cost communication, (iii) long battery life (up to 10 years), and (iv) low bandwidth communication. Among these technologies, LoRa and NB-IoT are those supported by many network operators and have been investigated in many research works. This section discusses the achieved results, possible improvements of these technologies and some real applications which may benefit from their use.

7.1.1 LoRaWAN

LoRaWAN is one of the first technologies defined for LPWAN applications. Its standardization was initiated by Semtech, an American company which owns the patent for the synthesizer used to generate the modulated signal, and later by the LoRa Alliance, an organization of many companies which shared the effort to define a new standard for these new applications.

Two independent studies aimed to characterize the link layer performance of LoRa modulation. In [FMB+16,FBV+18] two Semtech SX1272 modules were deployed at increasing distances and in LOS conditions, one module was on a 240 m high hill in Bologna and the other one, the transmitter, was deployed in different locations up to 10.8 km far from the receiver, the maximum distance reached in this experiment.

In [CLC+19], the ranging capabilities of LoRa modulation are evaluated in three distinct environments, i.e., coastal, forest, and urban with antennas at 1.5 m height. The SNR is boosted by increasing the Spreading Factor (SF). In addition to improving the SNR, the sensitivity is increased yielding a higher link budget. However, transmitting with a higher SF will not result in a higher Received Signal Strength (RSS). Hence, the SNR does not decrease to the same degree as the RSS, with increased distance. Even in the urban scenario, a good average SNR, w.r.t. the demodulation floor, is measured despite the low received signal strength. The experimental results indicate that the range is mainly limited by the SNR demodulation floor rather than the RSS. The transmit power, coding rate, spreading factor, and other parameters will determine the maximum coverage. The measurements show that the SNR sensitivity is reached before the RSS sensitivity for SF equal to 12. This confirms that the range is constrained by the SNR.

In [FMB+16,FBV+18] the orthogonality among transmissions with different SF was checked. Two devices transmitted simultaneously packets containing independent payload and using different SF. The packet success rate (P_s) was estimated by counting the number of received packets, while the Signal-to-Interference (SIR) was measured with a spectrum analyzer. In terms of capture ratio, defined as the minimum SIR allowing to guarantee a success rate of 50%, the experiment shows that when the SF of the two transmissions is the same it ranges between 1 and 2 dB, whereas when the SF of the two transmissions is different it ranges between -10 and -30 dB. This demonstrates that transmissions with different SF are not perfectly orthogonal, but in some cases the interference of concurrent transmissions should be taken into account.

In [FBV+18] the performance of a large network was evaluated through simulations, considering a square area with the LoRa Gateway (GW) located in the center and a variable number of nodes uniformly deployed. The network capacity is computed as the maximum number of packets per unit of time that a LoRa GW is able to process while guaranteeing a packet error rate lower than 10%. Fig. 7.1 (left) shows how the packet success rate changes when the offered traffic increases in a square area of 1 km^2, where all nodes can reach the gateway using any SF, and when 200 bytes of application data are transmitted. Different curves represent different settings of initial SF,[1] when using Acknowledgment packets and retransmissions (Confirmed) and when not (Unconfirmed). The simulations show that in case of a small area to cover, the best solution is to set SF equal to 7, in order to avoid collisions by keeping the packet time-on-air as small as possible. Moreover, when there is not much traffic in the network, the confirmed mode is the best choice to guarantee a high success rate; while when the network is more congested, retransmissions increase the collision probability. In this scenario a single GW is able to process up to 1.71 packets/s. Fig. 7.1 (right) shows how the packet success rate varies when the offered traffic increases in a larger area (46.5 km^2), where it is not possible to cover all devices with SF equal to 7 and only one GW. To overcome the coverage issue two solutions are represented in the figure, letting the device increase their SF when a confirmed message has not been acknowledged, or increasing the number of GWs deployed in the area avoiding confirmed transmissions. In the first case, the radio channel is utilized more intensively due to longer packet duration, a behavior that can lead quickly to the saturation of the network. The resulting network capacity is 0.46 packets/s. In the latter case, more GWs need to be deployed, but the average SF used by the devices, and thus the packet duration, is lower causing less collisions. The resulting capacity is 17.4 packets/s.

Some research has been devoted to possible enhancements of the LoRa technology [NCB18]. To limit the negative effect of collisions between transmissions performed with the same SF a solution is to utilize Successive Interference Cancella-

---

[1] SF is increased by one every two retransmissions.

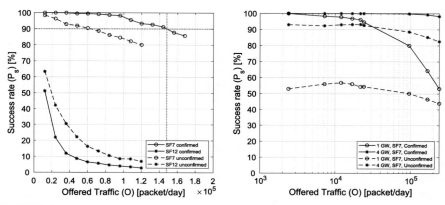

FIGURE 7.1

Packet Success Rate in a LoRaWAN network: One GW small area (left), Multiple GWs and large area (right).

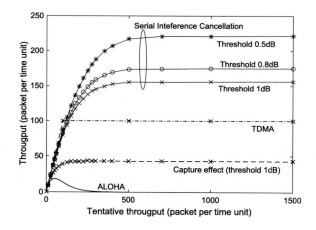

FIGURE 7.2

Performance comparison among different techniques employed to overcome collisions in LoRa.

tion (SIC), which is widely studied as a technique to enable receivers to demodulate multiple signals from their received sum. Fig. 7.2 shows the throughput of a LoRa based system affected by collisions as the network size, N, increases. The assumption is that all the nodes are using same SF and their packet generation rate is the maximum allowed by regional duty cycle limitations. It can be observed that SIC can significantly improve the network performance. This case study is in the sense of worst case scenario because no channel coding is used which would significantly improve the performance.

7.1.2 **NB-IoT**

NB-IoT is an access technology defined by 3GPP for massive Machine Type Communication (mMTC). NB-IoT implements several mMTC-oriented enhancements compared to other mobile technologies. Examples are: *(i)* narrow-band transmission and the exploitation of repetitions to reach devices in challenging positions such as basements or underground; *(ii)* differentiation of UE performance according to coverage areas by tuning parameters of the physical channels and network procedures; *(iii)* enhanced power saving mechanisms to improve the battery life; *(iv)* simplification of network procedures to reduce the UE complexity.

In [FCM$^+$18,FMPV17], the main characteristics of the technology are presented together with a mathematical model representing the network performance of a NB-IoT cell composed of multiple devices in different coverage conditions, transmitting small packets to the network. In a typical dense urban scenario, like the city of London, the eNBs are placed in an hexagonal grid with a variable Inter-Site Distance (ISD) each of them forming a three-sectorial site. The main network configuration parameters taken into account are the number of preambles per second available in each coverage class NB-IoT Physical Random Access Channel (NPRACH) channel (Z_c), the number of repetitions used in each physical channel (R_c, for simplicity common to all the channels), and the thresholds to determine in which coverage class each device belongs to based on the received power from the eNB (Th_c). The study aims to find the configuration which maximizes certain performance metrics. In general the model can be adapted to consider different performance targets, in this case the optimal configuration maximizes the network throughput provided that the success probability is more than 90%. This is achieved by generating random configurations and assessing the resulting network performance. Fig. 7.3 shows the result of this evaluation with ISD = 1732 m (left) and ISD = 3464 (right) where each point represents the resulting performance in a given configuration. With ISD = 1732 m the maximum throughput achievable is 52.4 kbps with a success rate of 96%. In this scenario the devices are split mostly in only two coverage classes with 1 and 2 repetitions respectively. A third class configured with 1 repetition is mostly unused showing that it is not needed, as coverage is not a major issue. Indeed, operators may decide to switch off the dedicated NPRACH in order to gain more resources for the NB-IoT Physical Uplink Shared Channel (NPUSCH) and accommodate more user data. With ISD = 3464 m the maximum throughput achievable drops to 10.4 kbps, but in this case the three coverage classes are configured with 1, 4, and 16 repetitions respectively and the devices are distributed evenly among them. In this scenario coverage is an issue that NB-IoT can solve by using the coverage classes concept and enabling a higher number of repetitions.

In [MPLC16] it is studied the possibility to enable mMTC applications by sharing the UHF spectrum with DTT. The proposed scenario considers a DVB-T2 network offering fixed rooftop reception as a primary service and NB-IoT network as a secondary service allocated to DTT white spaces. The results indicate that it is not feasible to allocate the NB-IoT carrier within the DVB-T2 channel, because of low power (between -38 and -36 dBm) that could transmit the small cell without interfer-

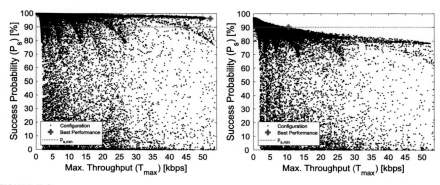

FIGURE 7.3

Random configurations of a NB-IoT network and best performance.

ing with the DVB-T2 channel. The NB-IoT small cell could transmit up to 15 dBm for adjacent channel with a 1 MHz guard band and there is no angular antenna discrimination. If it is considered, the improvement in the EIRP will be equivalent to the antenna discrimination (16 dB). The maximum allowable EIRP that the IoT devices can transmit varies between 9 and 14 dBm for the best case (Smart Parking) with a 2 MHz guard band. For the most restrictive case (Traffic Congestion) it is possible to transmit between 3 and 8 dBm with the same guard band, being this power enough to ensure a right operation.

In [FTC+19] it is considered another important scenario for mMTC communication, that is over-the-air firmware update. Given the presence of many devices per cell, a simple unicast update of the firmware generates a considerable amount of traffic which should be processed properly in order to avoid network saturation. In this work the transmission of one MBytes of firmware data is required in presence of normal application traffic. The performance using unicast transmission is compared to the one using Single Cell Point-to-Multipoint (SC-PTM), a feature introduced in Rel-14 of NB-IoT standard to enable multicast communication. The gains in terms of delivery time introduced by SC-PTM are quite obvious w.r.t. unicast. For unicast mode the delivery time varies from the order of hours to 1 day when increasing the ISD from 500 m to 1732 m, while it varies from the order of minutes to 1 hour for the SC-PTM. This indicates that the effective gains of SC-PTM w.r.t. unicast mode are strictly related to the location of UE. Nevertheless, it is worth emphasizing that while the delivery time is affected by the number of UE in the unicast case, the SC-PTM has a performance that does not vary with the number of UE being served. Thus, the choice of using either unicast or SC-PTM depends on the number of UE to be served and their coverage class.

A particular focus was given to Smart Grids application, which represent one of the target use cases that steered most the NB-IoT development.

In [PCF17] connectivity evaluations have been performed considering typical Smart Grid Wide Area Network (WAN) use cases in a real geographical zone of Italy

(Parma area), where Smart Meters can be deployed in different locations: outdoor, when the smart meter is located outside the building, indoor, when the meter is located in a propagation attenuation environment such as the first floor of the building, and deep indoor, when the meter is located in a deep propagation attenuation environment such as the basement of the building. The coverage evaluation were performed for both NB-IoT and LoRaWAN technologies assuming the same sites number and sites locations to enable a suitable comparison among the proposed solutions. Also it has to be considered both a single Mobile Network Operator (MNO) scenario and a roaming scenario. The evaluation highlights that the maximum percentage of uncovered area falls in the rural areas; indeed, considering only LTE sites, more than 40% of the rural areas is not covered by LoRaWAN, whereas 10% is not covered by NB-IoT.

In [NLZ18] various scheduling designs are compared with the aim of maximizing the transmission reliability. Use cases in the distribution segment include (on demand or periodic) remote meter reading, Real Time Pricing (RTP), and Object Relational Mapping (ORM). In the study ORM is considered to be the most demanding use case for the presented suitability assessment of NB-IoT technology in smart grids, meter reading is considered as background traffic. The network generally consists of a ring of substations (converting medium to low voltage), from where distribution feeders originate in a radial topology towards multiple households, each with a smart meter installed. The scheduler combining Earliest Due Date First and Shortest Processing Time First prioritization with Maximum Granularity Allocation subcarrier allocation achieves the highest reliability for nearly all outage percentages. We see a performance degradation as the granularity of the UL subcarrier allocation decreases (from Maximum Granularity Allocation to Least Granularity Allocation). Thus, due to the small packet sizes involved, increasing the granularity helps to decrease the waiting time of UEs which improves both the success rate and the 95th transfer delay percentile.

7.2 MAC and routing protocols for IoT

The design of efficient MAC and routing protocols in WSNs and IoT has been investigated for many years, with particular attention to design of distributed algorithms. However, some emerging applications, such as Industrial IoT (IIoT), are imposing strict requirements in terms of data transmission reliability, energy efficiency, throughput, and delay bounds, which cannot always be guaranteed by distributed solutions, like IEEE 802.15.4/Zigbee. Such applications demand new approaches to IoT protocols design, based on centralized solutions, where a central entity has a complete view of the network's characteristics and requirements, thus being able to compute a global schedule and paths. 6TiSCH is an example of centralized network and scheduling solution, aiming at providing deterministic and latency-bounded wireless communications.

This section presents main results of research activities related to centralized solutions, based on the IEEE 802.15.4e/6TiSCH protocol stack. One activity was devoted to the definition and experimental testing of a joint scheduling and routing protocol, while another one was dealing with the definition and testing of an energy-efficient routing protocol. The section concludes with a description of congestion control algorithm for IoT networks.

7.2.1 6TiSCH protocol stack

Time-Slotted Channel Hopping (TSCH) is a synchronous MAC protocol introduced in IEEE 802.15.4e [TSC12], as a recent amendment to the IEEE 802.15.4 standard [80211]. In TSCH network, channel hopping is used, while the superframe concept defined in IEEE 802.15.4 has been replaced by the concept of slotframes. A slotframe consists of a matrix of cells and each cell is defined by a pair of time slot (typically 10 ms) and channel offset. TSCH defines two types of cells: dedicated and shared. A dedicated cell is contention-free providing that only one transmitter can send a packet. If cells are shared among multiple nodes, TSCH defines a back-off algorithm to avoid the contention. One of possible scheduling approaches is that the overall communication is orchestrated by a centralized entity/controller. This entity (usually denoted as coordinator or sink) defines the action (transmit, receive, sleep) performed by each node in each time slot. The combination of time synchronization and channel hopping in IEEE 802.15.4e provides the reliable and efficient communication, which is robust against external interference and persistent multi-path fading. The 6TiSCH protocol stack [Thu18] presents the IEEE 802.15.4 standard at physical layer, TSCH at MAC layer, 6top for adding/deleting TSCH slots between neighboring nodes, IPv6 over Low-Power Wireless Personal Area Network (6LoWPAN) to adapt the transmission of Internet Protocol version 6 (IPv6) frames over TSCH, and Routing Protocol for Low-Power and Lossy Networks (RPL) as a proactive dynamic routing protocol (see Fig. 7.4).

OpenWSN[2] is an open-source implementation of the protocol stack shown in Fig. 7.4. In [VG16] an in-depth performance evaluation of this protocols stack using OpenMote-CC2538 devices[3] has been provided and further results can be found also in [GVV16]. Moreover, in [GPV19] Authors provide some preliminary results of an experimental campaign, aiming at better understanding the mechanisms of the scheduling and routing in 6TiSCH networks. The paper shows that a proper selection of 6TiSCH configuration parameters and synchronization mechanisms is needed for a stable and optimal coexistence of 6TiSCH and RPL protocols. This campaign was conducted using a new board, OpenMote-B, designed primarily for IIoT applications. OpenMote-B is a dual-band device allowing communications using 2.4 GHz, as well as 868 MHz, it is the first board that fully supports the IEEE 802.15.4g standard including MR-OFDM modulations for robust communications. [SGV18] presents

---

[2] See http://www.openwsn.org/.
[3] See https://www.openmote.com/.

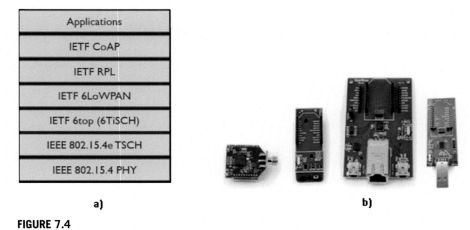

FIGURE 7.4

a) 6TiSCH Protocol Stack; b) OpenMote devices.

initial steps in porting RIOT[4] Operating System (OS) to OpenMote-B hardware platform.

7.2.2 Joint scheduling and routing protocols

The IEEE 802.15.4e standard does not specify details of the TSCH scheduling and resource allocation mechanisms, thus leaving many aspects to protocol designers. To bridge this gap, in [GVBV18] it was designed a Joint Scheduling and Routing Algorithm (JSRA), implemented on top of IEEE 802.15.4e. The algorithm jointly defines the set of paths connecting each node to the sink and the set of time slots they have to use for the communication (please, refer to [BV18b] for more details). Similarly to the Dijkstra's algorithm, which builds the tree iteratively by progressing from the root, at each iteration when a new link is added to the tree, a time slot is assigned according to the actual interference generated on the previously defined links. This ensures that packet losses are avoided, since the SIR (accounting for the sum of all possible interference) is kept above the capture ratio for all links.

As a second step, the JSRA algorithm has been integrated into the 6TiSCH protocol stack and implemented on the OpenMote-CC2538 platform. The testbed consists of ten nodes located into boxes on the walls of a corridor at the University of Bologna. Results, reported in Table 7.1, show the packet delivery ratio and throughput obtained by JSRA and 6TiSCH protocols. JSRA outperforms the standard solution in terms of throughput due to the possibility to assign the same slot to different links (see Fig. 7.5, where an example of routing and the scheduling outcomes for JSRA is shown), at the cost of a decreasing of the packet delivery ratio of 1%.

---

[4] See https://riot-os.org/.

Table 7.1 Comparison of JSRA and 6TiSCH protocols.

Protocol	Packet Delivery Ratio	Throughput [kbps]
6TiSCH	100%	12
JSRA	99%	15

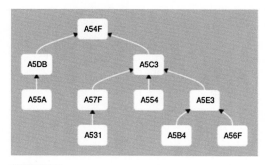

FIGURE 7.5

a) Routing topology obtained by JSRA and b) Scheduling obtained by JSRA.

Specific scheduling issues and challenges in 6TiSCH networks were also addressed in [GV18]. In this work a 6TiSCH simulator is used to demonstrate the performance of On-The-Fly (OTF) scheduling. The OTF algorithm is a distributed strategy for adapting the scheduled bandwidth to network requirements. The node negotiates the number of cells scheduled with its neighbors, without the intervention of a centralized entity. This report provides the simulation analysis based on various parameters including TSCH, RPL, OTF parameters, radio and propagation parameters, number of nodes and channels, node allocation area and deployment constrains, in order to measure performance, such as, End-to-End (E2E) reliability, E2E latency, the number of scheduled slots, etc. Simulation results show that OTF may achieve an E2E reliability up to 99%.

7.2.3 Routing protocols and congestion control

Routing in IoT networks has been extensively studied in the last decade, with the aim of reducing as much as possible the energy consumption, to increase the lifetime of battery-driven sensor nodes.

Authors in [DFGTV17] present a novel protocol, that is the combination of Source Routing (SR) and Minimum Cost Forwarding (MCF) protocols, which aims at reducing the energy consumption by communicating over paths with minimum cost. Source Routing for Minimum Cost Forwarding (SRMCF) is a reactive protocol, where nodes acquire routes on-demand and avoid saving information about the network topology. No information about the network topology is kept at nodes, but nodes always communicate over paths with minimum cost, independently of the traffic type. SRMCF has been implemented on Telos-B motes and experimental results in a real scenario has been performed over a small network, to compare SRMCF to MCF protocol. Results show that the proposed solution presents a 33% higher throughput and 24% less energy consumption than MCF. The impact of using SRMCF with two different MAC protocols, Berkeley-MAC and Contiki-MAC, has been also evaluated via simulations.

[DGGV17] aims at applying the SRMCF routing protocol into the IIoT context. In particular, the SRMCF protocol has been integrated into the 6TiSCH OpenWSN protocol stack and implemented on OpenMote-B devices. Preliminary results of this research are reported in [RVG18,RVG19], where the SRMCF protocol has been compared to RPL. An example of results is reported in Fig. 7.6, where it is shown the Round Trip Time (RTT) for the two protocols, defined as the time between the generation of a query at the sink, to be transmitted (via multi-hopping) to a specific node in the network, and the instant in which the reply, generated by the node, reaches the sink. The RTT is shown by varying the number of hops separating the sink and the node, and the payload size of the query/reply packets. From the results it is possible to observe the effects of slot frame size upon the performance of the protocol and how higher payload sizes generate higher values of RTT as the number of hops increases, mainly due to queue overflow at relays.

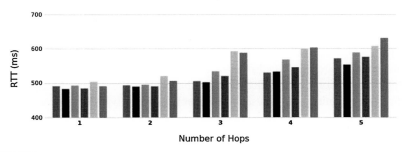

FIGURE 7.6

Average RTT as a function of the number of hops for RPL and SRMCF [RVG19].

We conclude this section addressing another important issues in IoT networks, that is related to traffic congestion, happening when a huge amount of information generated by different nodes needs to be transferred towards a single sink. Indeed, in case of congestion nodes queues may overflow and packets may be randomly dropped. As a consequence, valuable information can be lost on its way to the sink. To avoid this phenomenon, the network should be able either to avoid entering this state or to identify the congestion and recover (self-heal) in order to prevent loosing valuable information. Authors in [TSVG17] present a solution in which mobile

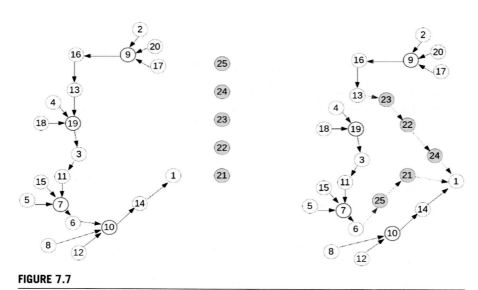

FIGURE 7.7

(a) Topology at congestion detection before Mobile-CC and (b) Topology after Mobile-CC node placement [TSVG17].

nodes are used to control and alleviate congestion and other impairments. In particular, they present a mechanism, denoted as Mobile-Congestion Control (CC), that can run above existing congestion control algorithms and employ mobile nodes in order to either create locally significant alternative paths towards the sink or generate direct paths of mobile nodes to the sink (see Fig. 7.7). A distributed congestion detection mechanism is also defined, such as a mechanism for deciding which nodes could not be used anymore as relays, because they are detected as congested. Simulation results show that the solution can significantly contribute in the alleviation of congestion in IoT. The same technique can be used to recover from other types of network faults as well.

7.3 Vehicular communications

The topic of vehicular communications has seen two main focus points. On the one hand, antenna design for V2X communications has been studied. This includes antenna types, patterns and MIMO deployments. Furthermore, the topic of antenna placement in recessed roof cavities has been researched. On the other hand, the topic of performance estimation and analysis in a new, highly dynamic environment was a strong focus. This included measurement campaigns as well as stochastic modeling approaches, and hybrid designs including hardware emulation and system level simulations. Both vehicular ad-hoc networks, as well as train communications were the focus of the analysis. In the following sections, the work of the two topics is presented.

7.3.1 Antenna design and integration

Applications in the IoT are usually thought of as small devices like sensor networks, but large vehicles such as trains and cars are now rethought as connected devices in the IoT. These connected vehicles can access infotainment via the IoT and share safety critical information directly by establishing ad-hoc networks. Thereby, vehicles become more communicative, which increases the demand for vehicular antennas. Especially, the exchange of safety related information requires reliable and redundant wireless communication. In principle automotive antennas should provide omnidirectional coverage, as the orientation of the car towards other stations will in general be unknown beforehand. However, once communication with a specific partner is established, it is desirable to have increased antenna gain towards this direction, e.g. by beam-steering.

Some research has been devoted to investigate the possibility to flush-mount pattern reconfigurable antennas inside chassis antenna cavities, see Fig. 7.8. Five pattern reconfigurable antennas were measured inside a chassis antenna cavity [AKMZ17a,AKMZ17b,AKMZ19].

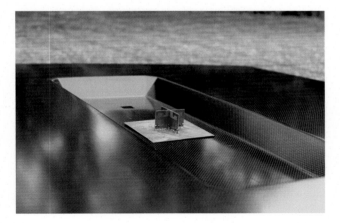

FIGURE 7.8

An antenna that can be reconfigured between four directions [AKMZ17b] is flush-mounted inside a chassis antenna cavity prototype made from carbon fiber reinforced polymer.

[AKMZ17a] compares the performance of three antennas that are reconfigurable between a front/back and a left/right state. All three antennas are built with two driven elements that can be switched to radiate in phase or with a 180° phase shift, but the antennas differ in the design of their radiating elements and switching mechanisms. Quarter-wavelength monopole antennas, inverted-L antennas and inverted-F antennas are explored as driven elements. It is shown that the antennas retain their reconfiguration capabilities when they are flush-mounted. Further, measurements show that such antennas can be designed with various switched gain differences, e.g. to reconfigure between states that are both quasi-omnidirectional and can provide a few additional decibel of antenna gain, or with large gain differences around 20 dB to reduce interference from vehicles in the front.

[AKMZ17b] prototypes and measures a flush-mounted antenna that can be reconfigured between four directions, e.g., front, back, left, and right. Such antennas can be used to selective transmit safety critical information in traffic. For example, a car that plans to change lanes on a highway can broadcast a message with increased gain towards its right side before starting to swerve. [AKMZ19] demonstrates an automotive antenna that can be hidden in a cavity and that is reconfigurable in 45°-steps towards eight directions. The antenna is designed as electronically steered parasitic array radiator and allows even finer beam-steering.

New concepts for automotive antennas have been explored in [GKDGH18], presenting a prototype of an antenna cavity that is built into the roof above the windshield. This concept is opposed to state-of-the-art shark-fin antenna modules that are located at the rear roof end. Measurements show that antennas at this location offer increased coverage towards low elevation angles in front of the car, which is paramount for communication with vulnerable road users.

FIGURE 7.9

An automotive antenna roof [AKGH] is prototyped and measured inside the anechoic chamber at VISTA, TU Ilmenau, Germany.

[AKGH] takes this concept a step further and proposes to use the whole car roof for antennas. Three antenna modules are introduced for flush-mounted antennas: A cavity above the windshield, a cavity in the roof center, and an antenna shelf at the rear roof end which elevates the antennas inside shark-fin modules to decrease shadowing from the roof curvature, see Fig. 7.9. The performed measurements show that quasi-omnidirectional coverage is possible from all three locations and that the increased distance between antennas increases the MIMO performance by lowering the correlation.

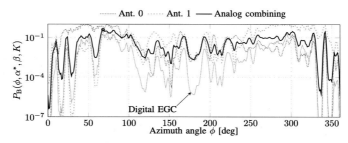

FIGURE 7.10

Burst error probability of using a single antenna (Ant. 0 or Ant. 1) compared with the proposed analog combining. As a reference, digital equal-gain combining (EGC) performance is also plotted.

A different take at robustness of the antenna patterns is shown in [NSB+18]. Here, the goal is to use L non-ideal antennas that can be purchased at low cost, and combine them in an efficient way to avoid error bursts due to bad antenna patterns. Crucially, the combination happens before the receiver, so that only one receiver is required. The authors choose minimizing the probability of error bursts as figure of merit, and aim to minimize the probability that K consecutive packets arriving from the worst-case angle-of-arrival are decoded incorrectly. To minimize complexity, the combining network does not estimate or use channel state information (complex channel gains, noise levels, etc.). The combining network consists of $L - 1$ analog phase shifters whose phases are affine functions of time. For a general L and the case when the packet error probability decays exponentially with the received SNR, the optimum slopes of the affine functions can be computed by solving an optimization problem that depends on the antenna far field functions. The authors provide an analytical solution for the special case of $L = 2$ antennas, which turns out to be independent of the antenna patterns. In an experimental setup consisting of two monopole antennas mounted on the roof of a Volvo XC90, the proposed combining method is shown to give significant performance gains compared to using just one of the antennas, see Fig. 7.10.

7.3.2 High mobility performance analysis and modeling

High mobility applications pose new challenges to the quality and reliability of data communications. The unique challenges of the outdoor channel with large delay spreads, coupled with the broad Doppler spectra associated with high mobility applications require thorough analysis of achievable data rates and latency. The full spectrum of tools has been employed to capture these conditions and move network connectivity analysis closer to reality.

The first presented problem is the coverage and throughput challenge associated with high speed trains. [BSOM17] presents two solutions to address this problem in cellular network services: On the one hand, amplify-and-forward Moving Relay

Nodes (MRNs) are studied. Furthermore, prototype windows are applied on on-board Wi-Fi enabled High Speed Trains (HSTs). The presented work applies the framework of hypothesis testing to extensive penetration loss and throughput tests. The tests are conducted with 3G/4G UEs located on-board the vehicles traveling throughout long-range geographical routes. This work is supported by real-world measurements conducted along Austrian railways such as from Vienna to Salzburg and from Vienna to Graz. The results are presented under the name SegHyPer, which works on route segmentation link quality parameters to enable micro-analysis in current and future cellular networks for mobile users on-board railjet HST.

In [BM18], the authors use vehicular highway throughput measurements such as the previous presented ones, and use them to estimate performance models. However, the focus of the work does not lie on throughput alone. Instead, the authors consider safety critical applications, where the burst properties of the transmission errors are just as important to estimate as the mean packet error rate. The presented approach takes its inspiration from the Gilbert-Elliot model, a two-state Markov chain. The model parameters are allowed to change to account for the non-stationarity of the channel. Based on this, maximum likelihood formulations for the model states are introduced. These formulations are then specified to estimate two different models. One model is parametrized on the measured 1-second mean SNR and uses the Baum-Welch algorithm to deduce the model parameters from recorded packet traces (called the mean SNR grouped estimator). The other approach takes the measured momentary SNR, and correlates it with packet loss events. This is used to estimate a momentary SNR threshold above which the packet will be transmitted successfully (the fading aware estimator). The results demonstrate that capturing the burst properties is essential, as the probability of burst errors is continuously high throughout the SNR range.

The previous work considered in-detail modeling of single link packet performance. However, in dense environments, the bulk of packet loss can be expected due to interference of neighboring nodes. This is the focus of [BBG+19]. Here, the goal is to use network simulations of dense city traffic, and analyze the resulting stochastic properties. To this end, SUMO is used as mobility simulator with a map of the city of Pristina in Kosovo. The resulting positions are used in OMNET++ with the VEINS extension, to generate VANET packet traffic. The traffic parameters are set to 10 packets per second and vehicle, with either 200 or 500 byte packets. The authors investigate the joint probabilities of number of neighbors in communication range and resulting packet loss due to hidden node interference. The first results show, that typical simple assumptions, such as Poisson Point Processes and Manhattan Grids do not reflect the results of the simulation. Hence, better assumptions are needed. Furthermore, the authors provide an estimate for the packet loss PDF due to interference, parameterized over the number of neighbors using the Gamma distribution. A similar analysis was carried out by the authors in [BLA+ed]. There, the Gilbert-Elliott approach for the single link was combined with the interference simulation approach, to evaluate an overall packet error probability. The result is shown in Fig. 7.11. The authors show that both aspects are equally important in this setting.

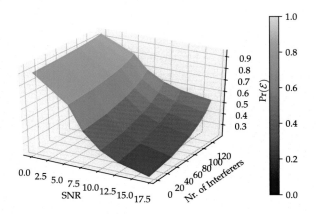

FIGURE 7.11

Combined packet error probability from single link analysis and interference modeling.

In these works, network simulators play an integral role for studying and evaluating vehicular communications. However, analysis is required on the accuracy of such simulations. An analysis on this has been conducted by [DJKed]. Specifically, the authors investigate the two most used network simulators, Veins (linked to OMNET++) or NS-3. The paper surveys the capabilities and employed propagation models. A head to head comparison of the two simulations reveals discrepancies in the results and caution to use them without further thought.

Finally, the last documents treat the challenging task of interfacing the real world with simulations. In the Veneris framework [ELLPGMGP19], OMNET++ is used as basis for a network simulator and generator. This is interfaced with the Unity game engine, to provide physically accurate simulation facilities. Through Unity, a 3D environment that reproduces real-world traffic dynamics, a ray-launching propagation simulator and bidirectional interfaces are implemented. It furthermore supports accurate driver behavior and vehicle models. The article discusses validation of the vehicle dynamics, the recreation of traffic flows, and the accuracy of the propagation simulations.

In the framework called Testing Environment for Vehicular Applications Running with Devices in the Loop (TRUDI) [MMC+ed] vehicles will be equipped with short-range wireless technologies with the aim to improve safety and traffic efficiency. Novel applications are thus being implemented for future cars and trucks, and one of the main issues is how to conduct tests and optimizations in an effective way, limiting the need to perform costly and time consuming experiments on the road. To cope with this issue, a simulator with hardware in the loop, called TRUDI, has been implemented. The aim of this new platform, illustrated in Fig. 7.12, is to provide a flexible solution to test V2X applications with real wireless communication devices in the loop. Starting from a logical separation of the communication device, called intelligent transport system station (ITS-S), from the processing and visualization unit,

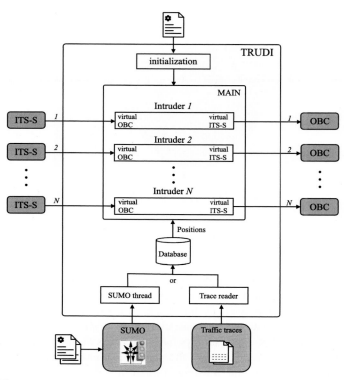

FIGURE 7.12

The internal structure of TRUDI.

denoted on-board computer (OBC), the idea is that TRUDI works in the middle of the two layers, with a man-in-the-middle approach. During the simulation, TRUDI modifies the exchanged information in order to simulate given positions and trajectories of the vehicles and to reproduce inaccuracies in the communication, such as message losses or errors in the localization. Once the application is tested, TRUDI is removed, the hardware is mounted on-board of the vehicle, and the application is ready to run. As an added value, a simulation of the ITS-Ss is also possible. As an example use case, an application for the intersection management has been implemented and tested, where the driver is warned of the presence and speed of other vehicles approaching the same junction. Once validated with TRUDI, the application was also verified on the road with real vehicles.

This approach allows for a very detailed analysis of the observed link-to-link performance. On the other hand, [BGB+19] aims at measuring in hardware the performance in a dense urban environment. The goal of the paper is to accurately assess the achievable performance for ad-hoc schemes in dense urban networks, where interference plays an important role as limiting factor. Historically, such analysis strongly simplifies the physical layer. Instead, the goal here is to keep the complexity of the physical layer, and simplify elsewhere. The paper approaches this problem by using mobility simulators to generate vehicular positions, and then model the communication between the vehicles through a graph. Hence, the authors simplify the graph while maintaining the essential communication behaviors and statistics. This simplification shows that the presented scenario can be modeled sufficiently accurate with a very low number of devices. Hence, three transceivers and two vehicular channel hardware emulators are connected to provide performance analysis of the presented network.

7.4 **Energy efficient/constrained solutions for IoT**

Recent advancements in wireless networking, microfabrication and embedded microprocessors have enabled a production of massive-scale wireless nodes suitable for a range of innovative applications. Size and cost constraints on individual wireless nodes result in corresponding constraints on available resources such as energy, memory, and computational capabilities. One of the main concerns is energy consumption, since distributed wireless nodes are mainly battery operated.

7.4.1 **Energy efficiency in IoT**

Many solutions for IoT are using the 2.4 GHz Industrial, Scientific, Medical (ISM) band (see, e.g., IEEE 802.15.4), which usually becomes crowded because of many radio technologies sharing this band, e.g., Wi-Fi, Bluetooth, and cordless phones. Interference impacts not only the QoS and reliability of communications, but also the energy consumption of the node, because packets need to be re-transmitted. [TIPV+16] reports about some experiments carried out to characterize the impact of interference on energy consumption. As primary network two IEEE 802.15.4 devices (WSN430 nodes, using the CC2420 radio), were considered; then up to other three devices, belonging to another network, have been added progressively to create interference. Experiments were conducted into an anechoic chamber (ideal case) and into an office (real case); in both cases the distance between the transmitter and receiver nodes was 3.5 meters, whereas the interfering nodes were deployed near the transmitter (at a distance of 20 centimeters), in the ideal case, and in different places between the two nodes, in the real case (see Fig. 7.13). Interfering nodes generated packets of 100 bytes of payload every 10 ms, causing serious interference on the channel. The average energy consumed at the transmitter side, as a function of the Receiver Signal Strength Indicator (RSSI) value measured at the receiver (being an estimation of the interference level), is shown in Fig. 7.14 for the ideal case. Results demonstrate that the average energy consumed varies according to the level of interference on the channel. Moreover, a packet transmitted with an interference level situated in the slice [-75, -70] dBm will consume 4.6 times more on average than a packet sent when the RSSI value does not exceed -85 dBm. As for the real case

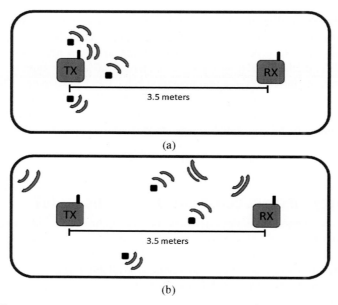

FIGURE 7.13

Scenarios of the experiment: a) in anechoic chamber, b) in office-type lab environment.

scenario (we refer to [TIPV⁺16] for the related results), a more significant variance of RSSI values than in the ideal case has been observed, due to presence of signal reflections and external sources of interference that could not be controlled.

It is well known that during a radio transmission, the most energy greedy part is the radio front-end. Methods to optimize the energy consumption by turning off this interface when unused have already been proposed and implemented. This could be directly managed by the hardware, adding a wake-up radio module in the transceiver [HKVG16]. The wake-up radio module has the energy consumption considerably lower than the one of the main radio front end. It will decide to activate the main radio front-end only if there is a communication demand. In Fig. 7.15, the global structure of the proposed wakeup radio receiver is presented. This receiver has two main paths. After the antenna, the signal is equally split in two branches by a power splitter. The upper branch, denoted direct path, is composed of a bank of N bandpass filters $D_1, D_2, \ldots D_N$ having the frequency shape identical to the identifier. It is also composed of an energy detector which provides V_{DC1} DC voltage, proportional to the V_{RF1} voltage at the output of the filters bank. The lower branch, denoted complementary path, is globally the same, but starting with a filter bank $C_1, C_2, \ldots C_N$ which is the exact complement of the one in the direct path. The V_{DC2} voltage at the output of the energy detector on the complementary path is subtracted from V_{DC1} and then the subtraction result is compared to a threshold by the means of a Schmidt trigger.

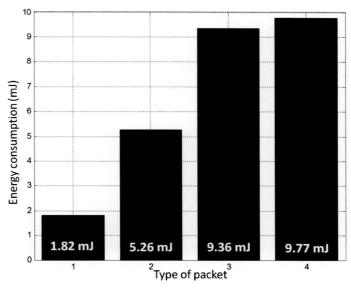

FIGURE 7.14

Energy consumption at the transmitter side versus average RSSI measured at the receiver.

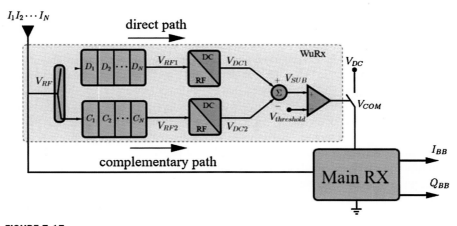

FIGURE 7.15

The wakeup radio architecture.

7.4.2 Energy harvesting aspects

Energy Harvesting (EH) has emerged as a technology capable of allowing network nodes to replenish their batteries using environmental energy sources such as, solar radiation, radio waves, or vibration. This, in turn, potentially allows the network nodes to operate for an infinite lifetime. However, the intermittent and random na-

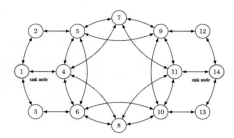

FIGURE 7.16

Example of a communications network.

ture of the energy supply makes it necessary to take a new approach to the design of communication policies. This has led to a great deal of research interest in EH-powered communication, with problems ranging from throughput maximization, source-channel coding, estimation, and others. A problem that often arises in IoT (and, in general, wireless) networks is that of routing information through the neighboring nodes. This problem was studied in [CAMR18] where the Authors proposed a joint routing and scheduling policy for data packets in an EH network. As Fig. 7.16 illustrates, each node independently generates traffic for delivery to a specific destination and collaborates with the other nodes in the network to ensure the delivery of all data packets. In this way, each node decides the next suitable hop for each packet in its queue (routing), and when to transmit it (scheduling). The solution to this problem when the nodes are not EH-powered is given by the BackPressure (BP) algorithm. By leveraging stochastic dual descent methods, the Authors proposed a generalization of the BP algorithm, referred to as EH-BP, which also accounts for the random nature of the energy supply. They also showed that the joint routing and scheduling satisfies the fundamental property of BP-type algorithms. Namely, if given data arrival rates can be supported by given energy arrival rates and some routing-scheduling policy, they can be supported by the EH-BP policy too. This extent is illustrated in Fig. 7.17 which shows a sample path of the total number of packets queued in the network at each time slot (it corresponds to the network scenario of Fig. 7.16, with nodes 1 and 14 acting as sink nodes). Average values are shown in dashed lines. As expected both the EH-BP and BP policies stabilize the queues. Nonetheless, an increase in the variance of the queue dynamics as well as the average number of packets in the network can be observed for the EH-BP policy. This is due to the random nature of the energy harvesting process. However, the average number of packets in the queues can be traded-off against the capacity of the battery (i.e., the larger the battery capacity, the lower the number of packets in the queues).

A different system scenario is addressed in [CFMAHR17]. In this work, the Authors focused on wireless networked control systems that, with the advent of Industry 4.0, are rapidly becoming prevalent. They are present in smart homes, robotic automation, smart transportation, industrial plants, and more. A critical component of these wireless control systems are the sensing devices. These sensor nodes measure

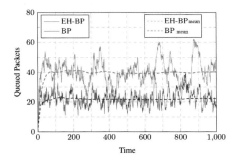

FIGURE 7.17

Number of packets queued in the network over time.

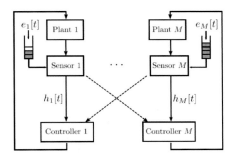

FIGURE 7.18

A wireless networked control system with EH sensors.

the state of the system and transmit their observations over a wireless channel (see Fig. 7.18). However, due to the uncertain nature of the wireless channel, the choice of communication policy critically affects the closed loop performance of the controlled systems. The sensors share the wireless communication medium and therefore one should aim for an efficient use of this resource (i.e., by avoiding packet collisions) in a way that meets the control performance requirements. Hence, the goal of this work was to design a decentralized random access policy such that all control loops satisfied their control performance requirements and the power consumption satisfied the energy causality constraints imposed by the EH process. The Authors proposed a simple dynamic threshold scheduling policy and, further, they computed the optimal scheduling policy by means of a stochastic dual method. The evolution of the control system performance (two-loop case) is shown in Fig. 7.19. Here we observe that the resulting random access policy is able to stabilize both systems since the averaged value of the state variables $x_i(t)$ asymptotically tends to a limiting value.

The work in [RDJK19] addressed the very complementary topic of Simultaneous Wireless Information and Power Transfer (SWIPT) in IoT networks. Specifically, the authors introduced a novel architecture called Modulation Split-based SWIPT. The

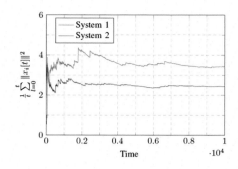

FIGURE 7.19

Evolution of the control system.

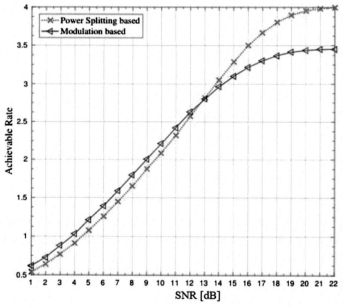

FIGURE 7.20

Achievable rate for the Modulation Split and PS schemes.

main idea here is to use specific constellation points for EH, and the rest for information transmission. This technique differs from traditional Power Splitting (PS) and time switching architectures which suffer from power losses at the symbol level. The achievable rate of Modulation Split approaches is expected to be lower than other PS schemes but QoS will not be compromised because, unlike in PS, the symbols allocated for IT do not carry power for EH. This extent is illustrated in Fig. 7.20 which reveals that Modulation Split outperforms PS in the (realistic) mid- to low-

SNR regime, whereas the situation is just the opposite for very high values of the SNR.

7.5 **SDN and NFV for IoT**

The use of SDN principles in IoT should enable the natural integration into future 5G wireless networks. Several research activities have been devoted to the design of SDN solutions for IoT. To validate their proposals, various prototypes have been developed and evaluated. In this section, we start by reviewing the efforts made in the definition of a SDN protocol architecture for IoT. In the second part, we present different proof-of-concept systems aiming at developing gateways allowing the communication between different IoT technologies; in these works both, the concept of SDN and SDR are exploited. We conclude the section with the topic of virtualization of IoT networks.

7.5.1 **Software-defined IoT networks**

A research activity has been dedicated to the design and test via experimentation of an SDN-based architecture for IoT networks.

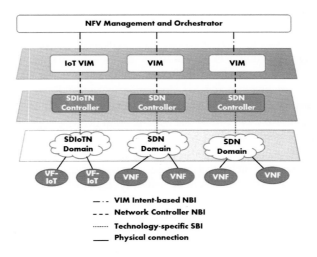

FIGURE 7.21

The general SDN-based architecture.

The proposed architecture is reported in Fig. 7.21 [CBC$^+$17,DCT$^+$18]. It is based on the principles of the ETSI NFV MANagement and Orchestration (MANO) framework, where high-level components, such as the orchestration layer, are in charge of managing the end-to-end network function life cycle and orchestrating the re-

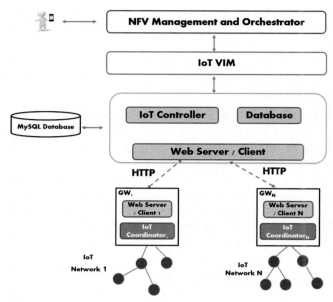

FIGURE 7.22

The Software-Defined IoT Architecture.

sources required to deploy a given service. The orchestration layer must then interact with domain-specific components, that in the case of the IoT domain include (see Fig. 7.22): i) *Management Plane*: IoT Virtualized Infrastructure Manager (VIM), which manages resources in the IoT domain; ii) *Control Plane*: SDN controller, responsible for proper traffic steering, contributing to the E2E service deployment across the domain; iii) *Data Plane*: IoT gateway and devices, representing the computing and networking resources to be managed and controlled by the other components.

The architecture also includes a database, containing the descriptors of all the IoT devices, including the IP addresses of the gateways enabling the connectivity of each IoT device, the services that they may provide and the related QoS that can be guaranteed. When a request comes from a user, the IoT VIM maps the incoming service request to the most suitable Virtual Network function (VNF); once having identified the best match, the controller will: i) program the selected IoT network ensuring the requested QoS requirements; and ii) forward the request to the gateway associated to the target VNF. The main novelty, w.r.t. previous works, such as [GMMP15] and [MR11], relies on the use of an IoT controller capable of programming the IoT networks responding to the QoS requirements specified by the end user.

In [DCT+18], the architecture is characterized in terms of the RTT performance metric. In particular, it is considered the setup shown in Fig. 7.23, where a user periodically asks for data measured by the three different sensor nodes connected to the

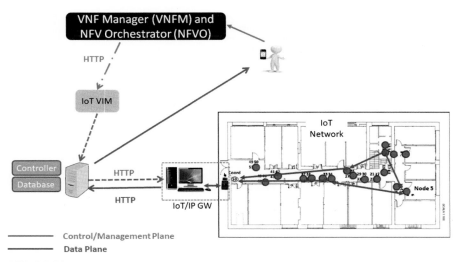

FIGURE 7.23

The Testbed Setup.

IoT network. These nodes are placed at one, two, and three hops from the coordinator, respectively. For each packet we measure: i) the RTT at the data plane, that is the interval of time between the arrival of the query coming from the controller, at the application layer of the coordinator, and the arrival of the reply from the target node, again at the application layer of the coordinator; ii) the RTT at the control plane, that is the time interval between the arrival of the query coming from the VIM, at the controller, and the arrival of the reply coming from the gateway, again at the controller; iii) the RTT at the management plane, calculated as the difference between the time stamps taken when a query arrives to the IoT VIM and when the response to the same query (if present) is sent back to the requesting user. Average RTT values are reported in Fig. 7.24. As expected the RTT slightly increases with the number of hops and when moving from data, to control, to management planes (each layer adds some processing). Results show the feasibility of the proposed solution, reporting end-to-end delays at the user in the order of 100 ms.

7.5.2 Integrating different IoT technologies

It is well known that nowadays there are many technologies available for the IoT, and there is not evidence of agreement toward one specific standard. This calls for the need of finding solutions allowing the integration of the existing technologies.

Authors of [GKAP18b] and [GKAP18a], proposed the implementation of an SDR based gateway for IoT, allowing the communication and data transfer among various networks. In particular, the gateway supports the most common short-range

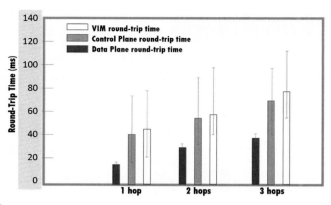

FIGURE 7.24

RTT at the different planes when considering different number of hops in the IoT network.

wireless protocols and standards, including IEEE 802.11, ZigBee, Bluetooth Low Energy (BLE), LoRaWAN, and satellite communication. SDR-based devices are able to dynamically switch between protocols and can also include spectrum sensing and packet sniffing capabilities. The first IoT gateway prototype was developed using Commercial Off-The-Shelf (COTS) radio modules. As can be seen in Fig. 7.25, four different technologies were included in the prototype. A second prototype was implemented based on the Ettus USRP310 SDR, which comprises a Zync System-on-Chip (SoC), an Field-Programmable Gate Array (FPGA), and an Advanced RISC Machine (ARM) processor. The latter prototype implemented the full protocol stacks of the IEEE 802.11g and IEEE 802.15.4 standards.

Another IoT gateway developed to create an ecosystem for overlay networks has been presented in [BRS17] and [BRS16]. The idea is to realize a mobile gateway (as shown in Fig. 7.26), enabling a reliable multi-homed communication, based on SDN concept. In particular, the gateway implements the Open Overlay Router (OOR) using the Locator Identifier Separation Protocol (LISP) protocol defined in [BRS17] and [BRS16]. The OpenDaylight SDN solution, one of the most popular SDN open-source solutions, is used to implement the controller functions. The prototype developed uses a Raspberry Pi implementation of OOR, that runs on a mobile WAN router that also includes 3G/LTE mobile connection. The mobile IoT gateway simultaneously connects different networks providing reliability and load balancing mechanisms. At the core of the WAN, the SDN controller is made available through the OOR LISPFlowMappings module. This approach opens-up new perspectives on the network management. The traffic can be steered on-the-fly based on the SDN controller OpenDayLight rules, allowing the implementation of different optimization techniques for accelerated delivery of applications or for policy enforcement, improving the IoT gateway functionality.

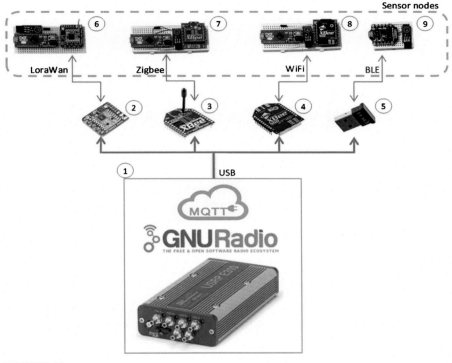

FIGURE 7.25

The architecture of proposed IoT gateway based on SDR and COTS elements.

7.5.3 Virtualization of IoT

The virtualization layer is very important for the future IoT software defined network. Thus, the authors of [DT18] examine file system performance for the native host OS performances, hypervisor-based virtualization (KVM), and container-based virtualization (Docker), based on their previous work in [TZJD17]. They have generated a workload through Filebench tool for: (1) webserver scenario dominated by random read components; (2) varmail scenario dominated by random read and random write components; and (3) file server scenario in which both random and sequential components are represented. The experiments are run for one, two, and three KVM/Docker Virtual Machines (VMs) or containers. As the number of instances increases, there is a significant drop in performance and this drop is constant on any hardware-software configuration. The Docker containers use the HostOS Fixed Service (FS) and have a small footprint and therefore have similar performances as the native host (in the case of one instance). In case of the instantiation of the container, certain performance fall will evidently occur, which is, depending on workload, slightly or strongly expressed.

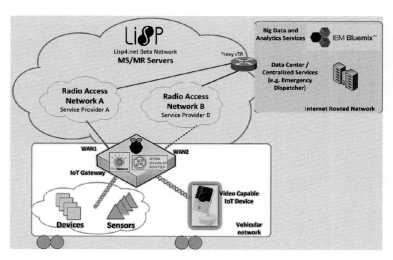

FIGURE 7.26

Scenario for a mobile IoT gateway with SDN management.

When it comes to KVM VMs, the performance drop is noticeable for the case of one VM and even more when applying larger number of VMs. Presented tests show that the Docker instance exhibits much better FS performance than KVM VMs.

An important characteristic of IoT elements is represented by the secure communication methods. Besides NFV and SDN, 5G networks envision also the implementation of Device-to-Device (D2D) communication. The ad-hoc and proximity nature of this communication introduces some very important security vulnerabilities. Key management, access control, privacy, secure routing, and transmission need dedicated signaling procedures and optimized implementation mechanisms that are appropriate for the mobile, low energy and low processing power environment. [BBS19] proposes security a mechanism for D2D communication involving the usage of physical unclonable functions (PUF) for unique key generation. Inspired by biometrics, PUFs provide a unique way to identify integrated circuits. Comparable in a simplistic way with a "unique fingerprint" of an integrated circuit, PUFs differentiate one integrated circuit from another (though apparently identical) [SCM16]. Besides the PUF unique key generation, also Elliptic-Curve Cryptography (ECC) and Diffie-Hellman Key Exchange (DHKE) are used for key management, while Salsa20/20 is the stream cyphering encryption method, suitable for confidentiality of the wireless transmissions. All these methods are implemented and tested on a SDR communication platform consisting of a Zync based SoC, complemented by RF daughter-boards from Analog Devices – an integration using hardware and software co-design.

7.6 Special applications of IoT

The IoT is the network of smart devices connected for various applications, such as process automatization and data gathering intended to simplify and secure our life. Ubiquitous sensor networks are present either in everyday life or dedicated for special applications, some of which are not seen from the consumer point of view. In this section we present some IoT solutions (either as a system or as technology components) thought for specific applications.

One of the most emerging application for the IoT is the Industry 4.0, also denoted as Factories of the Future (FoF). The concept of smart factories with advanced radio networks ensuring a reliable real-time communication was proposed at the end of 1980s [Rap89]. The main goal was to encourage commercialization and standardization of radio communication for the monitoring and controlling of machines. The shift from wired to the wireless networks dramatically helps to reduce huge capital expenditures of installation and maintenance of wires. However, the use of wireless into harsh industrial environments causes many propagation path loss issues, that must be included in the network operation analysis. The Horizon 2020 Clear5G project [htt18] envisions the FoF, where various radio technologies coexist and are converged to fulfill various QoS needs of different applications in the FoF [Zha17]. To achieve this goal, Clear5G has built a consortium with 7 partners from European Union and 5 from Taiwan, to further enhance research and international collaboration on how 5G may empower the FoF. Clear5G focuses on machine-type communications in the FoF environments employing both URLLC and mMTC services. The addressed major KPIs are consistent with the main stream of 5G development, more specifically including high density of connected devices, reaching 100 nodes per 1 m^3, low communication latency (down to 1 ms) and a high reliability (up to 99.999%) [htt18,PPP15].

Industry 4.0 scenario is also addressed in [CSM$^+$19], where it is designed a protocol for IIoT networks, based on the 2.4 GHz LoRa technology. The protocol is based on time slots scheduling, to allow high reliability, and incorporates energy harvesting issues, since assumes nodes are charges via wireless power transfer. Through proper configuration of the so-called Spreading Factor SF parameter defined by LoRa, nodes can trade transmission ranges with achievable bit rates. It must be noticed that a proper adjustment of the LoRa physical layer parameters is required in case of coexisting with an IEEE 802.11g network, for the sake of interference mitigation.

Another interesting application refers to smart buildings and smart living. In smart buildings, one of the main challenges is the adaptation of the building's infrastructure to the network installation and operation. Such situation occurs in most cases for already constructed buildings or old ones that cannot be significantly modified. For a large-scale building, the harsh character of the environment significantly disturbs the wireless connectivity or even sometimes makes it impossible due to the high value of the path loss exponent of the radio propagation. The difficulty of changing the physical structure of a building due to technical reasons, costs or law issues implies the investigation of alternative solutions. In [VFPK18] it was proposed to uses the ventilation network (known as Heating, Ventilation and Air-Conditioning (HVAC)). In most cases the HVAC ducts are made of metallic cylindrical elements. The propagation properties must be investigated to verify if the ventilation duct will act as cylindrical waveguide. This characterization for a typical metallic duct with diameter of 200 mm was investigated considering the following frequency bands: 433 MHz, 868 MHz, and 2.45 GHz. It was verified that the cut off frequency for the fundamental mode in such a cylindrical waveguide equals 879 MHz, which implies no occurrence of propagation modes for the 433 MHz or the 868 MHz frequency band. On the other hand, the 2.45 GHz frequency band was proposed as suitable band for applications in ducts with diameter down to 72 mm.

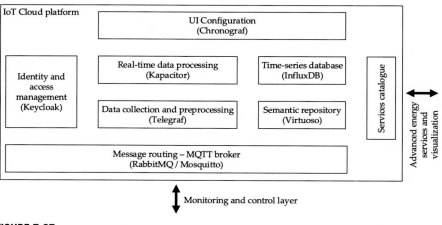

FIGURE 7.27

Cloud-based integration IoT platform for energy efficient living

Since the residential sector has been identified as one with the most demanding on energy, there exists a strong interest in exploitation of connected smart devices with the aim of improving energy efficiency, user comfort, and the overall quality of life. [BTB18] presents an IoT platform for monitoring user's energy consumption and habits with the aim of improving the energy efficiency of a household, as shown in Fig. 7.27. The platform ensures the provisioning and connection of sensors, smart meters, and actuators to the network and enables collection, processing, and exchange of data among the related components. In Fig. 7.27 the InfluxData Telegraf, InfluxDB, Chronograf, Kapacitor (TICK) Stack, composed of Telegraf (a server agent for collecting and reporting metrics), InfluxDB (a time series database for high write and query loads), and Kapacitor (a native data processing engine) were highlighted. These components can be configured and some tests can be performed in order to check sensor data acquisition, storage, and visualization through a testing Graphical User Interface (GUI). InfluxData TICK stack is an open source platform allowing users to manage metrics, events, and other time-based data.

A slightly technologically different but similar approach can be found on a Smart Campus of University of Malaga [FSRP+19]. The project Smart University of Málaga (SmartUMA) is based on four main pillars: deployment of measuring devices, setting a telecommunication networks and protocols, processing the data using the Artificial Intelligence (AI), and deploy a set of actuators. Its main goal is to improve the administration, management, and decision making at the university based on a precise knowledge of all the information that surrounds the campus.

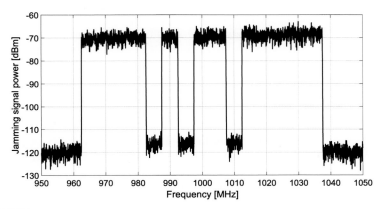

FIGURE 7.28

Spectrum of the 75 MHz width jamming signal with notched three protected bands.

Additional needs can be seen in the governmental sector. Officers of public entities like Police, Border Guard, and Army carry out their actions with left packages in different environments. Often there is a need to analyze unidentified parcels left on the airport or other public areas before their disposal. In many cases left packages are suspected to be Radio Controlled Improvised Explosive Devices (RCIEDs). To increase the security of the officers on duty and other bystanders usually some special device is used to jam radio signals that can be sent to detonate the explosives. Commonly this activity is named as setting an electromagnetic curtain. From a technical point of view, it can be understood as generating a radio signal with an occupied spectrum bandwidth of e.g. from a few MHz to hundreds of MHz. In most of the cases the jamming signal is generated in the communication bands of the cellular networks and ISM networks. In [RCM+19] such a concept is proposed, and an realization of a mobile device for generating an electromagnetic curtain is presented. The device is able to generate a wideband (with bandwidth of up to 1 GHz) jamming signal in the frequency range from 400 MHz to 2.7 GHz. The device was realized as a technological demonstrator with a size of a travel suitcase placed on a self-propelled platform. The jamming signal is generated digitally by using the software running on an ARM processor that is a part of the SoC Xilinx Zynq chip, integrating also the FPGA matrix. A unique feature of the device is that the jamming signal can be arbitrarily adjusted and generated on the basis of noise and chirp waveforms. Moreover, certain sub-bands in the jamming signal can be notched to maintain own communication during the operation of the jammer. The example of a jamming signal is presented in Fig. 7.28.

7.7 Conclusions

This chapter has provided an overview about some recent research activities developed in the framework of the Internet of Things. It appears clearly that there is not a winner Radio Access Technology (RAT) and the Mobile Network Operators will need to be able to offer services based on a multi-RAT approach, where the best technology will be selected depending on the specific application domain. Both, 3GPP solution like NB-IoT, and non-3GPP solutions, like LoRa and IEEE 802.15.4-based solutions (e.g., 6TiSCH), will be most probably included in the future 5G ecosystem. The problem of coexistence and interoperability among these different technologies, inevitably deployed in the field, is a future research challenge.

As for the architectural viewpoint, the application of SDN and NFV paradigms to IoT is still in its infant status, but seem very promising: the definition of reliable protocols to manage networks in a centralized way, both in terms of paths definition and scheduling decisions, are still open issues, together with the definition of cross-domain and multi-modal inference and analytics for managing large scale networks. In this framework, it is expected that new machine learning algorithms, data science, and analytics will play a key role to extract value from the next generation of IoT networks.

Finally, with reference to the application domains, Industry 4.0 seems to be a very challenging and interesting field for IoT, being characterized by very stringent requirements in terms of reliability and latency; vehicular applications are also attracting much attention from both academia and industry; finally, smart environments applications, mainly with reference to smart building, city, and agriculture, are another interesting field.

IoT for healthcare applications

**Kamran Sayrafian[a,b], Sławomir J. Ambroziak[c,b], Dragana Bajic[d], Lazar Berbakov[e],
Luis M. Correia[f], Krzysztof K. Cwalina[c], Concepcion Garcia-Pardo[g],
Gordana Gardašević[h], Konstantinos Katzis[i], Pawel Kulakowski[j], and Kenan Turbic[f]**

[a]*National Institute of Standards & Technology, Gaithersburg, MD, United States*
[c]*Gdańsk University of Technology, Gdańsk, Poland*
[d]*University of Novi Sad, Novi Sad, Serbia*
[e]*Institute Mihailo Pupin, Belgrade, Serbia*
[f]*INESC-ID / IST, University of Lisbon, Lisbon, Portugal*
[g]*Universitat Politècnica de València, València, Spain*
[h]*University of Banja Luka, Banja Luka, Bosnia and Herzegovina*
[i]*European University Cyprus, Nicosia, Cyprus*
[j]*AGH University of Science and Technology, Kraków, Poland*

The Internet of Things (IoT) has numerous applications in healthcare, from smart wearable or implantable sensors to remote monitoring of elderly, medical device networking, and in general creating a healthcare network infrastructure. IoT has the potential to create a pervasive environment for monitoring patients health and safety as well as improving how physicians deliver care. It can also boost patients engagement and satisfaction by allowing them to spend more time in the comfort of their residence and only interact with care centers and healthcare professionals whenever needed. A significant driver for the IoT-Health market is the increasing penetration of connected devices in healthcare. Wearable sensors have received a remarkable growth in recent years; however, a pervasive IoT-Health infrastructure is still a long way from commercialization. The end-to-end health data connectivity involves the development of many technologies that should enable reliable and location-agnostic communication between a patient and a healthcare provider.

This chapter summarizes IRACON contributions related to the application of IoT in healthcare. It consists of the following three sections. Section 8.1 presents the measurement campaigns and the related statistical analysis to obtain various channel models for wearable and implantable devices. In addition, the importance of physical human-body phantoms used for channel, Specific Absorption Rate (SAR), and Electromagnetic (EM) exposure measurements are examined. Methodologies to

---

[b] Chapter editors.

Inclusive Radio Communications for 5G and Beyond. https://doi.org/10.1016/B978-0-12-820581-5.00014-6
Copyright © 2021 Elsevier Ltd. All rights reserved.

improve the accuracy of these phantoms for various frequency bands are also discussed. Section 8.2 outlines methodologies to improve the medium access control (MAC) and networking layers of a body area networks along with possible architectures for remote health monitoring. Several applications such as localization, activity recognition, and crowdsensing and their corresponding technical challenges are also presented in this section. Finally, Section 8.3 introduces the concept of nanocommunications which can be considered as the nano-scale limit of the IoT technology spectrum. It provides an overview of the promising mechanisms that can establish data communication at molecular levels inside the human body as well as various interfacing techniques with macro-scale devices. It also highlights the revolutionary healthcare applications that could be enabled by this technology.

Remark

Certain commercial equipment, instruments, or materials are identified in this chapter in order to adequately specify the experimental procedure. Such identification is not intended to imply recommendation or endorsement by the respective organizations of the chapter editors.

8.1 Wearable and implantable IoT-health technology

For wearable and implantable sensors (or actuators), there could be several communication scenarios depending on the locations of the Tx and Rx with respect to the human body. These scenarios include wireless communication among devices inside the human body (in-body-to-in-body), between an implant and a wearable device (in-body-to-on-body), between two wearable devices (on-body-to-on-body), between a wearable and an external device e.g. an off-body Access Point (AP) (on-body-to-off-body), and finally between wearable devices located on different bodies (body-to-body). The characterization and statistical modeling of these communication channels is the central topic of this section. A thorough understanding of these channels is necessary and quite important for the design and optimization of the physical and medium access layers of any communication system that involves wearable or implant devices. In the following subsections, the IRACON channel models are presented for each communication scenario.

8.1.1 Channel measurement and modeling: On-body-to-off-body

In this communication scenario, either the transmitter (Tx) or the receiver (Rx) is a wearable device, while the other communicating node is at a fixed location away from the human body. Although this scenario may seem similar to use-cases including a mobile personal device, the impact of the human body on the antenna operation as well as the strong influence of the user dynamics on the communication link are among the distinguishing features that necessitate a dedicated channel model.

Analytical model

Turbic et al. have proposed a polarimetric geometry-based off-body channel model [TCB19] based on Geometrical Optics (GO) and Uniform Theory of Diffraction (UTD). The model also considers arbitrary antenna positions, orientations and radiation characteristics, i.e. polarization and gain. The transmission coefficient (h_{ch}) of a narrowband channel is represented as:

$$h_{ch} = \frac{\lambda}{4\pi} \sum_{n=1}^{N_m} \frac{1}{r_n} \, \mathbf{g}_r^H(\phi_r^n, \psi_r^n)_{[1\times 2]} \, \mathbf{H}_{p\,[2\times 2]}^n \, \mathbf{g}_t(\phi_t^n, \psi_t^n)_{[2\times 1]} \, e^{-j\frac{2\pi}{\lambda}r_n} \qquad (8.1)$$

where λ is the wavelength, N_m is the number of MPCs, r_n is the n-th path length, $\mathbf{g}_{t/r}$ is the polarimetric gain vector of the Tx/Rx antenna, $\phi_{t/r}^n$ and $\psi_{t/r}^n$ are the azimuth and elevation angles at the Tx/Rx respectively, and $\mathbf{H}_n$ is the path polarization matrix.

The polarization matrices differ for the LOS component ($\mathbf{H}_0$) and the reflected MPCs ($\mathbf{H}_n$) as follows:

$$\mathbf{H}_{0\,[2\times 2]} = \mathbf{Q}(\theta_{LOS})_{[2\times 2]} \qquad (8.2)$$

$$\mathbf{H}_{n\,[2\times 2]} = \mathbf{Q}(\theta_r^n)_{[2\times 2]} \, \mathbf{\Gamma}_r(\theta_i^n)_{[2\times 2]} \, \mathbf{Q}(\theta_t^n)_{[2\times 2]} \qquad (8.3)$$

where $\mathbf{Q}$ represents rotation matrix, $\mathbf{\Gamma}_r$ is the reflection matrix, θ_{LOS} is the polarization reference mismatch angle in the LOS direction, θ_i^n incidence/reflection angle, and $\theta_{t/r}^n$ is the mismatch angle between the polarization references associated with the Tx/Rx antenna and with the reflection/incidence plane. The angles θ_{LOS}, $\theta_{t/r}^n$ and θ_i^n are obtained from the scenario geometry as detailed in [TCB19, App. A].

To consider the human body dynamics, an analytical mobility model for the wearable antenna has been developed in [TCB18]. The model represents antenna motion as a composition of a linear forward motion at constant velocity, and a periodic component. This is illustrated in Fig. 8.1a. The antenna position over time (**r**) is therefore represented by:

$$\mathbf{r}_{[m]}(t) = \mathbf{r}_{0\,[m]} + v_{u\,[m/s]} t_{[s]} \, \mathbf{u}_v + \Delta\mathbf{r}_{[m]}(t) \qquad (8.4)$$

where v_u is the user's velocity, $\mathbf{r}_0$ is the starting point, $\mathbf{u}_v$ is the unit direction vector, and $\Delta\mathbf{r}$ represents the periodic displacement due to the changing posture. The corresponding orientation is represented by Euler angles, i.e. γ_1, γ_2, and γ_3, which specify a sequence of elementary rotations around the local coordinate axes Z-Y-Z, respectively. This will establish the orientation of the associated local coordinate system [TCB18, Eqn. 10-12]. The relation between the global and local coordinate systems is illustrated in Fig. 8.1b, where the rotation axes and the corresponding angles have also been indicated.

Due to the periodic changes in the human posture during walking or running, the periodic position component in (8.4) and the Euler angles are modeled by a Fourier series with up to two harmonics [TCB18, Eqn. 9,13]. The corresponding parameters are calculated from Motion Capture (MoCap) data [TCB18, Tab. 1].

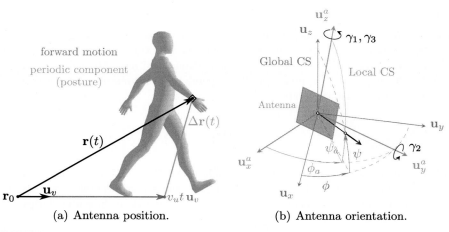

(a) Antenna position.   (b) Antenna orientation.

FIGURE 8.1

Wearable antenna mobility model.

The scattering environment surrounding the human body is represented by a simplified geometry i.e. scattering is assumed to occur on a circular cylinder centered around the body. This simplification is adopted to allow simultaneous consideration of multiple wearable antennas on the body. In other words, the scatterers are assumed to be fixed with respect to the human body; however, their impacts on different on-body antennas will depend on the relative positions of the antennas with respect to the scatterers.

The channel model was used to investigate the effects of different aspects of antenna motion in [TC20a]. A significant impact of user's motion on the polarization characteristics was observed, and the error in average cross-polarization ratio exceeded 23 dB when the antenna dynamics were neglected.

The antenna rotation has a dominant effect on the polarization matrix. However, the corresponding periodic displacement due to changing posture can be neglected. This allows for further simplification of the model. On the other hand, the antenna displacement has a significant impact on the small-scale fading characteristics. The fading dynamics are observed to vary over the motion period with distinct slow and fast phases [TC20b]. The latter is observed to result in 4 times higher Level-Crossing Rate (LCR) than the former when the antenna is located on the lower leg.

The channel model was validated against measurements, and a good agreement with the experimental data was observed. Using narrowband measurements data at 2.45 GHz [TAC17,ACK+16], the model's capability to reproduce polarization characteristics and temporal dynamics of the signal was demonstrated in [TCB19]. A good agreement with wideband measurements at 5.8 GHz [TACB18,TACB19] was also reported in [ATC19], and simulation error in the received power was shown to be within 3 dB. Fig. 8.2 shows the comparison between simulation (sim.) and measure-

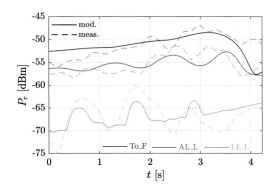

FIGURE 8.2

Comparison between the simulated and measured Rx power, for vertical off-body antenna polarization.

ments (meas.) for a scenario where the user is walking toward a vertically polarized AP antenna, while wearing antennas on the chest (To), wrist (AL), and lower leg (LL).

The wearable antenna rotation during motion was observed to strongly affect the antenna gain and polarization characteristics relative to the fixed off-body antenna, resulting in time-variant polarization losses [TCB17]. Therefore, the on-body antenna placement plays an important role on the off-body channel characteristics. Antennas on the arms and legs result in more severe Rx signal variations compared to locations such as the chest and head [TCB19].

Empirical models

The most common statistical path loss model based on narrowband measurements is[1]

$$L_{pl}(d,t)_{[\text{dB}]} = \overline{L_{pl}(d)}_{[\text{dB}]} + \Delta L_{ls}(t)_{[\text{dB}]} + \Delta L_{ss}(t)_{[\text{dB}]} \qquad (8.5)$$

where t denotes time, d is the Tx-Rx distance, $\overline{L_{pl}}$ is Mean Path Loss (MPL), ΔL_{ls} and ΔL_{ss} are the large- and small-scales fading components. A standard log-distance model is also typically adopted for the MPL component, i.e.

$$\overline{L_{pl}(d)}_{[\text{dB}]} = \overline{L_{pl}(d_0)}_{[\text{dB}]} + 10\,n_{pl}\log\left(\frac{d}{d_0}\right) \qquad (8.6)$$

where n_{pl} is the path loss exponent, and d_0 denotes the reference distance (e.g. 1 m). The fading components are modeled by random variables with distributions that are derived from statistical analysis of the measurements.

---

[1] By adopting the common practice in literature, the term path loss is used loosely herein. Due to the inability to de-embed the antennas from signal measurements, the measured loss actually corresponds to attenuation between antennas' terminals, i.e. system loss by definition [ITU16].

The data processing procedure used for the path loss component extraction and the estimation of model parameters is typically as follows. The small-scale fading variations are first eliminated from the instantaneous path loss by a moving average filter with an averaging distance of 10λ, the latter being a typical choice for indoor measurements [VLJ97]. After extracting small-scale fading, the log-distance model (8.6) is fitted to the remaining signal which includes a combination of MPL and large-scale fading.

With the path loss decomposed according to (8.6), a statistical analysis of large- and small-scales fading is performed for parameter selection. Commonly considered Goodness of Fit (GoF) tests for evaluation of different candidates are Akaike Information Criterion (AIC) [BA02], χ^2 and correlation tests [PP02]. One should note that the model parameters are usually provided separately for LOS, Quasi-LoS (QLOS), and NLOS conditions.

The large-scale fading is commonly reported to have a Lognormal distribution, while Rice, Nakagami-m, Rayleigh, and Lognormal distributions are the typical models used for small-scale fading. The latter is found to primarily depend on body-shadowing conditions and antennas' polarization. Moreover, the model parameters could vary with frequency, environment, antennas' radiation characteristics, and their on-body placement. The path loss model parameters reported to IRACON are summarized in Tables 8.1 and 8.2, and the corresponding experimental studies are briefly described in the following.

Table 8.1 Summary of MPL and Lognormal large-scale fading parameters in off-body channels.

n_{pl}	$\overline{L_{pl}(d_0)}_{[dB]}$	$\mu_{L[dB]}$	$\sigma_{L[dB]}$	$f_{[Hz]}$	Env.	Ref.
1.71	[32, 50]	0	[1.2, 3.0]	2.45 G	Office	[TAC17]
-	-	0	[1.4, 2.0]	2.45 G	Office	[WA19]
1.69	[25.2, 64.7]	0	[1.7, 6.5]	2.45 G	Ferry[*]	[KAS[+]18]
-	-	0	[1.2, 2.9]	2.45 G	Ferry[†]	[KACS19]
[0.16, 3.80]	[64.7, 76.2]	[-0.2, -0.4]	[2.0, 2.6]	868 M	Ferry[‡]	[CAR18a]
[0.13, 3.46]	[23.4, 30.7]	-0.3	[2.2, 2.8]	6.5 G	Ferry[‡]	[CAR18a]

[*]Dome-shaped discotheque; [†]straight corridor; [‡]L-shaped corridor

Narrowband measurements

The parameters in Tables 8.1 and 8.2 were obtained through several measurement campaigns, performed in different environments. Ambroziak and Turbic et al. [ACK[+]16,TAC17,ATC17] reported measurements results at 2.45 GHz, in a typical indoor office environment, while considering static, quasi-dynamic, and dynamic user scenarios. In static and quasi-dynamic scenario (i.e. with the user moving in place) different orientations of the user (leading to LOS, QLOS, and NLOS conditions) were considered. The dynamic scenario had the user walking towards and away from the off-body antenna, over a straight path.

Table 8.2 Summary of small-scale fading parameters in off-body channels.

Dist.	Parameters		$f_{[Hz]}$	Env.	Ref.
Nakagami[1]	$m_{Nk} \in [0.9, 19.5]$	$\Omega_{Nk} \in [1.0, 2.0]$	2.45 G	Office	[TAC17]
Rice[2]	$s_{Ri} \in [0.8, 1.0]$	$\sigma_{Ri} \in [0.5, 0.7]$	2.45 G	Office	[WA19]
Nakagami[1]	$m_{Nk} \in [0.8, 1.5]$	$\Omega_{Nk} \in [1.5, 2.1]$	2.45 G	Ferry[*]	[KAS+18]
Nakagami[1]	$m_{Nk} \in [0.8, 0.9]$	$\Omega_{Nk} \in [1.9, 2.2]$	2.45 G	Ferry[†]	[KACS19]
Lognorm.[3]	$\mu_{Ln\,[dB]} = -0.4$	$\sigma_{Ln\,[dB]} \in [2.1, 2.2]$	868 M	Ferry[‡]	[CAR18a]
Lognorm.[3]	$\mu_{Ln\,[dB]} = -0.3$	$\sigma_{Ln\,[dB]} \in [1.5, 1.7]$	6.5 G	Ferry[‡]	[CAR18a]

[*]*Dome-shaped discotheque;* [†]*straight corridor;* [‡]*L-shaped corridor*
[1]m_{Nk} *(shape) and* Ω_{Nk} *(scale);* [2]s_{Ri} *(noncentrality) and* σ_{Ri} *(scale);* [3]μ_{Ln} *(log-mean) and* σ_{Ln} *(log-standard deviation)*

The measurements were repeated with co-polarized [ACK+16] and cross-polarized antennas [TAC17]; therefore, providing a data set for channel characterization and estimation of model parameters in orthogonal polarizations. An alternative approach to estimate the MPL model parameters was proposed in [TAC17,ATC17]. The authors calculated the path loss exponent for the LOS case in the co-polarized channel, and estimated the intercept term $\overline{L_{pl}(d_0)}$ in (8.6) for each scenario.

The large-scale fading was reported to follow a lognormal distribution, while Nakagami-m was the best overall statistical model for small-scale fading. Parameters of the Nakagami distribution were found to be considerably different in the co- and cross-polarized channels. The distribution was closer to Rice and Rayleigh, respectively in the former and latter case.

Another set of measurements in the same environment and at the same frequency were performed by Wiszniewski and Ambroziak [WA19]. However, the scenario involved the user passing by an off-body antenna that was placed at a fixed distance from the user's walking path. Various on-body antenna placements such as chest, back, wrist, and head were considered in this measurement. Lognormal and Rice distributions were reported as the best model for large- and small- scale fading components respectively.

In order to avoid MPL model parametrization for each user orientation, Turbic et al. [TAC18a] have introduced an additional term to (8.5) to account for the orientation-dependent body-shadowing loss, i.e. shadowing pattern (S_{sh}), given by:

$$S_{sh}(d, \phi)_{[dB]} = S_m(d)_{[dB]} \frac{1}{2} \left\{ 1 + \cos\left[\frac{2\pi}{\Delta\phi}(\phi - \phi_0) \right] \right\} \qquad (8.7)$$

where ϕ is the azimuth angle of arrival/departure at the wearable antenna, S_m is the distance-dependent maximum body-shadowing loss, ϕ_0 is the azimuth angle of maximum loss, and $\Delta\phi$ is the shadowing pattern angular width. The model parameters were obtained from the indoor measurements at 2.45 GHz [TAC18a, Table I], with the user rotating at different distances from the off-body antenna. Fig. 8.3 shows the shadowing pattern (8.7) that is fitted to the measurement data. While the general

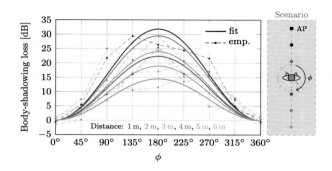

FIGURE 8.3

Body-shadowing loss as a function of user orientation.

model is applicable for on-body-to-off-body and body-to-body channels, additional measurements are required to estimate the parameters for the latter case.

In addition to typical indoor environments, measurements were also conducted in a passenger ferryboat [KAS+18,CAR18a,KAC17], where various metal structures could result into strong signal reflections. Authors in [KAS+18] have performed narrowband measurements at 2.45 GHz inside a dome-shaped discotheque within the ferryboat. They considered a scenario where the user walks towards and away from an AP at a fixed position. The measurements were obtained in two orthogonal polarizations, and repeated for wearable antennas located on the arm, chest, and head.

While the authors adopted the same MPL model (8.6), an additional term (L_{pa}) was introduced in (8.5), in order to take the effects of the wearable antenna height and orientation into account, i.e.

$$L_{pa\,[\text{dB}]} = 10\,a \log |\Delta h_{[\text{m}]}| + b_{[\text{dB}/\degree]}\mu_{\phi\,[\degree]} + c_{[\text{dB}/\degree]}\sigma_{\phi\,[\degree]} \qquad (8.8)$$

where Δh is the difference in Tx and Rx antennas' heights, a, b, and c are the model parameters, μ_ϕ is the mean angle in between the walking direction and the maximum on-body antenna radiation, and σ_ϕ is the corresponding standard deviation (Table 8.3). The statistical analysis performed by the authors shows that the best models for large- and small-scales fading in this environment are Lognormal and Nakagami-m distributions, respectively,

Table 8.3 Model parameters for L_{pa} [KAS+18].

a	b	c	Δh	μ_ϕ	σ_ϕ
[-1.06, 0.79]	[0.0, 0.1]	[-3.3, -1.4]	[0.05, 0.45]	[-97.0, 69.9]	[2.6, 5.2]

Considering the same scenario and settings, the authors also performed additional measurements at 2.45 GHz in a straight corridor within the same ferry [KACS19]. The same statistical models (with different values of parameters) were reported to fit this environment as well (Table 8.2).

Authors in [CARC17,CAR18a] have performed measurements in an L-shaped corridor with the user walking toward and away from an AP placed in one leg of the corridor. LOS and NLOS conditions were distinguished depending on whether the user and the AP are in the same leg of the corridor. Using a custom-developed dual-band measurement stand, the data were simultaneously recorded at 868 MHz (narrowband) and in the 6489 MHz UWB channel. The embedded UWB radio module was additionally used to associate each sample with the corresponding Tx-Rx distance. The measurements were repeated for the same wearable antenna placements as in [KAS+18], i.e. the head, chest, and wrist. An example of the measured path loss for this scenario is shown in Fig. 8.4, which also illustrates LOS/NLOS classification and shows the MPL model fitted to the data.

The authors have reported the Lognormal distribution as the best fitting model for large-scale fading. This also agrees with other studies presented here. On the other hand, in contrast to the other reports, the authors found that the small-scale fading also follows a Lognormal distribution.

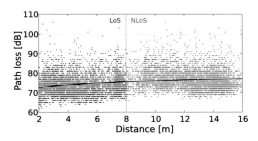

FIGURE 8.4

MPL model fitted to measurements in a ferry (L-shaped corridor).

In [KAC17], Kosz et al. have performed measurements with the user sleeping in a passenger cabin. In their study, narrowband measurements at 2.45 GHz were performed simultaneously with two wearable antennas. The wearable antennas were placed on the chest and back side of the user in one configuration, and the chest and wrist in the other. The data was obtained with two types of wearable antennas, namely FlexPIFA and FlexNotch. The authors reported the Rx mean power and standard deviation. The mean power was found to vary between 56 dB and 70 dB over the considered scenarios, while the variation of the standard deviation was between 3.7 dB and 8.0 dB. A strong dependence of these parameters on the user's orientation, type of the wearable antenna and its placement was observed; however, the height at which the user slept inside a bulk bed showed little impact on the results.

Wideband measurements

While most of the empirical contributions to IRACON considered narrowband fading channels, authors in [TACB18,TACB19] reported the results of dual-polarized CIR measurements at 5.8 GHz, with 500 MHz bandwidth, conducted in an indoor environment The measurement campaign included a number of different scenarios designed to investigate the influence of depolarization, user dynamics, body-shadowing from the user or another person, as well as the scattering impact of people in the environment around the user [TACB18]. The measurements were obtained simultaneously with orthogonally polarized antennas, and repeated for antenna placement on the chest, wrist, and lower leg. Results show strong influence of the user dynamics and body-shadowing on the CIR parameters, i.e. number of paths, Rx power, and delay spread; therefore, leading to an effectively non-stationary channel [TACB19]. The ratio of the received powers in the orthogonal polarization was found to vary up to 21.3 dB when the user walks toward or away from the antenna.

Another set of CIR measurements over the UWB frequency range 3.8-10.2 GHz was reported by Wilding et al. [WMW19], and provided an insightful analysis of the body-shadowing effect and off-body channel characteristics under obstructed LOS conditions. A significant pulse distortion and widening due to attenuation and body-diffracted waves was observed. The attenuation of the LOS component due to the body-shadowing is characterized by introducing the effective energy pattern of a wearable antenna, i.e. Rx power distribution over azimuth angles relative to the maximum radiation direction. Fig. 8.5 shows this pattern for different antenna placements indicated in the figure, with the free-space antenna scenario being also provided for reference.

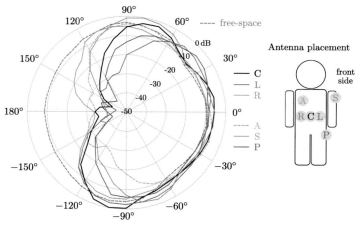

FIGURE 8.5

Wearable antenna effective energy pattern.

Millimeter-wave measurements

While all of the empirical studies discussed so far consider frequencies below 6.5 GHz, two off-body channel measurement campaigns at centimeter- and millimeter-waves were also reported to IRACON [SPCN18,ZGL+17]. Sana et al. [SPCN18] reported wideband off-body channel measurements at 60 GHz, obtained by using a channel sounder developed at Durham University (UK). While only the initial results based on visual inspection of the measured CIR were presented, the authors also investigated the achievable diversity gains by employing a 2×2 MIMO antenna, and considering three different combining techniques.

Zhao et al. considered outdoor mobile off-body channel at 15 GHz and 28 GHz [ZGL$^+$17]. The impact of the human body on the radiation characteristics of mobile terminals was investigated by evaluating the radiation efficiency reduction and gain pattern distortion due to blockage. The study was carried out for three types of antennas: notch, slot, and edge-patch antenna. According to the reported results, the radiation efficiency reduction due to body proximity is up to 4 dB lower at 15 GHz and 28 GHz, than at frequencies below 6 GHz. By using the measured mobile antenna radiation patterns in ray-tracing simulations, the coverage areas at 15 GHz and 28 GHz were compared for an urban pedestrian scenario. The latter frequency was reported as lass favorable due to higher path loss and body-blockage losses.

Finally, the empirical studies reported to IRACON also addressed the body-to-body communication scenario as a special case of the off-body communication. A narrowband measurement campaign at 2.45 GHz performed in indoor and outdoor environments was reported in [ACT16a]. The experiment considered scenarios where the users walking toward and away from each other, as well as walking in parallel. The large- and small-scales fading were found to follow Lognormal and Nakagami-m distributions, respectively, similar to the off-body channels with fixed APs [TAC18b]. Body-shadowing was observed to strongly affect the small-scale fading characteristics. The average amount of fading (fading merit) was up to 2.56 times higher when LOS is obstructed. The environment was also found to have a significant impact as well. The small-scale fading was observed to be more severe in the indoor environment i.e. 1.88 times higher average fading and 3.81 dB lower K-factor.

8.1.2 Channel measurement and modeling: On-body-to-on-body

For body surface communication, antenna orientation and polarization with respect to the environment is an important issue that is sometimes ignored in path loss calculation. This issue could be particularly significant in dynamic scenarios where the person wearing the antenna is in motion.

Authors in [Aoy16] took antenna directivity into consideration for Body Area Networks (BANs) using millimeter or terahertz frequencies. In their research, variation in the direction of on-body antennas and the frequent occurrence of body-shadowing in dynamic scenarios have been highlighted. The temporal variation of the azimuth and zenith angles of the on-body antennas and shadowing rate is investigated using simulation. Antenna placements considered in their study included possible wearable applications such as abdomen, chest, head, upper arm, ankle, thigh, and hand. Results show that in some scenarios, azimuth and zenith angles of the antenna exhibit a large variation during a person's walk. Therefore, high speed beamforming mechanisms should be considered if the on-body antenna has high directivity.

8.1.3 Channel measurement and modeling: In-body-to-on-body and in-body-to-off-body

As mentioned earlier communication with an implant device could include three scenarios i.e. communication between two implants (in-body-to-in-body), between an implant and an on-body device (in-body-to-on-body), and between an implant and an off-body device (in-body-to-off-body). Radio wave propagation from an implant device is highly affected by the dielectric properties of the human body tissues on its path. These properties impact both the antenna characteristics (e.g. matching, radiation pattern) and the propagation channel. In the following subsection, implant communication channels in the human body or an animal are discussed.

Human body implants

Two different MPL models are proposed for these channels: a log-linear model similar to Eq. (8.5) but with reference distance d_0 of 1 cm, and a linear model as the following:

$$\overline{L_{pl}(d)}_{[\text{dB}]} = \overline{L_{pl}(d_0)}_{[\text{dB}]} + \alpha_{pl\,[\text{dB/cm}]}\, d_{[\text{cm}]} \qquad (8.9)$$

where α_{pl} is the slope, and $\overline{L_{pl}(d_0)}$ is the MPL at zero distance.[2]

The path loss variations around the MPL are found to exhibit Lognormal distribution similar to the off-body channel.

The parameters of the models are summarized in Tables 8.4 and 8.5 for all three channel scenarios. The parameters values were derived from numerical full-wave simulations, wideband measurements using physical phantoms and *in vivo* experiments. For the liquid phantom experiment [AGF+16], one antenna was fixed at a specific location and the other was moved over a spatial grid in order to collect sample measurements. The in-body-to-off-body channel was investigated based on *in vivo* and phantom-based measurements reported in [GFA+16]. There are several studies contributing to the in-body-to-on-body channel model parametrization [AGF+16,GFA+16,AGC+18,PSAGP+19]. The phantom-based measurements were reported in all studies, and the *in vivo* measurements have been reported in [GFA+16,PSAGP+19].

The measurements obtained with the liquid phantoms are observed to generally overestimate the path loss, specially when the distance between the antennas increases. Therefore, a correction model can be applied to the set of available phantom-based measurements for more realistic path loss estimates [GFA+16]. The distance-dependent correction factors (C_{pl}) for the linear (8.5) and log-linear (8.9) MPL models are given by (8.10) and (8.11) respectively:

$$C_{pl}(d)_{[\text{dB}]} = 1.1 + 7.4 \log(d_{[\text{cm}]}) \qquad (8.10)$$

$$C_{pl}(d)_{[\text{dB}]} = 5.8 + 2.2\, d_{[\text{cm}]} \qquad (8.11)$$

---

[2] This is a control point of the model without a meaningful physical interpretation, as the model is not applicable for co-located antennas.

Table 8.4 Log-linear MPL model parameters and standard deviations in channels with implants.

Ch.	$\overline{L_{pl}(d_0)}_{[dB]}$	n_{pl}	$\sigma_{Ln\,[dB]}$	$d_{[cm]}$	$f_{[GHz]}$	Ref.
in2off	[70.4, 71.5]	[0.7, 1.4]	-	4 - 50	3.1 - 8.5	[GFA+16]
in2on	47.8	1.98	1.2	5.5 - 20	3.1 - 8.5	[AGF+16]
	[-12.2, 35.8]	[5.8, 9.3]	[5.0, 5.7]	2.8 - 8	3.1 - 5.1	[AGC+18]
	[-29.7, 26.2]	[5.4, 10.3]	[2.0, 4.6]	2.8 - 8	3.1 - 5.1	[PSAGP+19]

Table 8.5 Linear MPL model parameters and standard deviations in channels with implants.

Ch.	$\overline{L_{pl}(d_0)}_{[dB]}$	$\alpha_{pl\,[dB/cm]}$	$\sigma_{Ln\,[dB]}$	$d_{[cm]}$	$f_{[GHz]}$	Ref.
in2in	45	4.6	4.3	3 - 8	3.1 - 8.5	[AGF+16]
in2on	[30.8, 36.6]	[5.2, 7.4]	-	3 - 11	3.1 - 5.0	[GFA+16]
	[14.8, 53.4]	[4.5, 7.4]	[5.0, 5.7]	2.8 - 8	3.1 - 5.1	[AGC+18]
	[13.8, 41.6]	[4.3, 6.8]	[2.0, 4.6]	2.8 - 8	3.1 - 5.1	[PSAGP+19]

Animal implants

In addition to health monitoring applications in humans, implants can also be used in animals, for example tracking the health conditions of dairy cows and facilitating herd management. Benaissa et al. in [BPN+19] presented the in-body-to-off-body path loss between an implanted device inside a cow and an external node. The proposed model was obtained based on measurements conducted with several cows in a 6 m × 18 m barn, housing seven fistulated dairy cows. The authors adopted the statistical path loss model (8.5) and fitted the MPL model (8.6) to the measurement data. The large- and small-scales fading components (ΔL_{ls} and ΔL_{ss}) were considered jointly and assumed to have a zero-mean Gaussian distribution with standard deviation $\sigma_{[dB]}$. The reported MPL model parameters are summarized in Table 8.6.

Table 8.6 MPL model parameters for in-body-to-off-body channels with animal implants.

$\overline{L_{pl}(d_0)}_{[dB]}$	n_{pl}	$\sigma_{[dB]}$	$d_{[m]}$	$f_{[MHz]}$	Ref.
[48.3, 98.5]	[1.9, 2.1]	[1.8, 4.9]	1 - 20	433	[BPN+19]

Using the proposed channel model, the authors also calculated the expected transmission range for LoRa technology. Depending on the Tx power and the desired data rate, it was shown that a transmission range of up to 100 m could be achieved.

8.1.4 Human body phantoms and SAR measurement

Analysis of the exposure of the human body to electromagnetic waves is an important field of research for IoT applications in healthcare. The goal in this research is to evaluate the impact of the electromagnetic fields on the human health. RF exposure tests focus on confirming that the maximum allowable absorption rate is not exceeded by the regulatory limits. To carry out such tests, it is also important to know the electromagnetic properties of the human body tissues.

The electromagnetic properties of the body tissues are expressed in terms of their permittivity and permeability. Since the human body is a non-magnetic medium, the permeability of its tissues is equivalent to that of the air. In contrast, the permittivity varies considerably for different body tissues. This is especially the case at Gigahertz frequencies and above. The relative permittivity which is the absolute permittivity normalized to that of the air, ε_r^*, is a complex frequency-dependent property. The real part of the relative permittivity (ε_r') is the dielectric constant, and its imaginary part (ε_r'') is the loss factor. Therefore, relative permittivity can be expressed as: $\varepsilon_r^*(f) = \varepsilon_r'(f) - j\varepsilon_r''(f)$. In some cases, the imaginary part of the permittivity is given as the dielectric conductivity that can be deduced from the loss factor i.e. $\sigma(f) = 2\pi f \varepsilon_r''(f)\varepsilon_0$, where ε_0 is the relative permittivity of the vacuum.

Both dielectric constant and loss factor (or conductivity) define the behavior of the electromagnetic waves traveling through different human body tissues. Therefore, it is important to consider the values of the complex permittivity in the study of implant propagation channels. The most widely used repository of such values was given by C. Gabriel in [Gab96] more than 20 years ago. It provides the values of the dielectric constant and loss factor as a function of frequency (from several kHz to GHz). However, researchers should take into account that Gabriel's repository was produced from experiments in different animal species using various measurement techniques.

In all IoT use-cases involving implants, laboratory measurements and/or software simulations are necessary in order to evaluate the performance of the wireless link in the human body environment. Researchers should use hardware or software models that replicate the complex permittivity of the tissues involved in the study. These kinds of models are referred to as *phantoms*. They are intended to accurately emulate electrical properties of different body tissues.

Dielectric properties of body tissues

There are several methods for measuring the dielectric properties of any material. Open-ended coaxial probe, transmission lines, resonant cavities or parallel plates are the most known methodologies reported in the literature. The requirements of the measurement, the type of material and the frequency range determine which kind of methodology is the most useful in each case. In particular, the open-ended coaxial probe is the most used technique for measuring the electromagnetic properties of body tissues. It allows broadband measurement for liquids, gels or semisolids with high accuracy. This approach is based on a rigid coaxial cable with a flat cut end (open-ended coaxial) submerged into the liquid or posed over the surface of the gel or semisolid. The probe is connected to a Vector Network Analyzer (VNA) which is responsible for measuring the values of the reflection coefficient (scattering parameter $S_{11}(f)$) for all frequencies under analysis. Next, the $S_{11}(f)$ values are translated into their corresponding dielectric constants i.e. $\varepsilon'_r(f)$, and loss factor, $\varepsilon''_r(f)$.

Considering the open-ended coaxial technique, an adequate prior calibration of the probe is the key in order to obtain accurate values of both dielectric constant and loss factor. In the literature, several calibration procedures have been addressed so far. The most common procedure consists of measuring the reflection coefficient of at least three different elements (known as reference standards) with well-known complex permittivity. The most used reference standards are open circuit, short circuit, and water (hereinafter called "typical calibration") since their dielectric properties are well reported in the literature. However, in [FLGPC+17], authors demonstrate that these standards are not the most appropriate when measuring high water-content body tissues such as muscle, heart, stomach, and liver. For these cases, the lowest uncertainty of the measurements was achieved by adding methanol to the three calibration standards used in the typical calibration, i.e. by using an open circuit, short circuit, water, and methanol. This is due to the fact that the complex permittivity of methanol is at the same order of magnitude as that of the high water-content tissues. This effect can be observed in Fig. 8.6 where authors measure a liquid with permittivity similar to that of the high water-content tissues i.e. Dimethyl sulfoxide (DMSO). Authors also considered the typical calibration as well as methanol, ethanol, and 2-propanol as additional calibrators. The lowest values of the measurement uncertainty for the complex permittivity values was achieved by adding methanol to the calibration standards. For low water-content body tissues, authors in [FLGPC+17] reported that ethanol is also the best additional calibrator and provides the lowest uncertainty.

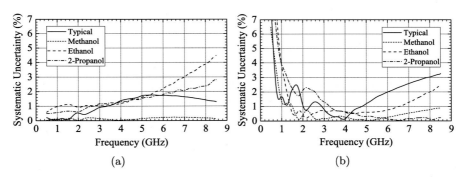

FIGURE 8.6

Systematic error (%) in the measurement of the DMSO (equivalent to high water-content human tissues) with different calibration setups: a) dielectric constant, b) loss factor.

As mentioned before, the Gabriel's database [Gab96] is the most widely used reference for test and evaluation of wireless devices operating in the vicinity of the human body tissues. However, one major issue with this database is that the reported results are obtained from different experiments, animal specimens, and measurement techniques. Therefore, direct comparison between different tissues is not very suitable. Furthermore, this database only considers the average values of the measurements, and not the variability among different measurements. Authors in [FLGPF+18] analyze this issue by reporting the mean and the standard deviation X2 (2SDM) of the complex permittivity values for several tissues at the gastrointestinal area i.e. muscle, colon serosa, fat, and skin. It was shown that the higher heterogeneity of the tissue translates into wider variability of the complex permittivity. This can be clearly observed in Fig. 8.7 where the fat tissue shows a higher variability than skin or muscle tissues. This is because the fat tissue has different content of water depending on the measurement point; thus, increasing the heterogeneity of the tissue. Such variability of the complex permittivity can have a negative impact on the communication link. For example, the antenna matching and its radiation pattern can be affected since the real values of the permittivity of the surrounding tissues are quite different than the values used during its design.

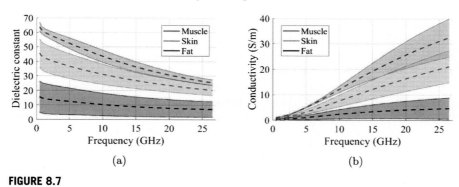

(a)   (b)

FIGURE 8.7

Mean (dashed line) and variability (shadowed area) of the dielectric properties of muscle, skin, and fat tissue.

Electromagnetic phantoms for radioelectric measurements

As mentioned before, phantoms try to replicate the electromagnetic properties of different human body tissues by using the values of the dielectric constant and loss factor given in [Gab96] or other similar databases such as [FLGPF+18]. Software or computational phantoms are computer-based models of various body tissues that can be used for electromagnetic simulations. In contrast hardware phantoms are physical materials used for experimental measurements.

Hardware phantoms (hereafter referred to as phantoms) can be solid, semi-solid, or liquid depending on the requirements of the test. Furthermore, use of phantoms is only valid for certain frequency range in which they accurately reproduce the complex permittivity of the target tissue. In the literature, there are a number of formulas for producing phantoms for many tissues and several frequency bands especially the microwave. However, it is really hard to find a common formulation for systems operating at ultra-wideband frequencies. This is due to the complexity of mimicking not only a certain value of permittivity but also its trend versus frequency. The approach reported in [CPGPFL+16] aims at overcoming this constraint by using acetonitrile-based liquid aqueous solutions. These solutions mainly use water, acetonitrile, and salt in different proportions to replicate both the dielectric constant and the loss factor for a variety of body tissues within the 0.5 GHz to 26.5 GHz frequency band as observed in Fig. 8.8 [CPGPFL$^+$18]. These broadband phantoms are specially appropriate to carry out physical measurements for implant scenarios. The liquid nature of the phantom allows flexibility in positioning the antenna at the desired location.

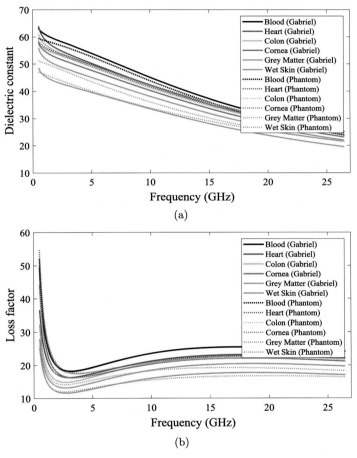

(a)

(b)

FIGURE 8.8

Relative permittivity of several phantoms (dashed lines) [CPGPFL$^+$16] in comparison with the target values [Gab96] of their corresponding tissues (solid lines). a) Dielectric constant, b) Loss factor.

Testing radiation exposure as a result of using mobile devices is another important application of phantoms. Most of the phantoms used for this purpose are liquids applicable in a narrow frequency band. This makes testing mobile devices that use multi-frequency technology quite difficult as multiple phantoms will be required to evaluate all frequencies under consideration. Even then, the real radiation effect would not be accurately evaluated for novel technologies such as carrier aggregation or simultaneous wireless connection. This limitation will become a real constraint for future 5G mobile systems operating in the mmWave band. In [CPGPC$^+$18], the authors investigate a novel phantom formulation for 5G mobile communications. Due to the low penetration depth of millimeter waves, the most important tissue for the phantom is the skin. The authors analyzed propanol and methanol and concluded that propanol aqueous solutions are the best for mimicking the skin tissue for both dielectric constant and loss factor at mmWave. This level of matching can be observed in Fig. 8.9, where the shaded area depicts the ±10% deviation with respect to the mean value of the target skin tissue given in [Gab96]. The inherent heterogeneity of the tissues and variation among specimens leads to a variability in the measured values as explained before.

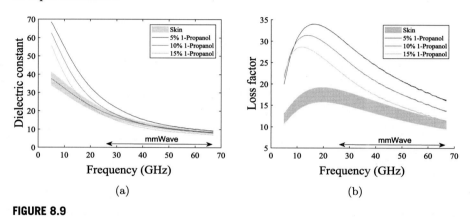

FIGURE 8.9

Relative permittivity of the k-carrageenan gels. a) real and b) imaginary part.

Human exposure to EM fields

It is known that long term exposure to EM sources could lead to several health problems [who]. Because of that there are restrictions on the maximum EM radiation enforced by national and international regulatory bodies. The EM exposure is typically limited in terms of the maximum electric field strength/power density incident on the human body or in terms of the SAR. The latter is a measure of the amount of energy absorbed by the human body when it is exposed to an EM field. SAR is expressed in watts per kilogram. The exposure limits and SAR also depends on the dielectric properties of the human tissues; and therefore, on the frequency under study.

The evaluation of the SAR resulting from a Personal Wireless Communication Device (PWD) is a challenging task due to its operation in the close vicinity of the human body. Accurate computational estimation of the SAR for such devices will be critical to ensure adherence to national or international safety limits. Therefore, near-field validation of the numerical models used for SAR computation becomes necessary. In [HBPH17], authors present a novel validation technique based on the comparison of measured and simulated one-port characteristics of the PWD antenna while the near field is systematically perturbed by a dielectric control object near the PWD. In particular, the authors use the input impedance Z as a single-port parameter and propose a formula for the change of the antenna impedance in the near field, as well as a formula that relates the error of such changes to the near field values. Using these formulas, authors numerically verify the equivalence of the two near-field validation techniques by comparing measured and computed validation data for a Planar-Inverted F Antenna (PIFA) as observed in Fig. 8.10. After applying a deconvolution directly to the values of the impedance changes and considering Tikhonov regularization of the convolution kernel, the reconstruction of the 3D electric field with reduced error will be possible. Further technical details can be found in [HBPH19].

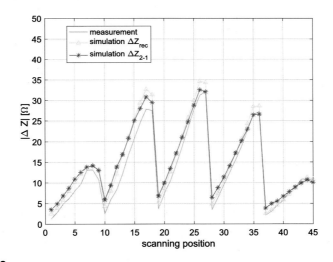

FIGURE 8.10

Comparison of the measured (red solid line; mid gray in print version), numerical impedance (green solid line with triangles; light gray in print version) from the electric field, and numerical input impedance change (blue solid line with stars; dark gray in print version).

Determination of the real exposure levels is a critical task when planning mobile communication services. In [HLHI18], authors present a large set of exposure measurements performed in different places of Kosovo. Measurements were taken for Global System for Mobile Communications (GSM), Universal Mobile Telecommunications System (UMTS), and LTE mobile technologies in DL using a commercial and calibrated spectrum analyzer. The results show that even after the implementation of LTE 1800 system (Re-farming), the main contributor to the EM exposure is

GSM 900 followed by UMTS and GSM 1800. The lowest levels of downlink base station emissions are captured from LTE. All measured values in outdoor and indoor areas (LOS and NLOS positions) are well below the International Commission on Non-Ionizing Radiation Protection (ICNIRP) reference levels [icn]. Looking at different scenarios, the highest values of the total electric field exposure were obtained in public transportation vehicles such buses, followed by coffee shops and outdoor environments. The lowest values were captured in home and office environments.

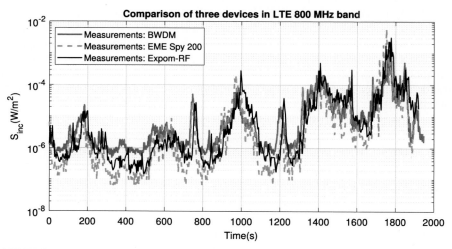

FIGURE 8.11

Incident power densities measured during the walk for several Personal Exposure Meter (PEM).

The measurement of the incident power density is conducted by the measuring (i.e. meter) device. In [ATA+18], the authors design a Multi-Band Body-Worn Distributed (BWDM) Radio-Frequency Meter as a Personal Exposure Meter (PEM). The BWDM consists of 22 textile antennas integrated in a garment covering the back and torso of the human subject. It was calibrated on 6 different human subjects and for 11 different telecommunication frequency bands. The BWDM meter has an improved uncertainty of up to 9.6 dB in comparison to commercially available personal exposure meters. Furthermore, as illustrated in Fig. 8.11, commercial PEM underestimate the exposure to EM fields when considering real measurements on a outdoor route.

8.2 IoT-health networking and applications

Future communication networks involve a plethora of 'smart' wireless devices that can exchange information in real time. These devices are expected to be part of our daily lives, interfacing not only with humans, but also with other devices; therefore, leading to the new paradigm known as "machine to machine" communications [RRKS14]. Connected medical devices (including wearables and implants) are expected to form a smart environment capable of exchanging health information i.e. IoT-Health. The communication links in this environment can be characterized by polymorphic requirements such as latency, throughput, reliability, security, etc. [PB10]. Remote physiological monitoring is a critical application of IoT-Health which requires connectivity of wearable or implantable sensors to local or wide area networks. Remote monitoring of large numbers of patients requires sufficient bandwidth, reliable communication links and high quality of service (QoS). As part of the future 5G networks, smart medical or personal health devices operating within such environment are expected to have the capability to observe and understand the relevant physical and social parameters of their operating area. This section reports on the IRACON research activities related to IoT-health networking and applications.

8.2.1 Networking and architectures

A Body Area Network (BAN) is formally defined by the IEEE 802.15.6 standard [IEE12] as a communication and networking protocol for wireless connectivity of wearable and implantable sensors (or actuators) located inside or in close proximity of the human body. A BAN typically includes several sensor and actuator nodes along with a controller also known as coordinator. Sensors are mainly used to monitor a physiological signal while actuators apply a signal to the body or cause an operation to take place inside or on the surface of the human body. A BAN can operate as a stand-alone network or as part of a larger infrastructure. There are still several technical challenges involving implementation and integration of BANs that need to be addressed in order to achieve high reliability or Quality of Service. One such challenge is mitigating radio interference from coexisting wireless networks or other nearby BANs.

When several BANs are within close proximity of each other, inter-BAN interference may occur since no coordination across multiple networks exists in general. Authors in [BSA15] investigate the performance impact of the Energy Detection (ED) threshold within the IEEE 802.15.6 CSMA MAC protocol when the system is comprised of several co-located BANs. They have shown how the static value of this threshold can lead to starvation or unfair treatment of a particular node(s) when there are potential interferers in the vicinity. To demonstrate this, they implemented a simplified CSMA/CA protocol as outlined by the IEEE 802.15.6 standard. Same authors in [BSA16] have proposed adaptive schemes that can be used to adjust the ED threshold in the transmitting nodes of a BAN. The objective is to fairly allow channel access to all nodes regardless of the level of interference that they are experiencing. Simulation results indicate benefits of the proposed strategy and demonstrate improvement in the overall performance.

Another challenge in heterogeneous Body Area Networks is the development of methodologies for the allocation of multiple data streams with simultaneously operating radio interfaces. In [CAR+18b], authors have proposed a novel bit-rate adaptation method for data streams allocation in heterogeneous BANs. Block diagram of the proposed adaptive method is shown in Fig. 8.12. A dedicated simulator has been developed using the results of measurements in a realistic environment. Using this simulator, the efficiency of the proposed algorithm was compared with other known data stream allocation algorithms. It was shown that using transmission rate adaptation based on radio channel parameters can increase the efficiency of resource usage compared to fixed bit-rate transmissions and other algorithms.

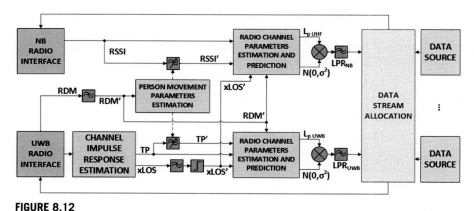

FIGURE 8.12

Block diagram of the proposed novel adaptive method for data stream allocation [CAR+18b].

Architectures for remote health monitoring

A Remote Health Monitoring (RHM) system aims to provide real-time or near real-time monitoring of patient's vital signs, thus enabling direct medical care or treatment. This technology can be used to help patients with a number of medical conditions; improving their quality of life and general well-being. It can also assist with medical and drugs' database administration. Technological advances in microelectronics, wireless communications and low-cost medical sensors are setting the stage for a 24/7 connected healthcare environment. Current medical sensors can provide a variety of physiological signals such as heart-rate, electrocardiogram, blood pressure, blood glucose levels, oxygen saturation, etc. If such sensors can be assigned a unique IPv6 address, then health information (i.e. sensors' readings) can be collected and transferred to other IP end-devices or to the Cloud for further processing and decision making.

Authors in [GFT+18] proposed the heterogeneous IoT-based architecture for remote monitoring of physiological and environmental parameters using Bluetooth and IEEE 802.15.4 standards (Fig. 8.13). The RHM system consists of a BAN with Shimmer physiological sensors (communicating using Bluetooth radio), and OpenMote environmental sensors communicating using IEEE 802.15.4 radio. This architecture enables data collection from multiple wearable and environmental sensors in order to process and extract useful information about the current state of the patient, as well as the environment in which the patient resides. Measurements are collected in a relational database on a local server (i.e. a Fog node for fast data analysis) as well as on a remote server in the Cloud.

Processing of the data generated by various health monitoring sensors and extracting valuable information is a complex task which requires significant resources. Authors in [TGR16] proposed a cloud-based solution to deal with this challenge. Their solution includes personalized medical devices as the source of sensor data and the cloud delivering various healthcare services over the Internet. The medical cloud hosts specially developed applications that process the data from medical devices and communicate the results to caregivers or medical institutions.

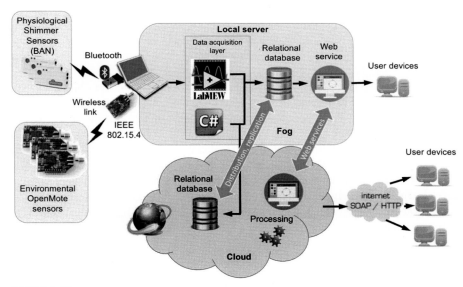

FIGURE 8.13

Heterogeneous IoT-based architecture for remote monitoring of physiological and environmental parameters [GFT+18].

Furthermore, authors in [KJD17] have provided a review on how 5G can be used as an underlying technology to provide remote health monitoring. They have also presented a novel architecture shown in Fig. 8.14. The proposed system employs White Space Devices (WSD), along with IEEE 802.22 (WiFAR) standard to provide seamless connectivity for the end-user devices. Initial findings indicate that the proposed communication system can facilitate broadband services over a large geographical area taking advantage of the freely available TVWS.

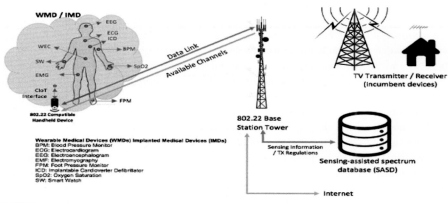

FIGURE 8.14

Proposed architecture connecting WMDs and IMDs to the internet using WiFAR (IEEE 802.22) [KJD17].

8.2.2 Applications

Localization

Localization has been mainly considered as a technology to locate objects or people in outdoor or indoor environments. However, there are new applications of this technology to locate and track indigestible electronics or capsules inside the human gastrointestinal (GI) tract. Specifically, authors in [BPG+18] explain how using UWB radio frequency transmission could be beneficial in applications such as capsule endoscopy localization. Performance analysis of RF-based localization is typically conducted through simulations using computational human body models or through experimental measurements using homogeneous phantoms. One of the most common methodologies in RF-based localization is using the received signal strength (RSS) to estimate the position of the transmitting node. Laboratory measurements using a customized multi-layer phantom test-bed [BPG+18] as well as in-vivo experiments were conducted [BGN+18] to evaluate the performance of RSS-based localization. The experimental laboratory measurements were conducted in the 3.1 GHz to 8.5 GHz UWB frequency band, and the results were used to perform two-dimensional (2D) localization. Fig. 8.15 illustrates how authors envisaged real scenarios into the phantom model that was used to conduct a number of experiments [BPG+18].

A magnetic sensor was attached to the in-body and on-body antenna so that the tracker could precisely evaluate the distance between antennas as well as their respective coordinates. Five on-body antenna positions with a separation of 2 cm along the y and z axes were considered on the outer edge of the fat phantom layer. The in-vivo measurements were conducted in a living porcine at the Hospital Universitari i Politècnic la Fe in Valencia, Spain [BGN+18] and the results were used to perform three-dimensional (3D) localization. Same antennas and equipment (Vector Network Analyzer (VNA) and magnetic tracker) were used for both in-vivo and laboratory

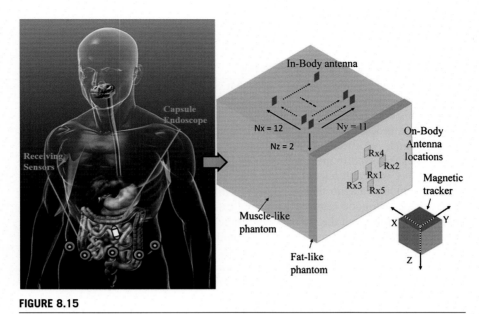

FIGURE 8.15

A multi-layer phantom-based model.

experiments. The in-body antenna was placed in three different positions inside the small bowel through laparoscopy procedure. For each in-body location, the on-body antenna was placed on the abdomen of a porcine model, and in direct contact with the skin. Measurements were taken in the 3 GHz to 6 GHz UWB frequency band, considering a resolution of 1601 points in frequency. Preliminary results on the estimation of the in-body antenna coordinates showed that RSS-based localization can achieve an average accuracy of (0.5-1) cm (assuming a limited range of distances between the in-body and on-body antennas).

Localization techniques can also be employed to locate dairy cows in a farm. In preparation for that objective, authors in [PTB+17] studied the cow's on-body and the off-body channels in both outdoor and barn environments. Bluetooth Low Energy (BLE) and UWB wireless technologies were used in that study. On-body measurements were performed in a large area of about 6 m × 12 m, so that reflections from the walls could be ignored. Path-loss measurements were performed using two ZigBee (XB24-Z7WIT-004) motes. On-body, off-body and location tracking measurement results have been presented in [BPT+16a], [BPT+16b], and [TPMJ17] respectively. The propagation path-loss for different on-body communication links on a dairy cow (ear to neck, hind leg to neck, front leg to neck, and udder to neck) has been characterized by both measurements and simulations. Measurements on a dairy cow in a multipath environment (i.e. barn) have also been performed to validate the simulations conducted using a cow body phantom. The path-loss results obtained from the simulations showed good agreement with the values derived from the measurements. A log-normal path-loss model has been constructed from the measurement and simulation data for the whole cow's body.

Remote monitoring and crowdsensing

The concept of mobile crowdsensing (MCS) can be described by volunteers, equipped with sensors, who extract and share information for personal or common benefit. This concept aligns with the four pillars of the internet of everything (IoE): people, data, process, and things [SJKB19]. Initial crowdsensing typology included environmental, infrastructural, and social applications only [CCLG16], but health applications followed immediately [JS17], [PRLS15], [PRH+15], [MMA+18]. The large number of health related information, captured on a daily basis, constitutes a valuable source of diagnostic and prognostic information waiting to be explored. The hypertension is a good crowdsensing application example [JJS+18]: it exhibits no obvious symptoms and the patients tend to behave as if they were healthy. Crowd-sensing can collect the location, weather condition and the cardiovascular features in order to form a large database of cardiovascular parameters in different environmental circumstances for further analysis. Patients are motivated to participate in MCS through feedback, advising them to continue or pause their current activity. The feedback is generated using the Random forest [Bre01] machine - learning algorithm, and the mean decrease impurity method for the feature importance assessment [Bre01]. The confusion matrix and the feature importance are shown in Figs. 8.16 a and b respectively. An official recommendation for MCS [GYL11] to transmit the processed features and not the complete signal initiated the development of alternative processing tools that are insensitive to artifacts in raw data. An example is Binarized entropy that estimates the approximate entropy from binary differentially coded time series [SMM+17].

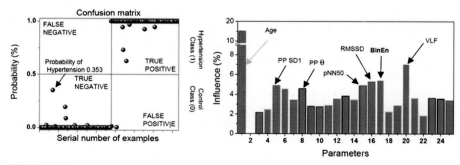

FIGURE 8.16

a) Confusion matrix; an arrow points the new patient; b) Feature importance; features are typical for cardiology; [JJS+18].

Activity recognition and motion analysis

Traditional approaches to measure a person's activity include attaching special measurement devices on predefined locations like hip and ankle. Data from those devices are typically recorded in an internal memory for analysis at a later stage [RT12]. Many of the epidemiological and clinical studies still use this method in their research [MPGRS18]. With the advances in wearable technology, sensor measurements can be directly sent to the users' smartphone for real-time analysis. In [OR16], a system architecture for activity recognition using smartphones and smartwatches as sensor devices has been proposed. The system also includes a remote cloud which is in charge of the training and improving the neural network models used for activity recognition. Accuracy of the algorithm(s) has been evaluated to identify which sensor combination gives the best results for activity recognition. Results are shown in Fig. 8.17. The bars represent the sensors combination that have been used as input to the models. It can be seen that in general the accuracy is higher when the smartphone sensors' data is used.

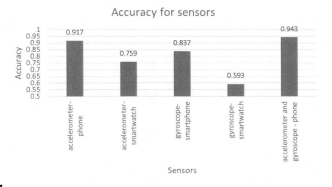

FIGURE 8.17

Accuracy for each sensor combination.

In [BPMS17], a system for remote monitoring of patients with movement disorders is presented. In addition to the system architecture, the authors describe an Android application aimed for recording of the patient's neck movements i.e. TremorSense. Initial results demonstrating the effect of botulinum toxin therapy on patients with neck tremor are shown in Fig. 8.18. The modulus of angular velocity for patient P1 before (top figure) and after (bottom figure) receiving Botulinum toxin therapy is presented in the Figure. As observed, the Botulinum toxin therapy has significantly decreased the neck tremor. This also supports the patient statement regarding an improvement on the order of 30%.

In [CAN18], authors present a low-cost portable system to capture gait signals and propose a novel method for automatically obtaining gait phases (swing and stance) using wavelets and Microsoft's Kinect RGB-D sensor. Human gait patterns are characterized by a basic gait cycle that is composed of two phases: stance and swing. Stance phase represents the state where the heel remains in contact with the ground and the swing phase represents the state where the

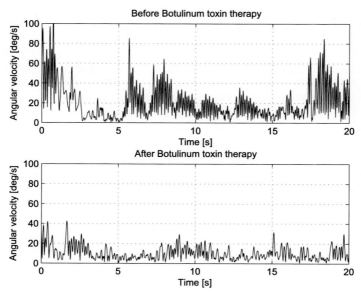

FIGURE 8.18

Angular velocity over time for patient with neck tremor.

heel moves without any ground contact. Using human gait analysis, authors derived values of spatiotemporal variables such as the walking speed, cadence, and stride length from these two phases. Table 8.7 shows how these variables clearly and quantitatively distinguish patients with Parkinson's Disease (PD). Authors explain how this approach can be used to provide an objective metric for evaluating PD progression. Furthermore, they recommend PD clinical diagnosis should include complementary gait analysis using the proposed technology.

Table 8.7 Average spatiotemporal variables and standard deviation obtained for PD and non-PD volunteers.

Variables	PD patients		non-PD patients	
	Left	Right	Left	Right
Stance Time (s)	2.24 (0.31)	2.17 (0.23)	0.91 (0.10)	1.06 (0.10)
Swing Time (s)	1.33 (0.14)	1.33 (0.18)	0.76 (0.09)	0.76 (0.06)
Step Number	10 (0.55)	9.67 (0.19)	6.83 (0.36)	6.17 (0.29)
Test Duration (s)	3.7 (0.41)	3.65 (0.32)	1.72 (0.07)	1.89 (0.09)
Speed (m/s)	0.63 (0.06)	0.65 (0.05)	1.20 (0.05)	1.04 (0.07)

8.3 **Nanocommunications**

The next revolutionary phase in the Internet-of-Things in healthcare is the development of nanomachines for use inside the human body. Nanomachines will enable new mechanisms for gathering information about the human health, and will open the door to innovative medical diagnosis and treatments dealing with serious infections and heart attacks [KWAC⁺19], vascular system diseases [KWTC19], remote surgery on an extremely small scale, drug delivery, tissues regeneration, etc. Due to their very small dimensions, nanomachines have very limited capabilities when working as a single device. This necessitates efficient cooperation of a large group of nanomachines which in turn requires exchange of information or communication among them. This is referred to as nanocommunications.

8.3.1 **Nanocommunication mechanisms**

In recent years, research on nanocommunications is making considerable progress along the following main three approaches: (a) EM-based, (b) molecular, and (c) FRET-based.

The first approach is based on the idea of continued miniaturization of the existing microelectronic devices. These miniaturized devices, which are expected to operate in the THz band (0.1 THz to 10 THz), are made of new materials such as carbon nanotubes or graphene. The wavelength of the EM waves in the THz band is in the micro-meters range. This allows for the development of antennas with proper dimensions. To power EM-based nanomachines, solutions based on nanowires made of zinc oxide have already been proposed. The nanowires are able to generate electric voltage if bent or reshaped, e.g. in an environment with fluid flow. The dimensions of graphene-based devices are generally in the micro-meter range. Since their operation is still based on EM communication, they can be easily integrated with existing wireless networks in the macro scale. Consequently, they could perform as a gateway between the macro and nano devices.

The second approach to nanocommunications is based on the communication mechanisms in biology. Here, the information is not carried through the transmission of an EM wave, instead a group of molecules convey the message. These mechanisms are commonly described as molecular communication (Fig. 8.19). One example of molecular communication is calcium wave propagation using diffusion (Brownian motion) [NSM⁺05]. This is the mechanism commonly used for signaling between living cells. Another example is using diffusion for broadcasting larger particles (such as polymers) that carry information coded in their properly modified structure [UPA13]. In contrast to passive diffusion, some molecular communication mechanisms are based on active transportation of encoded information i.e. data can be encoded in the DNA chain of a plasmid located inside a bacterium or attached to a catalytic nanomotor [GA10]. Finally, molecular motor is another mechanism that allows carrying information encoded in the RNA (or a sequence of peptides in a vesicle), and traveling along protein tracks called microtubules [EMSO11].

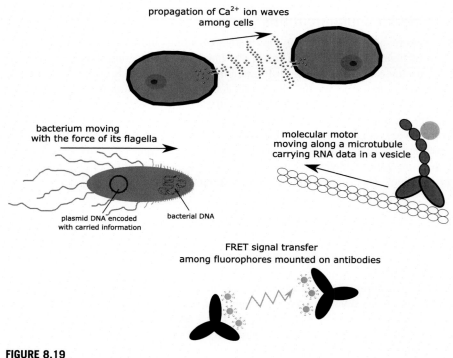

FIGURE 8.19

Examples of nanocommunication mechanisms.

The third approach to nanocommunications is based on a phenomenon called Förster Resonance Energy Transfer (FRET). This mechanism allows for non-radiative energy passing between two molecules. The first molecule, excited by an external radiation or a chemical reaction, operates as a nanotransmitter. The nanotransmitter that is in a high energy state may pass its energy to a neighbor molecule, operating as a nanoreceiver (Fig. 8.19). The phenomenon of FRET occurs only between spectrally matched molecules, i.e. the nanotransmitter emission spectrum should overlap the nanoreceiver absorption spectrum. ON-OFF modulation is achieved through this energy transfer between the two molecules. This means that sending an information bit '1' is realized by a FRET transfer while sending an information bit '0' does not involve any transfer of energy. FRET operates in nano scale as its communication range is usually limited to (5-15) nanometers. However, the communication efficiency is strongly dependent on the distance between the transmitter and the receiver [KKWJ17]. Similar to traditional wireless communications, FRET efficiency and communication reliability might be increased by using diversity techniques where multiple molecules operate at both sides of the communication channel i.e. MIMO-FRET. Designing routing techniques suitable for FRET-based nanonetworks is a major challenge. In [KSW17], new routing mechanisms have been proposed based on biological properties of specific molecules like photoswitchable fluorophores, quenchers, proteins with changeable shape, and ATP synthases. FRET nanonetworks can be possibly constructed from fluorescent molecules like dyes or fluorophores, as well as, quantum dots and photosynthetic molecules of plant origin like carotenes and chlorophylls [OFK18]. Nanomaterials based on carotenes and chlorophylls might also be considered as parts of nanostructures for intelligent fabrics and materials.

8.3.2 Interface with micro- and macro-scale networks

As the communication mechanisms discussed in the previous section do not share the same physical medium, it will be really challenging to transfer data between nanomachines and other traditional wireless networks. Several solutions to this problem were investigated in [KTC19]. One such solution involves the use of channelrhodopsins molecules as nanoreceivers. Channelrhodopsins are able to create naturally occurring ion channels. After an excitation, i.e. absorption of a photon or FRET, a channelrhodopsin opens itself for at least 10 ms, creating a pore where cations (positive ions, like Ca^{2+}, K^+, Na^+) flow through. Consequently, the electrical potential behind the channelrhodopsin changes and can be measured with an electrode [KWKJ18]. This property means that such a channelrhodopsin can serve as an energy-to-voltage nano-converter. This is extremely useful for the purpose of reading FRET signals by electrical devices. Channelrhodopsin molecules might also be embedded into a nerve cell membrane replacing the neuroreceptors [KKWJ18]. When a FRET signal is received at a channelrhodopsin, cations which are flowing through the channelrhodopsin, open the pore and cause an action potential to propagate in the nerve cell. Since a single nerve cell can be over one meter long, this technique enables conversion of the FRET to electrical signals and its transmission over relatively large distances to other macro scale electronic devices (Fig. 8.20).

Transmission in FRET nanonetworks may also be initiated by devices outside the body, e.g. wearable devices in a BAN (Fig. 8.20). It can also be triggered by optically-enabled micro-size devices built from graphene. FRET nanonodes such as fluorophores can be bio-engineered molecules with distinct absorption spectra ranging from ultraviolet (380-400) nm up to infrared (700-750) nm. These molecules could be used for selective reception of optical signals coming from other micro/macro devices.

Communication between different types of nanonetworks (namely FRET and molecular), although challenging, is possible as well. Molecular motors such as kinesins or dyneins may not only carry information encoded in the RNA, but also fluorophores (Fig. 8.20). Depending on the actual location of the carried fluorophore, such mobile fluorophores could forward a signal to different remote FRET networks. The signaling mechanism in FRET typically starts with a nanotransmitter excited by a photon. However, a chemical reaction can also initiate the delivery of the required energy. This process is referred to as Bioluminescence Resonance Energy Transfer (BRET). For this chemical reaction to occur, luciferin (substrate), luciferase

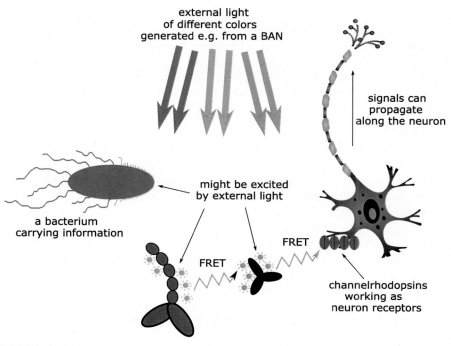

FIGURE 8.20

Communication mechanisms between the molecular/FRET networks and external devices.

(enzyme), oxygen, and ATP (energy source) are required. The process of energy delivery may also be controlled by molecular mechanisms. For example, the oxygen flow could be regulated or ATP could be transported in vesicles or molecular motors.

Finally, micro-size mobile nodes using THz communication can be used to interface with mm-scale static gateways using classical wireless communications [KWAC+19]. However, this connection is limited to distances of several millimeters due to large signal attenuation by the human tissues as well as possible regulatory limitation on the transmit power.

Methodologies discussed in this subsection show alternative mechanisms that can be used to create efficient interfaces between nano-, micro-, and macro-communication networks. These interfaces are clearly essential to realize the full potential of such integrated networks and support future technological advances in various IoT-Health applications.

Localization and tracking

Klaus Witrisal[a,b]**, Carles Anton-Haro**[c,b]**, Stefan Grebien**[a]**, Wout Joseph**[d]**,
Erik Leitinger**[a]**, Xuhong Li**[e]**, José A. Del Peral-Rosado**[f]**, David Plets**[d]**,
Jordi Vilà-Valls**[g]**, and Thomas Wilding**[a]

[a]*Graz University of Technology, Styria, Austria*
[c]*Centre Tecnològic de Telecomunicacions de Catalunya, Barcelona, Catalonia, Spain*
[d]*Ghent University, Ghent, Belgium*
[e]*Lund University, Lund, Sweden*
[f]*Universitat Autònoma de Barcelona, Barcelona, Spain*
[g]*Institut Supérieur de l'Aéronautique et de l'Espace, Toulouse, France*

The use of radio signal for position tracking (positioning) and navigation has a long tradition. Most notably, global navigation satellite systems (GNSS), which were originally conceived for military purposes, are now widely applied for civilian applications ranging all the way from tracking of cargo containers to gaming. Nevertheless, GNSS suffers from limited coverage in dense urban areas and indoors, as well as from a limited position accuracy. These limitations rule out most location-based applications that are concerned with the (natural) interaction of humans with their immediate surroundings, the physical environment in which we live, work, and spend our free time. Besides, GNSS also falls short in terms of fulfilling the stringent requirements associated to location-awareness services needed, for instance, for autonomous driving applications.

Besides, the fifth generation (5G) of mobile communication systems along with the advent of the Internet of Things (IoT) networks will provide a range of new, advanced wireless technologies. A white paper was written within the COST-IRACON action in April 2019 [WAHdPR⁺18], to highlight the expected benefits of these new wireless technologies for the purpose of improved positioning.

This chapter starts with a discussion of the initial consideration made in the white paper (Section 9.1)[1] followed by four technical sections, focusing on measurement acquisition and modeling (Section 9.2), position estimation (Section 9.3), multipath-assisted positioning (Section 9.4), and system-level studies (Section 9.5), which summarize the technical contributions presented within COST-IRACON. Fi-

---

[b] Chapter editors.

[1] The chapter editors would like to acknowledge the contributions by the authors of the white paper, in particular the contributors to Sections II–IV, cf. [WAHdPR⁺18].

Inclusive Radio Communications for 5G and Beyond. https://doi.org/10.1016/B978-0-12-820581-5.00015-8
Copyright © 2021 Elsevier Ltd. All rights reserved.

nally, a range of experimental facilities are described in Section 9.6, which have been used by members of the experimental working group Localization and Tracking, of the COST-IRACON action.

9.1 Introduction

Here, we provide a summary of the most promising future application scenarios for high-accuracy positioning (Section 9.1.1), followed by a discussion on the technical challenges arising from those applications (Section 9.1.2), and on the expected features and limitations of 5G and IoT wireless systems with respect to positioning (Section 9.1.3).

Since the scope of this chapter is necessarily limited, the interested reader is referred to a number of recent tutorial and review articles where further information can be found. Specifically, [WML+16] envisions mmWave based, single-anchor positioning systems for indoor applications, [WSGD+17] identifies the key properties of 5G as they relate to vehicular positioning, [CTJ+17] is a survey on robustness, security, and privacy in location-based services for the future IoT, [dPRRLSSG17] is a survey of cellular mobile radio localization methods, from 1G to 5G, and [FFCM17] is a survey of reliable and accurate indoor positioning system for emergency responders.

9.1.1 New application scenarios, user requirements

Intelligent transportation systems: In the context of ITS, positioning is required, both from the point of view of individual vehicles and from a vehicle network perspective. With increased levels of automation, more precise positioning is needed to ensure vehicle stay within planned trajectories. Vehicles sense their environment and connect to the vehicle network perspective, where both absolute and relative position information of vehicles is needed for completing cooperative maneuvers. In all these scenarios, due to the possibly high mobility, position information must be both timely and accurate [5GP15]. Requirements based on typical velocities were detailed within the EU H2020 HIGHTS project [S+16], EU H2020 TIMON project [TIM17], and the more recent EU H2020 5GCAR project.

Aerial vehicles: Unmanned aerial vehicles (UAVs) will play a significant role in both civilian and military applications ranging from remote sensing, search and rescue, to environmental monitoring, to aerial communications and networking [GJV16]. UAV navigation and control requires precise localization and tracking. Traditionally, this is realized with data fusion approaches that exploit inertial sensors and global navigation satellite systems (GNSSs) [ST18,PT16]. However, the availability and reliability of GNSS-based position estimates cannot be always guaranteed, for instance in an urban or sub-urban environment because of severe multipath conditions, blockage of the satellites' signals and, also, due to the increase of non-intentional or intentional interferences like jamming and spoofing [NA10]. Both ad-hoc solutions, e.g., based on ultra-wideband (UWB) ranging [GQM+16], or opportunistic solutions [PT15,MVR09] that exploit existing radio transmissions, e.g., in WiFi or cellular networks, for trilateration or triangulation with antenna arrays, can be developed.

Industrial applications: Industrial applications were classified in [3GP16] mainly in three groups: industrial control, factory automation, and process automation. The requirements in these groups vary between applications: industrial control and factory automation requires precise positioning, low latency, and the need to support very high reliability and availability. For example robot positioning applications will require [3GP16] that the positioning accuracy is within 10 cm and the latency is less than 15 ms. Industrial process automation targets monitoring and tracking operations of IoT devices (for example tracking valuable tools, containers, or electronic equipment). Here, a massive number of devices is foreseen; in addition, the battery lifetime of the nodes shall operate over multiple years.

Retail and assisted living: In retail, precise indoor navigation can be beneficial for consumers to easily navigate to their desired products, and also for the management to track consumer movement habits and to perform optimal product placement. Moreover, identifying consumers behavior and in-store movements in big shopping malls or airports, can be used for proximity marketing and advertisements [YBF+14,HKC17]. For ambient assisted living, location discovery is crucial for context-aware service provisioning, ranging from home entertainment and home automation, to activity detection and elderly monitoring for medical tele-care solutions [ADMO+16] or inside hospitals [HPC+16]. The common requirements of those applications are position accuracy of about the human range, minimal setup efforts, easy usage and low complexity of the algorithms, full coverage, adaptiveness to the environment, low power consumption and scalability, while they tolerate latencies of a few seconds.

3GPP approach towards positioning use cases and requirements for 5G: 3GPP as the joint standardization body of mobile network operators, equipment supplier, and chipset manufacturers has been addressing the definition of next generation use cases for 5G from the onset. The SMARTER study (Feasibility Study on New Services and Market Technology Enablers) [3GP16] collected already during 3GPP Release 14 about 74 use cases grouped into five categories: Enhanced Mobile Broadband, Critical Communications, Massive Machine Type Communications, Network Operation, and Enhancement of Vehicle-to-Everything. Many of those use cases include the need for positioning. Indications on future requirements already make clear that the targets are high, for example sub-meter accuracies and low latency values below 15 ms for positioning results. Requirements have been further studied for example in the "Study on Scenarios and Requirements for Next Generation Access Technologies, TR 38.913" [3GP17a]. This chapter, in Section 9.2, provides an outlook on the actual realization of positioning in 5G systems.

9.1.2 Technical challenges

End-user requirements dictate the performance levels needed for any technical system. This section describes a number of technical challenges associated to emerging location-aware services.

Heterogeneity: Arising from the wide variety of potential applications, the challenge of heterogeneity has to be discussed. Unlike GNSS and cellular networks, where a single technology platform is capable of supporting an extremely wide range of application scenarios, a *diversity of (wireless) technologies* will be needed to support location-aware electronic systems with the performance requirements outlined above.

Multipath propagation: For radio-based positioning systems, multipath propagation is considered the key physical challenge hindering the implementation of 10 cm-level position accuracy. The presence of multiple propagation paths on top of the line-of-sight (LOS) path severely degrades the performance of localization algorithms (see Section 9.2). However, if the propagation paths can be resolved, the presence of multipath can become an asset for localization (Section 9.4).

Line-of-sight availability: The continuity of service is strongly related to the availability/visibility of infrastructure such as beacon radio signals. In case of radio systems targeting at the 10 cm-accuracy level, even one blocked line-of-sight (LOS) connection may be sufficient to severely degrade positioning accuracy.

Time synchronization: Radio transceivers derive their internal timing reference from independent local oscillators. Synchronizing the timing reference of independent radio transceivers is therefore an important prerequisite for systems that use propagation delay estimation.

Hardware complexity in large antenna array systems: As discussed in Section 9.2 ahead, large antenna array systems can significantly improve the precision of angle-of-arrival based localization systems, [SLZ15]. However, a number of challenges need to be tackled both at the hardware domain and at the signal processing domain. Firstly, fully digital architectures require the realization of a prohibitively high number of RF-to-Baseband chains [RPL+13]. Furthermore, appropriate antenna configurations have to be studied to meet stringent physical space limitations [LETM14]. Denser antenna arrays introduce mutual coupling, unequal gains, and phase response effects that require improved design and calibration techniques [IT12,RPL+13].

Latency and control overhead of random access/MAC protocols: The need to support a massive number of always connected devices that communicate sporadically (e.g. machine-type communications or IoT) is of particular importance for certain types of future networks. Traditional centralized MAC solutions for cellular systems are based on orthogonal access schemes which cannot support a massive number of devices due to the large control overhead and severe latency. A number of random access and MAC protocols have been proposed for different low power wide-area networks (LP-WANs) such as LoRa with a reduced control overhead and capable of achieving low latencies. To support accurate localization methods, however, future decentralized MAC schemes must also include additional features, such as time guaranteed services to support the positioning anchors' transmission in a reliable and exclusive way.

9.1.3 Expected features and limitations of 5G and current IoT technologies: Impact in positioning

Current IoT technologies are aimed at providing communications to a massive number of low-power and low-cost devices, with emphasis on mesh topologies with simple protocols. As a consequence, the positioning capabilities of IoT systems are limited. On the contrary, the expected positioning accuracy in 5G wireless networks is expected to be below 1 m in urban (and indoor) scenarios and less than 2 m in suburban scenarios where the vehicle speeds are up to 100 km/h [For15]. This section discusses the main features of 5G and IoT communication technologies in relation with positioning, along with their main shortcomings and limitations.

Features and limitations of current IoT technologies

Different IoT applications impose different, and often conflicting, goals in terms of range or coverage, rate or throughput, power consumption, and latency. The current IoT landscape includes both proprietary and standardized solutions, that try to satisfy different requirements. These IoT solutions can then provide positioning capabilities, which have different limitations depending on the system design.

Long range: Long-range IoT applications include asset tracking, smart city, smart metering, smart farming, smart retail, and logistics, etc. Currently, there are two main competing IoT proprietary solutions to achieve *long-range* communications: LoRa and Sigfox. Long Range Wide Area Network is a low-power proprietary technology operating in the ISM band, based on chirp spread spectrum modulation [RKS17,PMHI16]. Sigfox relies on the so-called Ultra Narrow-Band (UNB) modulation (i.e., bandwidths below 1 kHz) which helps achieve long ranges, as the ultra-narrow receiver filters most of the in-band noise. These proprietary solutions do not have any positioning features specified or included in the standard. Thus, classical positioning approaches, such as fingerprinting based on signal powers, can be applied primarily. However, the performance of these methods is significantly degraded due to the long-range communications. Timing-based schemes can be also implemented, but their performance is greatly reduced for these narrowband signals [FP17].

Furthermore, there are two main standard technologies for long-range IoT solutions, which are IEEE 802.11 Long Range Low Power (LRLP) and the 3GPP narrowband technologies, i.e., LTE-M, LTE NB-IoT, and EC-GSM-IoT. The IEEE 802.11 LRLP is a relatively new interest group (established in 2015) within IEEE 802.11 [P80] for IoT, M2M, energy management, and sensor applications. 3GPP narrowband technologies are evolving really fast with a major support from the industry. In addition, their positioning capabilities are under standardization, as it is described in [LBG+17]. The main advantage of such standardized technologies with respect to proprietary solutions is the use of licensed bands to reduce interference, and the dedication of network resources for positioning, such as pilot signals and positioning protocols. Following the trend of LTE NB-IoT, the 5G standard plans to include dedicated pilot signals and protocols for positioning, in order to improve the limited positioning capabilities of current IoT technologies.

Short range and narrowband: Short-range IoT applications, such as smart home, wearables, health and fitness, etc., are typically provided indoors. Examples of short-range IoT technologies are the Radio Frequency Identification (RFID), Bluetooth Low Energy (BLE), and in particularly BLE mesh, ZigBee, ANT and ANT+, DASH7, etc. [ASAAA17]. These technologies have a low bandwidth, but thanks to the short-range communications, they are able to achieve high-accuracy positioning. The main techniques are based on fingerprinting or proximity techniques, where meter-level accuracies can be achieved, such as with RFID in [LLY+16]. Advanced techniques can be based on the use of angle measurements [DDGG16]. However, to achieve a high precision, they require large antenna arrays at the receiver side (e.g. readers in RFID technologies), resulting in an increase on the complexity and cost of the IoT system [DDGG16], while multipath propagation remains as a limiting factor [WGMW18].

Short range and wideband: Short-range IoT communications can also be based on wideband technologies, such as the IEEE 802.15.4a UWB standard. This standard is based on the direct sequence UWB modulation, which offers spectral efficiency, robustness at low transmit powers, and precise ranging. Thus, thanks to its larger bandwidth, the IEEE 802.15.4a UWB standard is able to achieve centimeter-level accuracies [DCF+09]. In addition, these short-range wideband signals result in practically deterministic MPCs, thus they can also be exploited for multipath-assisted localization (Section 9.4), as it is shown in [KHG+17] with off-the-shelf devices. Using these MPCs, single-anchor localization can be implemented reducing the infrastructure required for the IoT positioning system.

Wideband and UWB signals can also be implemented in RFID-based systems, as discussed in [GKG+17,DDGG16] (see Section 9.6). The use of wideband backscattered signals allows precise ranging measurements, while preserving the system complexity and cost. The main limitations of the UWB-RFID technology are poor energy efficiency, short range, and costly ad-hoc infrastructure.

Low power and low throughput: Currently, the vast majority of IoT solutions have been designed for low-power low-rate applications, that is, for planned battery durations of several years. Current solutions in IoT range from few years of the battery life on the IoT sensor (e.g., 6LoWPAN/6LPWA, BLE) up to more than 10 years battery life (e.g., NB-IoT, LoRa, Sigfox). Typically, IoT solutions have very low throughputs from few tens of bits per second (bps) up to few hundreds of kbps.

The low-power and low-rate features have a strong impact on the positioning techniques applicable to IoT technologies. Since the transmission time is reduced, the number of positioning measurements is reduced, leading to simple positioning protocols with minimal assistance. Furthermore, the positioning operations on the IoT device need to have a very low complexity to achieve a low power consumption.

Features and shortcomings of 5G wireless networks

5G systems come with a number of disruptive features which have direct impact in positioning capabilities [WSGD+17]. These advanced features include network densification, operation in mmWave bands and massive MIMO [ABC+14], and device-to-device communication [BHL+14]. The impact of such features on positioning capabilities is discussed next.

Higher bandwidth and new frequency bands: 5G systems include operation in new frequency bands in the mmWave spectrum (above 24 GHz), where more bandwidth is available than in the crowded below 6 GHz bands. This has a dual effect on positioning. On the one hand, larger bandwidths allow for a higher degree of delay resolution, so that individual multipath components can be estimated and tracked. On the other, higher carrier frequencies lead to more optical-like propagation, with reduced diffraction and very limited trough-wall penetration [RMSS15]. Hence, only few paths will be present and each path has a geometric connection to the physical propagation environment. In the sub-6 GHz band, large bandwidths will also be available with carrier aggregation, but more dominant propagation paths will be present.

More antennas: With higher carrier frequencies and shorter wavelengths comes the opportunity to pack more antennas into a given area [SRH+14]. Above 24 GHz, planar arrays with hundreds of antennas are feasible. Using more antenna elements provides the opportunity to increase the resolution of the channel in the spatial domain (angle of arrival and angle of departure, in azimuth and elevation), providing a new way to separate multipath components (other than in the delay domain). Hence, combined with higher carrier frequencies and large bandwidths, using more antennas leads to a high degree of resolvability and high accuracy of estimating multipath components, each with an associated delay, angle of arrival, and angle of departure. This implies that absolute positioning with respect to a single reference transmitter is possible, as well as identification of the sources of each reflected or scattered path (cf. Section 9.4).

Network densification: Increased area spectral efficiency is obtained from a denser deployment of base stations with a reduced coverage area and aggressive spectral reuse. This requires sophisticated solutions for interference and mobility management [BLM+14]. For positioning, ultra-dense networks are beneficial since the distance to reference transmitters is reduced. The reason is two-fold. Firstly, with ranges from 5-50 meters, the probability of having a line-of-sight connection is upward of 50% [Li16], which in turn is highly beneficial for localization as the delay and angle information from the optical line-of-sight path provide the most location-relevant information of all multipath components. Secondly, with a dense radio-access network seamless multi-connectivity can be established between terminals and base-station via multi-beam communication. This increases the robustness to line-of-sight obstruction and improves the localization accuracy.

Device-to-device communications: Since LTE Release 12, D2D communication is considered a candidate technology for proximity detection. In 5G, D2D will be native, for high-rate links between nearby users, benefiting from low path loss, low transmit powers, and extremely low latency [TUY14]. Such D2D links also provide an additional source of positioning information, where well-positioned users can serve as (noisy) location references for users out of direct coverage of base stations. Moreover, D2D links can provide relative positioning information as well as a mean to develop efficient cooperative positioning schemes for achieving higher accuracy.

9.2 Measurement modeling and performance limits

The process of localization via radio signals involves a Tx transmitting a set of signals which are received by a Rx (possibly multiple Tx and Rx). The received signal can be used for direct localization methods or localization is achieved in a two-step process: first position-related measurements are extracted which are in turn used by dedicated localization algorithms. By analyzing the received signal, position-related measurements can be observed such as RSS, Time Of Arrival (TOA), AOA, Doppler frequency, or combinations thereof.

To be able to characterize the measurement, and thus the acquisition process of a measurement system, knowledge of the signal and channel models is necessary. This radio channel knowledge can be exploited to derive theoretical performance bounds on the observed parameters, predict channels for fingerprinting localization, and perform simulations on a link-level or system-level. The accuracy of the acquisition process is strongly dependent on channel conditions and, in particular, on the presence of MPCs. To be able to mitigate/exploit these MPCs, a channel model capturing the dependence between position and channel properties is necessary.

A radio signal and channel model is introduced in Section 9.2.1 which is subsequently utilized in Sections 9.2.2 to 9.2.5 to describe the different position-related measures to be acquired.

9.2.1 Signal and channel model

Radio localization depends heavily on knowledge of the interplay between the observed environment, radio propagation, and the measurements used for localization. This knowledge is usually leveraged by means of a radio channel model. A Tx at position $x_{Tx}(t)$ transmits a radio signal $s(t)$ over a linear time-variant channel $h(t, \tau; x_{Tx}(t), x_{Rx}(t))$ to a Rx at position $x_{Rx}(t)$. The noisy received signal at the Rx is found via the convolution as

$$r(t; x_{\text{Tx}}(t), x_{\text{Rx}}(t)) = \int s(t-\tau)h(t, \tau; x_{\text{Tx}}(t), x_{\text{Rx}}(t))\,d\tau + n(t), \qquad (9.1)$$

where $n(t)$ represents Additive White Gaussian Noise (AWGN) attributable to the receiver. The channel impulse response can be written as

$$h(t, \tau; x_{\text{Tx}}(t), x_{\text{Rx}}(t)) = \sum_k a_k(t, \tau; x_{\text{Tx}}(t), x_{\text{Rx}}(t)) \times$$
$$\times \delta(\tau - \tau_k(x_{\text{Tx}}(t), x_{\text{Rx}}(t))) \qquad (9.2)$$

where $\tau_k(\boldsymbol{x}_{\mathrm{Tx}}(t), \boldsymbol{x}_{\mathrm{Rx}}(t))$ is the delay of the kth MPC from Tx to Rx at time t and $a_k(t, \tau; \boldsymbol{x}_{\mathrm{Tx}}(t), \boldsymbol{x}_{\mathrm{Rx}}(t))$ captures its complex amplitude.

Propagation in real environments, especially in urban and indoor scenarios, is characterized by a multitude of MPCs. These MPCs can be split up in SMCs and a DMC [Mol05]. The SMCs relate to a few specular reflections whose position-related parameters are well defined and depend on the environment including the position of Tx and Rx. This geometrical dependency opens up the possibility to exploit SMCs for positioning [WML$^+$16]. The DMC stems from a multitude of MPCs which can not be resolved by the Rx. This DMC is usually modeled statistically and needs to be estimated as nuisance parameters during the measurement acquisition process. Otherwise, the estimates of the position-related parameters will be degraded.

Ways of exploiting propagation over multiple geometric paths (multipath propagation) will be covered in Section 9.4. The remainder of this section deals with ways of extracting various position related measurements from the signals received at a single antenna or at an antenna array by spatially sampling the impinging wavefield.

9.2.2 Received signal strength

A simple way of getting distance related information is by exploiting the relation of the RSS or RSSI with the distance, often modeled by a path-loss model [Mol05].[2] Even though RSS values are strongly influenced by fading, shadowing and blockage, and antenna radiation characteristics, they still allow for achieving a positioning accuracy in the range of meters to tens of meters [DUM$^+$16,RS18] with a strong dependence on the propagation environment and system setup.

On-body nodes suffer greatly from the strong attenuation due to the human body. Differential RSS values between (at least) two on-body sensors are shown in [DUM$^+$16] to allow estimating the body orientation jointly with the position, resulting in a higher robustness with respect to body shadowing, which can be especially interesting in crowded indoor scenarios. The resulting accuracy for the orientation estimation is in the range of $25 - 30°$ and $9 - 15°$ in the worst and best cases respectively, and a median position error of 1 m for joint estimation.

The influence of different channels and multiple antennas employed for measuring RSS values is investigated in [RS18], showing that spatial and frequency diversity improve the stability of the RSS values.

9.2.3 Time of arrival and time difference of arrival (TOA/TDOA)

Time-of-flight-based positioning (often termed ranging, Time Delay Estimation (TDE) or TOA estimation) is maybe the most popular method for positioning, due to the achievable high accuracy and ease of implementation. In contrast to RSS, the relationship between time measurements and the underlying distance between spatial locations is a linear one, connected via the speed of light, hence a constant accuracy

---

[2] In (9.2) the relation is hidden in the complex amplitudes $a_k(t, \tau; \boldsymbol{x}_{\mathrm{Tx}}(t), \boldsymbol{x}_{\mathrm{Rx}}(t))$.

is achieved. The synchronization requirements between transmitter and receiver can be a main limiting factor in the achievable performance, depending on the protocol used for obtaining the distance estimates, e.g. one-way-ranging, two-way-ranging, or round-trip-time measurements. Using Time Difference Of Arrivals (TDOAs), the differences between several TOA measurements are exploited, which allows to relax the need for tight synchronization [Qui17].

It is well known from fundamental accuracy bounds, e.g. the Cramér-Rao lower bound (CRLB) [Kay93,SW10], that in an AWGN channel, the achievable ranging accuracy (REB, or Ranging Error Bound) depends primarily on the employed signal bandwidth and the SNR. For Diffuse Multipath (DM) channels, e.g. encountered in indoor scenarios, it was shown in [HLM+16,WGMW18] that the dependence on the signal bandwidth is even stronger, due to the large number of unresolvable or overlapping, diffuse multipath components that are often encountered. This is quantified by a closed-form solution of the achievable variance of range estimation var[$\hat{d}$]:

$$\text{var}[\hat{d}] \geq \frac{c^2}{8\pi^2 \beta^2 \, M \, \text{SINR}_\tau}, \tag{9.3}$$

allowing an intuitive interpretation of the role of various signal and system parameters. The employed bandwidth is represented via the mean-square signal bandwidth $\beta^2 = \int f^2 |S(f)|^2 \, df$, M is the number of independent measurements or antennas, and SINR_τ is the effective SINR that is encountered for ranging,[3] quantifying the influence of diffuse multipath (9.3). The connection between AWGN and DM channels can be quantified by the fact that

$$\text{SINR}_\tau \leq \text{SNR}. \tag{9.4}$$

The SINR can be interpreted as the ratio between the power of a specular component to the power of the diffuse components plus noise interfering with it. By modeling the DMC with a delay power spectrum, a closed-form solution can be found for the SINR_τ [HLM+16,WLHM16,WGMW18,KWA+07]. In Fig. 9.1(a), the CRLB is depicted over a wide range of bandwidths for a LOS plus DM scenario. Furthermore, two estimators, a naive MF and a ML estimator are included. For the ML estimator a stochastic description of the DM component is needed by means of the delay power spectrum, which can be applied as a whitening filter. This proper treatment of the DM allows the ML estimator to follow the CRLB down to smaller bandwidths. However, the parameters of the delay power spectrum have to be included in the estimation process as nuisance parameters, leading to a more difficult estimation task. An increasing antenna number leads on the one hand to a higher accuracy at a certain bandwidth, and on the other hand to a detection gain as the estimators are able to follow the bound to smaller bandwidths.

---

[3] For ranging the DM acts as an interfering term; another interpretation is a 'per component' SINR_τ.

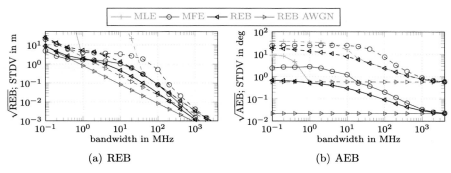

FIGURE 9.1

Ranging (left) and Angular (right) Error Bounds for different pulse bandwidths $1/T_p$, and performance results for ML and MF estimators. Simulations were performed for $N_r = 10\,000$ realizations for Uniform Linear Arrays (ULAs) with $M = 2$ (dashed lines) and $M = 16$ (solid lines) elements and inter-element spacing of $\lambda/2$ and an AOA $\phi = 0°$.

The closed-form solution in (9.3) is accurate for non-overlapping paths, but becomes overly optimistic when path overlap between MPCs increases [WGMW18]. This is the case especially indoors for lower bandwidths and in corners and close to walls or reflective surfaces where many specular MPCs overlap. The theoretical bound for the achievable ranging accuracy is validated with measurements in [HLM+16].

With the bandwidth being an important factor in the ranging accuracy, narrowband systems are inherently suboptimal for range-based positioning. A possibility to overcome the drawback of small bandwidth is the use of Frequency-Hopping (FH) to increase the overall bandwidth, which is investigated in [dPRLSSG17], focusing on current 4G and future 5G systems. The main limitations of FH compared to native wideband systems are the synchronization between the different frequency measurements, phase offset, and group delay that can occur at each hop as well as the changing transmission channel due to movement of the mobile agent with which ranging measurements are performed.

For a TDOA-based positioning scheme, an overall positioning accuracy below 50 m has been achieved by two frequency hops over a system bandwidth of 10 MHz [dPRLSSG17]. Especially the contiguous allocation of the positioning reference signal has a positive effect as it limits the impact of receiver impairments. Fig. 9.2 shows the effect of increasing the number of RB for EPA and ETU multipath channels via simulations, with the FH scheme consisting of two hops equal to an in-band transmission of 6 RB or 50 RB or 1.08 MHz and 8.94 MHz respectively. It shows that for the EPA channel an error below 50 m can be achieved in the high C/N_0 region, whereas for 50 RB FH positioning reference signal remains largely independent of the C/N_0 around 10 m.

A system study for uplink-TDOA based positioning-schemes is performed in [dPRBAZ+17a], investigating the effect of uplink pilot placement on the achievable

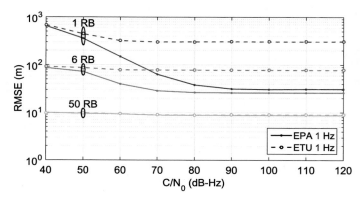

FIGURE 9.2

Root-mean-square-error (RMSE) over the carrier-to-noise ratio (C/N_0) for different channels and number of resource blocks (RB).

channel estimation and positioning accuracy (e.g. in terms of CRB), efficiency of power amplification (e.g. in terms of PAPR) and multiplexing schemes, scheduling of pilot resources and signaling overhead that is required for resource allocation. More details on related system-level studies can be found in Section 9.5.

As LOS or NLOS propagation in general has a severe effect on the overall positioning accuracy, it is of interest to neglect NLOS measurements from ranging results. This possible field of application is especially interesting for neural networks which can be used to learn NLOS-characteristics from channel impulse responses and allows classifying range measurements into LOS and NLOS [BHM16].

9.2.4 Angle of arrival (AOA)

When employing transceivers with multiple antennas, information regarding the direction from which the direct path (and possibly reflections) arrive becomes available. Especially at low bandwidth where the ranging accuracy is limited, AOA based positioning schemes can achieve better results in terms of positioning accuracy [MLV07]. Jointly estimating AOAs and TOAs of one or multiple paths can be desirable as it enables single anchor positioning schemes as well as making full use of the positioning information in the received signal, a topic which will be covered in more detail in Section 9.3.

In a similar fashion as for distance estimation, fundamental performance bounds for AOA estimation are often investigated using the CRLB due to the ease of derivation and the possibility to find a closed-form solution under certain assumptions on the signal, system, and environment [Kay93,vT02]. In [WGL$^+$18,WGMW18] such a closed-form solution was derived for the CRLB var$[\hat{\phi}] = \sigma_\phi^2$, termed the Angulation Error Bound (AEB), for AOA estimation of a LOS signal in DM and AWGN, showing the additional influence of DM in comparison to the standard AWGN channel

[MLV07,SW10,HSZ$^+$16]. The closed-form solution for the AEB is given as

$$\text{var}[\hat{\phi}] \gtrsim \frac{1}{8\pi^2 D_\lambda^2(\phi) M \, \text{SINR}_\phi} \tag{9.5}$$

where SINR_ϕ is the effective SINR for angulation and $D_\lambda^2(\phi)$ is the normalized, squared array aperture. This array-geometry-dependent factor is defined as $D_\lambda^2(\phi) = \frac{1}{M} \sum_{m=1}^{M} \frac{\delta_m^2}{\lambda^2} \sin^2(\phi - \varphi_m)$ with $\lambda = c/f_c$ being the employed carrier wavelength and δ_m and φ_m the distance and angle of the mth array element at position $\boldsymbol{p}_m$ from the array reference point $\boldsymbol{p} = [x, y]^\mathsf{T}$ in polar coordinates. The most favorable choice for the reference point $\boldsymbol{p}$ was shown to be the center of mass of the array, as it allows finding the closed-form solution of the AEB as given in (9.5) in [WGMW18] or in [vT02,MLV07,HSZ$^+$16] for general AWGN channels. Comparing (9.5) with (9.3) the different effective SINRs for ranging and angulation are found to follow the relation

$$\text{SINR}_\phi \le \text{SINR}_\tau \le \text{SNR}. \tag{9.6}$$

The results in Fig. 9.1(b) show that especially at low bandwidth the dependency $\text{SINR}_\phi < \text{SINR}_\tau$ is usually fulfilled, as the whitening gain is observable for ranging only. For AOA estimation the SINR_ϕ converges to the Rician K-factor, the ratio of the LOS to all other MPCs [WGMW18]. This behavior is attributable to the fact that even in UWB systems the bandwidth is much smaller than the center frequency, i.e., $\beta \ll f_c$ is usually valid. In contrast to the Ranging Error Bound (REB) (9.3), the AEB (9.5) exhibits an additional dependence on the actual array geometry, apart from the number of antenna elements M. Array elements that are farther away from the reference point contribute more to the overall achievable accuracy. The higher accuracy at higher carrier frequencies only comes to effect when the array geometry is kept fixed, e.g., when using an aliased array[4] resulting in a lower AEB due to the more narrow main lobe, at the expense of ambiguities due to grating lobes in the array response [vT02]. The possible use of the increased angulation accuracy achievable with aliased arrays for positioning was shown in [WGL$^+$18] by using information contained in MPCs in combination with a known environment geometry. Practical algorithms for AOA estimation are beamforming, beamsteering, or parametric methods [KV96,vT02], but also by beamswitching, for example using directional antennas [AHM19a].

The possible use of standard BLE signals for AOA based positioning is examined in [MNP$^+$18] using the MUSIC algorithm. The experimental setup consists of a single-antenna agent sending out standard advertisement packets to two anchors, each consisting of 2 Universal Software Radio Peripherals (USRPs) with 2 antennas, i.e., 4 antennas in total at each anchor. As the USRPs introduce a different hardware

---

[4] We refer to an aliased array when two adjacent array elements do no longer spatially sample the received signal within one carrier period, i.e., when the inter-element spacing $d > \lambda/2$.

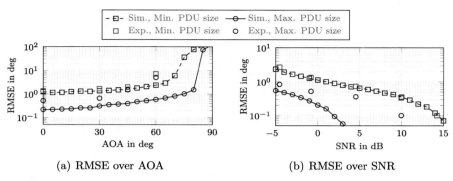

FIGURE 9.3

RMSE of estimated AOA for different SNRs and different AOAs for SNR $= 0$. The plots show a comparison of simulations and measurements with different PDU sizes.

phase shift during signal acquisition, they need to be phase calibrated. A comparison of the AOA RMSE obtained from simulations and measurements is shown in Fig. 9.3 for different SNRs and AOAs for a single-antenna agent and an anchor with a ULA with $M = 4$ antennas, showing that a larger Packet Data Unit (PDU) generally achieves a higher accuracy for AOA estimation.

When physical arrays are not practical, e.g., due to application constraints, an alternative approach is the use of synthetic aperture arrays, where virtual arrays are formed by movement of a single-antenna device and accurate relative position information. In [QGZT16] the achievable accuracy of such an approach is investigated using a mobile device equipped with and Inertial Measurement Unit (IMU) in two different motion scenarios for AOA estimation. For an arc shaped synthetic array, the standard deviation of the AOA error standard deviation was shown to be smaller for a larger movement radius R, which is in line with the array-geometry-dependence of the AEB shown in (9.5). For a radius of $R = 0.3$ m the AOA error standard deviation over 10 measurements was $\text{std}[\hat{\phi}] = \sigma_\phi \approx 12$ deg and $\text{std}[\hat{\phi}] = \sigma_\phi \approx 5$ deg for $R = 0.5$ m.

9.2.5 Joint measurements

In many circumstances, one type of position related measurement can be either not accurate enough or not sufficient to solve the positioning problem. An example thereof would be single-anchor localization without map information using either angle or range measurements only. Due to the high achievable accuracy, a popular combination for joint parameter estimation are angles and ranges, provided an antenna array is available.

A straightforward method is to jointly solve either a deterministic maximum likelihood problem [VPP97,FTH+99,Ric05], use subspace method (e.g., MUSIC) with the corresponding extension to delay estimation [VPP97], or perform joint estimation based on least squares methods [TDM11]. Incorporating uncertainty of the parame-

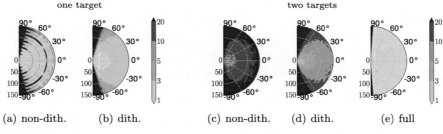

one target   two targets

(a) non-dith.   (b) dith.   (c) non-dith.   (d) dith.   (e) full

FIGURE 9.4

Position error in meters for Monte Carlo simulations and $M = 512$ measurements with one target and 1-bit quantization (non-dithered 9.4(a) and dithered 9.4(b)) and for two targets with $M = 512$ and 1-bit quantization (non-dithered 9.4(c) and dithered 9.4(d)) and $M = 256$ and 32-bit quantization (9.4(e)).

ters to estimate allows the use of Bayesian methods described in [SF11] or [LGW19]. More recently, line spectral estimation was extended to joint delay and angle estimation [HBFR14].

Radar target tracking [HLS93,BSL95] is investigated in [FXVJ18] for joint range and angle estimation using a Frequency Modulated Continuous Wave (FMCW) 2-element radar array. By using compressed sensing to enforce sparsity in the discretized range profile while estimating the AOA from the continuous phase in combination with 1-bit quantization, the effects of dithering on the resulting positioning accuracy are investigated, based on joint AOA and TOA estimation. The accuracy is investigated by comparing simulations and measurements for a K-band radar at 24 GHz with a bandwidth of 250 MHz. The results in Fig. 9.4 show that dithering improves the positioning accuracy by resolving angle ambiguities, whereas a tradeoff between the number of measurements and quantization to a certain number of bits is observable.

9.3 Position estimation, data fusion, and tracking

This section deals with location estimation and tracking methods using measurement data and network-level input. Table 9.1 lists an overview of the different works described in this section, indicating the parameters used for localization, possible additional info that is exploited, and whether the works consist of simulation or show experimental results.

9.3.1 Tracking and sensor fusion for moving persons and assets

Indoor tracking

For indoor tracking the following techniques and combinations will be discussed (also Table 9.1): (i) hybrid RSS and AoA [TBDG17]; (ii) sensor fusion with inertial

Table 9.1 Comparison of the different localization and tracking techniques in indoor and outdoor environments (IMU: inertial measurement unit, RSS: received signal strength, ToF: time of flight, AoA: angle of arrival, OSM: Open Street Maps).

Environment	Ref.	Parameters used				Additional info		Experiment	Simulation
		ToF	RSS	AoA	phase	context	IMU		
indoor	[TBDG17]		x	x					x
	[RSS17]	x					x	x	
	[LLO+19]	x		x	x				x
	[RBZ17]		x			building		x	
	[MMdHB18]		x					x	
	[VRL+17]	x	x	x	x				x
	[HLMW16]		x	x					x
	[dHBGV19]		x					x	
	[PDT+19]	x	x	x		building			x
	[RCS19]	x						x	
	[BNMJ19]	x				building			x
outdoor	[TPS+19]	x	x			OSM		x	
	[PPT+18]	x				OSM		x	
	[QHD17]	x						x	
	[NC17]		x	x					x
	[PMQ18]		x	x				x	x

navigation [RSS17]; (iii) the combination of time of arrival (TOA) of multipath components and phase information from a MaMIMO tracking framework [LLO+19]; and (iv) the combination of RSS, AoA, and Time-of-Flight (TOF) information [PDT+19].

In [TBDG17], target tracking is realized by extracting received signal strength (RSS) and angle of arrival (AoA) information from the received radio signal, in the case where the target transmit power is considered unknown. By combining the radio observations with prior knowledge given by a target transition state model, a maximum a posteriori (MAP) criterion is applied to the marginal posterior PDF proposed. However, the derived MAP estimator cannot be solved directly, so it is approximated for small noise powers. The target state estimate is then obtained at any time step by employing a recursive approach (Bayesian method). Simulations confirm the effectiveness of the proposed algorithm, offering excellent estimation accuracy in all considered scenarios. Fig. 9.5 shows the performance of the algorithm of [TBDG17].

[RSS17] proposes a method to increase position estimation precision by an inertial navigation algorithm on top of a Time-of-Flight based Ultra-Wide Band (UWB) system, when less than three anchor nodes are available. Experimental tests were carried out to verify the precision of the position estimation of a moving person in an indoor environment. The root-mean-square error (RMSE) of the position estimates reduced from 1.01 m to 0.5 m, when applying the algorithm (50% improvement in precision). Also [RCS19] considers UWB experiments for person tracking in a ferry environment, which is a harsh metallic environment. It is possible to monitor the po-

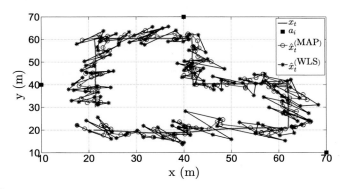

FIGURE 9.5

The true target trajectory versus the estimated trajectory for a hybrid RSS and AoA estimation, where the Maximum a Posteriori (MAP) approach outperforms the classical Weighted Least Squares (WLS) approach [TBDG17].

sition of a moving person in the passenger ferry environment using UWB ToF with measurement errors below one meter by ensuring connection with at least three reference nodes. However, the required spacing of the UWB nodes should be kept small: e.g., 15 m from the measuring device placed on the monitored person.

A robust phase-based positioning framework using a massive multiple-input multiple-output (MIMO) system is proposed in [LBÅ$^+$17]. The phase-based distance estimates of MPCs together with other parameters are tracked with an Extended Kalman Filter (EKF), the state dimension of which varies with the birth-death processes of paths. The iterative maximum likelihood estimation algorithm (RIMAX) and the modeling of dense multipath components (DMC) in the framework further enhance the quality of parameter tracking by providing an accurate initial state and the underlying noise covariance. The tracked MPCs are fed into a time-of-arrival (TOA) self-calibration positioning algorithm for simultaneous trajectory and environment estimation. Throughout the positioning process, no prior knowledge of the surrounding environment and base station position is needed. The performance is evaluated in Fig. 9.6 with the measurement of a 2D complex movement, which was performed in a sports hall with an antenna array with 128 ports as base station using a standard cellular bandwidth of 40 MHz. The positioning result shows that the mean deviation of the estimated user equipment trajectory from the ground truth is 13 cm. In summary, the proposed framework is a promising high-resolution radio-based positioning solution for current and next generation cellular systems.

[PDT$^+$19] proposes a software positioning framework for jointly using three parameters that are relevant for positioning: the RSS, AoA, and ToF of the wireless signals. Based on empirically obtained measurement accuracies of available non-hybrid systems, positioning errors were modeled for different possible hybrid localization systems. Simulation results are shown in Fig. 9.7. For the considered hardware (Chronos for ToF and SpotFi for AoA), ToF yields the most accurate re-

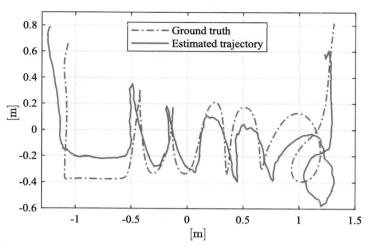

FIGURE 9.6

The ground truth (dashed-gray) and the estimated user equipment UE positions (blue; gray solid line in print version) [LBÅ+17].

sults, with errors below 1 m with only 3 access points in a 27 m x 41 m office building. ToF hardly benefits from adding RSS and/or AoA information. RSS and AoA have a similar performance, with errors around 2 m for 13 deployed access points. A hybrid RSS/AoA system achieves this error with only 7 access points.

Outdoor tracking

Current outdoor tracking solutions rely on GPS technology. Although providing accurate real time location updates, GPS has a disadvantage of consuming a fair amount of power at the mobile device. The availability of telecommunication infrastructure [TPS+19,PMQ18] and the advent of public sub-GHz network deployments (LORA, Sigfox) [PPT+18], provides the possibility to perform an alternative way of outdoor geolocalization. [PMQ18] proposes the use of unmanned aerial vehicles (UAVs) for ground user localization.

[TPS+19] uses 3G signal measurements, the topology of a mobile cellular network, and enriched open map data to locate a mobile phone. Location updates are based on timing information, signal strength measurements, base station configurations, mode of transportation estimation, and advanced route filtering. All calculations are completely performed on the network side without any required adaptations on the mobile side. The proposed AMT (antenna, map, and timing information based tracking) location tracking algorithm is experimentally validated in urban and rural environments near Ghent and Antwerp, with trajectories on foot, by bike, and by car. Fig. 9.8 shows the estimated positions with the proposed location tracking algorithm as red dots for a cycling trajectory in the city center of Ghent, Belgium. It is shown that both the mode of transportation, base station density, and environment

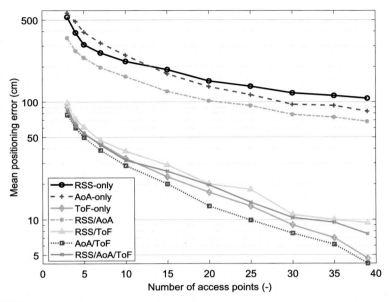

FIGURE 9.7

Simulated positioning accuracies for hybrid and non-hybrid configurations in a
27 m x 41 m office building.

impact the accuracy and that the proposed AMT is more robust and outperforms existing techniques with relative improvements up to 88%. Best performances were obtained in urban environments with median accuracies up to 112 m. [RCBO19] gives a brief overview of localization using Orthogonal Frequency Division Multiplexing (OFDM) such as 4G LTE networks.

Localization performance in a public LoRa network using Time Difference of Arrival (TDoA) is investigated in [PPT+18]. Like in [TPS+19], the processing load is entirely on the backend side of the network, making it much more suitable for localization of constrained devices. Six LoRa nodes at 868 MHz were configured with respective (fixed) SFs between 7 and 12, in a public LoRa network in and around Eindhoven, the Netherlands. One-hour 'walking' (4.4 km, maximum speed of 5 km/h) and 'driving' (38.7 km, maximum speed of 130 km/h) routes were defined. Two important performance metrics are the achieved location update rate (SF8 was selected in [PPT+18] as it provides the highest number of location updates) and the positioning accuracy. In order to further improve the positioning accuracy, a road mapping filter is applied to SF8's raw location outputs [TPMJ15]. Fig. 9.9 compares the TDOA results for SF8 without ('raw') and with ('v = $v_{allowed}$') road mapping filter. When it is known a priori that the node will be moving at walking speed, upper speed limits of the route mapping filter can be set to 5 km/h instead of using the (higher) road speed limits (v = 130 km/h). Fig. 9.9 shows that this significantly

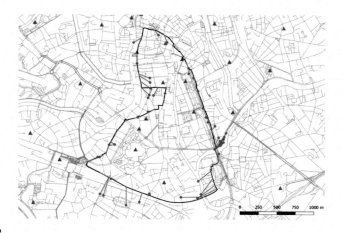

FIGURE 9.8

Trajectory for cycling in the city center of Ghent (black lines), estimated positions (red dots), error between estimation and ground truth by GPS (red lines), and NodeBs (gray triangles).

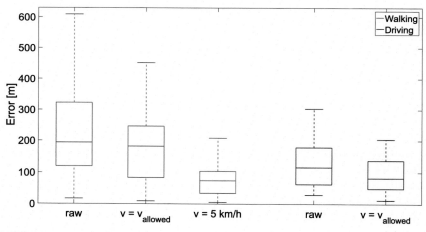

FIGURE 9.9

Positioning error distribution for SF8 of raw TDOA data and after applying road mapping filter using road speed limits ($v = v_{allowed}$) and using mobility speed limits ($v = 5$ km/h, when walking).

further improves tracking accuracy, in this case to a median error of 74 m (62% reduction).

[PMQ18] investigates multiple techniques to detect a ground user with a UAV without the need for expensive hardware or software processing but, rather, by leveraging on the movement of the UAV. Using raytracing simulations (CloudRT), the ground-to-air channel is investigated and it is shown that the multipath components (MPC) dispersion is relatively small in the angular domain indicating that, for ground-to-air-channels, most scattering occurs nearby the ground. Then, three Direction-of-Arrival (DoA)-based methods are presented for ground user localization, with both simulations and experiments. The methods actively exploit the UAV movement, which can effectively be controlled and leveraged to obtain DoA estimates or to obtain multiple independent estimates of the user's DoA. Fig. 9.10 shows the result of six test flights for one of the methods, showing that the UAV moves towards the source in an almost straight line.

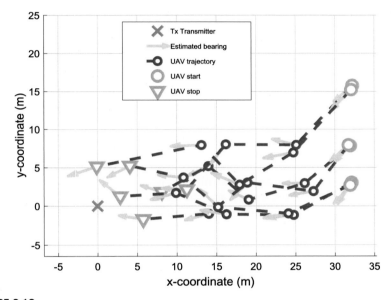

FIGURE 9.10

Test flight trajectories from UAV to Tx ground user using a weighted RSS algorithm.

9.3.2 Fingerprinting and ray tracing for localization

Fingerprinting, whereby the received signal strength (RSS) of multiple beacons is compared to those contained in a previously-constructed database, can deliver accurate positioning results in indoor environments. A drawback is that these reference fingerprints are usually obtained via extensive and time-consuming measurement campaigns inside the environment of interest, unless the fingerprinting is done based on available propagation models. Different types of fingerprinting are explored here (Table 9.1): model-based RSS fingerprinting via raytracing [RBZ17] or via a measurement-based propagation model construction [MMdHB18,dHBGV19], MaMIMO channels combined with convolutional neural networks (CNNs) [VRL+17], or power difference of arrival (PDOA) fingerprinting [NC17]. In [BNMJ19], ray-tracing is used to simulate accuracies of ToF measurements.

In [RBZ17], ray-tracing (RT) simulations are proposed as a fingerprinting technique to deterministically predict the radio wave propagation mechanisms, identify the main channel characteristics, and construct the RSS-based fingerprint database offline. To validate this approach, the RSS fingerprints simulated by RT are compared to those measured in an indoor office environment in which three Bluetooth Low Energy (BLE) beacons are distributed. Good agreement is obtained between the measured and simulated RSS fingerprints, especially at positions benefiting of a strong LOS component dominating the radio wave propagation. On the other hand, significant discrepancies were observed in NLOS situations (e.g. when the LOS is obstructed owing to human shadowing). The susceptibility of BLE to fast fading and the impact of human shadowing on RSS fingerprints are also demonstrated, and a model is proposed to describe the fluctuations of the RSS fingerprints in the environment.

RSS fingerprinting information is used in [MMdHB18] with the purpose of particle-filter based localization. The RSS estimation using a log-distance path loss model is improved by robust model fitting. A spatial statistical analysis algorithm accounts for the estimation of the RSSI and key parameters used in the signal propagation model. Basically, the approach consists in dividing the environment in several zones and determining the path loss coefficient factor value for each combination of the access point and environmental zone. The division is based on structural changes like the presence of corridors or rooms. In [dHBGV19], the authors expand the work of [MMdHB18] by improving the calibration of WiFi-based indoor tracking systems for smartphones. An in-motion calibration methodology with three different signal propagation representations is proposed. The use of the motion-aware calibration mechanism is shown to improve the localization accuracy.

[BNMJ19] proposes a ray tracing-based simulation approach for evaluating tracking algorithms based on ToF measurements. It combines a model for the generation of a random human walk with a communication protocol and a calibration of the anchor links' real ranging performance. With the proposed system, the performance of a simple least square localization algorithm and with an extended Kalman filter is evaluated. The work of [BNMJ19] shows that TDMA two-way ranging tracking systems are suitable for applications where position update frequency is low and tracked agents are mostly stationary.

Fingerprint-based positioning using measured massive MIMO channels and the application of convolutional neural networks (CNNs) are proposed in [VRL$^+$17]. When represented in appropriate domains, measured massive MIMO channels have a sparse structure which can be efficiently learned by CNNs for positioning purposes. The positioning capabilities of CNNs tend to generalize well to most propagation scenarios thanks to their inherent feature learning abilities. State-of-the-art CNNs with channel fingerprints are generated from the COST 2100 channel model with a rich clustered structure. Moderately deep CNNs can achieve fractional-wavelength positioning accuracies, provided that an enough representative data set is available for training. RMSE errors of 0.6 wavelengths only were obtained. Further research aspects are how to achieve a robust CNN design that is able to deal with impairments such as measurements and labeling noise, or channel variations that are not represented in the training set, or how to design complex-valued CNNs that perform well and are robust during optimization.

[NC17] presents simulation results using a game engine ray based tool for an outdoor scenario. The location of a transmitter in Non-Line of Sight (NLOS) with two receivers is investigated using one AOA/PDOA receiver and one PDOA receiver. Results at 900 MHz show an error of near 60 meters compared to the real position of the emitter.

9.3.3 Advanced localization techniques

The difficulty of estimating TDOA lies in the fact that the network of receiver nodes needs to be synchronized to obtain accurate results. [Qui17] presents a technique for estimating TDOA where no explicit synchronization between nodes is required. By forwarding the baseband signal from a first receiver node to a second receiver node, the time offset between the nodes is effectively canceled out, and the TDOA between the two nodes can be easily estimated. The only requirement is to know the propagation delay between the two receiver nodes, and to ensure that the clock skew is relatively low. The proposed method is implemented on a software-defined radio testbed and validated in a controlled lab environment. The method is also tested in an outdoor line-of-sight environment with a network of four receiver nodes. Fig. 9.11 shows an experimental result for the wireless outdoor setup with three relay nodes and one receiver. It can be seen that successive measurements might yield slightly different results, and the different hyperboles intersect close to the transmitter's real location. The architecture permits to have a TDOA error that is lower than when using GPS-synchronized nodes, and can achieve errors below 10 ns in controlled environments, and errors with zero-mean and a standard deviation of 30 ns in outdoor environments.

9.4 Multipath-based localization and mapping

Instead of mitigating the influence of multipath propagation on the estimation of the line-of-sight (LOS) path, multipath-assisted localization systems turn the multipath effect into an advantage by exploiting geometrical information, i.e., delays/distances, AOAs and AODs, of deterministic MPCs extracted from the radio channel. In detail, each estimated deterministic MPC is assumed to be associated with an environment feature, which can be a physical anchor (PA), one of the virtual anchors (VAs) which represent the mirrored positions of the PA w.r.t. planar surfaces in the propagation environment (see Fig. 9.12), or one of the point/line scatterers. Currently, most of the multipath-based localization approaches only consider the reflection model, meaning

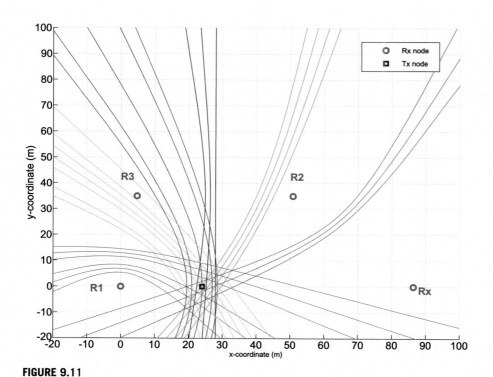

FIGURE 9.11

[Qui17]. Hyperboles corresponding to the TDOAs measured with three relay nodes and one receiver. The blue lines correspond to the TDOA between Rx and R1, the red lines between Rx and R2, the green lines between Rx and R3, the cyan lines between R1 and R2, the magenta lines between R1 and R3, and the black lines between R2 and R3.

that only PA and VAs are modeled (we only consider specular reflections originating from flat surfaces but not the ones originating from scatter points). The deterministic MPC parameters are estimated by using a parametric radio channel estimator. The estimates are subsequently used in a localization algorithm if the feature positions are assumed to be known, given prior knowledge of the environment floor plan (map), or in a simultaneous localization and mapping (SLAM) algorithm where the positions of agent and geometric features are jointly estimated.

9.4.1 Signal model and geometry model

In general, at time instant n, the baseband radio signal $s(t)$ is exchanged between the jth PA located at position $a_k^{(j)}$, $j \in \{1, \ldots, J\} = \mathbb{J}$, and a mobile agent at position p_n, the channel impulse response is given in (9.2). In real propagation environments, the channel impulse response is described by two separate components: deterministic components and dense multipath component (DMC) [LMTW17], yielding the

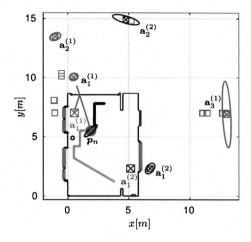

FIGURE 9.12

Illustration of the environment map obtained by the multipath-assisted SLAM algorithm [LMTW17]. Two physical anchors at $\boldsymbol{a}_1^{(1)}$ and $\boldsymbol{a}_1^{(2)}$ represent the infrastructure. The gray dotted-line represents the measured trajectory along which the mobile agent is moving. Gray squares indicate geometrically expected VAs, blue and red plus markers with uncertainty ellipses represent discovered VAs. An agent tracking result is shown in black with a error ellipse corresponding posterior covariance.

received signal given as

$$r^{(j)}(t) = \sum_{k=1}^{K_n^{(j)}} \alpha_{k,n}^{(j)} s(t - \tau_{k,n}^{(j)}) + s(t) * \nu_n^{(j)}(t) + w(t). \qquad (9.7)$$

The first term describes a sum of $K_n^{(j)}$ deterministic MPCs with complex amplitudes $\alpha_{k,n}^{(j)}$ and delays $\tau_{k,n}^{(j)}$ with $k \in \{1, \ldots, K_n^{(j)}\} = \mathbb{K}_n^{(j)}$. The number of MPCs $K_n^{(j)}$ is equivalent to the number of visible features at the current agent position. The delays $\tau_{k,n}^{(j)}$ correspond to the distances between the mobile agent position $\boldsymbol{a}_k^{(j)}$ and the jth PA (for $k = 1$) or the virtual anchors (VAs) of the jth PA for $k \in \{2 \ldots K_n^{(j)}\}$, which represent the mirrored positions of the PAs w.r.t. the planar surfaces in the propagation environment (Fig. 9.12). Thus the delays are geometrically formulated as $\tau_{k,n}^{(j)} = \| \boldsymbol{p}_n - \boldsymbol{a}_k^{(j)} \|/c + \epsilon^{(j)}$, where c is the speed of light and $\epsilon^{(j)}$ represents the clock-offset due to clock asynchronism between the jth anchor and the mobile agent. The second term denotes the convolution of $s(t)$ with DMC $\nu_n^{(j)}(t)$. The last term $w(t)$ is the additive white Gaussian noise process (AWGN) with double-sided power spectral density $N_0/2$. The power ratio between useful deterministic MPCs and impairing DM plus noise, i.e., $\text{SINR}_{k,n}$, is used as a reliability measure of the estimated MPCs, which is also related to the MPCs' range and angle estimation uncertainties given in (9.3) and (9.5). At each time n, a parametric radio channel estimator is used to estimate MPC parameters along with DM and noise statistics. It is possible that some deterministic MPCs are misdetected and/or clutter components which did not originate from any features are produced during the estimation process. The estimates are further used as noisy measurements in the localization or SLAM algorithm.

9.4.2 Technical challenges

Estimation of multipath component parameters

The key step of multipath-assisted algorithms is that they require accurate extraction of location-related parameters of MPCs, i.e., distances/delays and angles. The estimation quality of MPC parameters in turn determines the localization performance. In [UG19,ULG18], the authors use the Kalman enhanced super resolution tracking (KEST) algorithm to estimate MPC parameters. KEST uses snapshot-based signal parameter estimates obtained from the Space-Alternating Generalized Expectation-Maximization (SAGE) algorithm and tracks these parameters over time with Kalman filters. At the same time, it keeps track of the number of detected MPCs. The estimation of signal parameters via rotational invariance technique (ESPRIT) algorithm is used in [CZT19] to estimate MPC parameters in each snapshot, the estimates in consecutive snapshots are associated by using a global nearest neighbor tracker.

To improve the resolvability of MPCs, ultra-wideband (UWB) signals have been used in many multipath-assisted algorithms, e.g., [KHG+17], [LMTW17], due to their superior time-resolution. Under the limited bandwidth condition, the work in [LBÅ+17] utilizes large-scale antenna arrays to resolve closely spaced MPCs by exploiting the spatial sparse structure of the multipath channel. Also, the proposed work continuously tracks the phase changes of individual MPCs between consecutive snapshots using an Extended Kalman Filter (EKF)-based parametric channel estimator, which makes it possible to resolve the centimeter-level distance changes between consecutive time instances.

Data association

Another fundamental issue with the multipath-assisted algorithm is how to find robust associations between the estimated MPCs and the environment features i.e., PAs/VAs or scattering points. In most of the current works, one estimated MPC is assumed to be uniquely associated with one geometric feature. Furthermore, if an additional MPC is detected, it would be beneficial to know if it is from a new or a legacy environment feature. The data association (DA) problem becomes more challenging in situations where path-overlap of MPCs occurs or many clutter components are estimated. The work in [CZT19] performs the DA based on a global nearest neighbor (GNN) tracker with a Kalman filter. In detail, each potential target within the gating region is evaluated, then the best one to incorporate into tracking is chosen. The authors of [LMTW17] introduce a probabilistic DA approach in a belief propagation (BP) algorithm based SLAM framework. The factorization of the joint posterior probability density function (PDF) makes the algorithm scalable with the changing deterministic MPC number, hence frequent birth and dead processes of tracked features can be captured. The work in [UGJD17] extends a channel-SLAM algorithm by a multi-hypothesis tracking (MHT) based DA approach. Specifically, two MHT schemes aimed to decide for associations on a particle basis are presented and compared, i.e., the maximum likelihood (ML) method and data association sampling (DAS). The result shows that in case of measurement ambiguities, the DAS scheme is expected to outperform the ML method, but with higher computational complexity.

9.4.3 Localization approaches

In the cases that the geometric feature positions can be assumed to be known given prior knowledge of the environment floor plan, the estimated MPC parameters and known feature positions can be jointly used to localize the agent. For example, in [KHG$^+$17] the authors proposed a single-anchor localization solution (a maximum likelihood position estimator) using the CIR and range measurements from DecaWave's DW1000 chip incorporated into the Pozyx hardware platform. The DecaWave DW1000 chip is an UWB transceiver that provides range measurements and allows retrieving the measured channel impulse response (CIR), which is necessary to exploit deterministic MPCs. Using the range estimate only with known anchor position a, the position estimate of the agent p_n is uniformly distributed on a circle around the anchor. The additional position-related information contained in the CIR can be utilized to further identify the agent's position. This work is of particular interest due to the low cost of the hardware infrastructure. For two IEEE802.15.4 channels, the ranging error standard deviation in challenging indoor scenarios is shown to be below 5 centimeters (i.e., very low positioning error).

9.4.4 Localization-and-mapping approaches

If the environment geometrical information is not available or time-variant (due to e.g., people walking or interior changes), the positions of the agent and a map of geometric features should be jointly estimated using a localization and mapping algorithm. For example, in [CZT19], a Rao-Blackwellized, particle-filter-based SLAM algorithm is used to estimate a vehicle trajectory and positions of the base station and virtual transmitters, with the MPC delays estimated from the long term evolution (LTE) cell specific reference signal. Analysis based on data logged from a measurement campaign with a single base station shows the performance of the proposed approach with a maximum agent position error of 1.5 meters.

In [LMTW17], a multipath-assisted feature-based SLAM algorithm is proposed based on probabilistic data association and belief propagation (BP) message passing. This method jointly applies data association and infers the states of the mobile agents as well as the environment features. Factorization of the joint posterior probability density function (PDF) makes the algorithm scalable with the changing deterministic MPC number.

The authors in [JHFP18] propose to use the physical scatterers as VAs in the environment instead of image points of the receiver w.r.t. the map, hence the environment geometrical information is no longer needed (map-free approach). The locations of the scatterers in the environment are estimated with respect to the prior known antenna array location. The estimated scatterers are then used as VAs to locate the mobile agent through trilateration. The influence of people walking or interior changes are implicitly accounted for in the changes of the virtual anchor locations. Hence, the proposed method is adaptive to dynamic environments. The performance of the proposed method is also assessed with measurements conducted in a cluttered room using ultra-wideband large-scale array systems at high frequency bands including both LOS and NLOS scenarios. The results show that the error of the estimated target location increases with the distance between the transmitter and the receiver but, most of the time, the estimation error is below 30 cm.

In the localization and mapping approaches mentioned above, a *single agent* scenario was considered. In the presence of *multiple mobile agents*, it would be beneficial to share the map of observed surrounding geometrical information, e.g., positions of features, either directly between different agents or via some local entity, e.g., a base station in a cellular network. The map that has been estimated by other users already in the scenarios is denoted as a prior map, and the map estimated by a user itself is called a user map. To make the prior map useful to a user, few considerations should be taken into account. First, in practical scenarios, different mobile agents experience different clock offsets $\epsilon^{(j)}$ from the PA/PAs, which leads to different distance estimation offsets. Second, SLAM algorithms give the position estimates in their local coordinate systems, which means the user map and a prior map are in different coordinate systems with an unknown rotation and an unknown translation. Third, the correspondences among estimated anchor positions should be found. Only a reliable match between maps can ensure that the user benefits from the cooperative solution instead of introducing additional ambiguities. The work in [ULG18] firstly finds correspondences among the transmitters in the two maps and subsequently the corresponding rotation and translation of the coordinate systems. This map matching process is performed in parallel to the actual Channel-SLAM algorithm. Results indicate that the map-sharing approach improves the performance of channel-SLAM in terms of both accuracy and computational complexity. Another relevant approach in [UG19] proposes a variant of the random sample consensus (RANSAC) algorithm for the estimation of relative rotation and translation between the coordinate systems when exchanging the maps between agents. Simulations carried out in an indoor scenario reveal that RANSAC increases the accuracy of map matching significantly compared to the standard least squares method. It also helps decrease the convergence time of SLAM.

9.5 System studies and performance limits

This section summarizes a range of system-level studies of positioning systems, covering the system design and system-level performance analysis. The focus will be placed first on indoor systems and then shifted towards vehicular localization outdoors. This section also addresses aspects of network optimization, leveraging the concept of location-awareness [TMR$^+$14].

9.5.1 Indoor localization systems

For indoor positioning, a wide range of radio-based positioning strategies can be found in the literature. Severe multipath propagation conditions as well as shadowing by objects are considered to be the main impairments on positioning performance. Such systems, for example, perform positioning based on received signal strength (RSS), time-(difference)-of-arrival (ToA/TDoA), angle-of-arrival (AoA), or various hybrid variants thereof. Several alternatives to traditional indoor positioning have been recently explored, the most advanced techniques relying on antenna arrays.

Useful CRBs for array-based positioning in presence of dense multipath are derived in [WGMW18]. This provides the relationship between the ToA and AoA information in multi-anchor positioning scenarios and the influence of the anchor-agent placement. The different measurement types can complement one another. AoA information is more useful at close range, allowing for accurate single-anchor positioning at close range, while, at far range, only the ranging information remains accurate. Interfering specular multipath components have a maximum detrimental influence near the walls.

In [HLMW16], authors explore the achievable ranging and positioning performance for two design constraints in a radio frequency identification (RFID) system: (i) the bandwidth of the transmit signal and (ii) the use of multiple antennas at the readers. The MIMO RFID system for the monostatic, bistatic, and multistatic scenarios is shown in Fig. 9.13. The ranging performance is derived for correlated and uncorrelated constituent channels by utilizing a geometry-based stochastic channel model for both downlink and uplink. The ranging error bound is utilized to compute the precision gain for a ranging scenario with multiple collocated transmit and receive antennas. The positioning error bound is then split into a monostatic and bistatic component to analyze the positioning performance of the MIMO RFID system in dense multipath scenarios. Simulation results indicate that the ranging variance is approximately halved when utilizing uncorrelated constituent channels in a monostatic setup, and that the bandwidth and the number of antennas reduce the error variance roughly quadratically and linearly, respectively.

Another interesting antenna array-based approach has been proposed in [WGL$^+$18], using a single anchor equipped with an antenna array to be able to exploit information contained in specular multipath components (SMCs). This allows to increase the positioning accuracy of the system and to reduce the required infrastructure, using a-priori information in form of a floor plan. Authors analyze through the CRB the benefits of spatial aliasing of antenna arrays on the achievable angular resolution. Then, they assess the array-based multipath-assisted positioning accuracy. Results show that SMCs help reduce ambiguities arising due to the aliased array which, in turn, allows to benefit from the large array aperture while reducing the complexity compared to a full array of similar size.

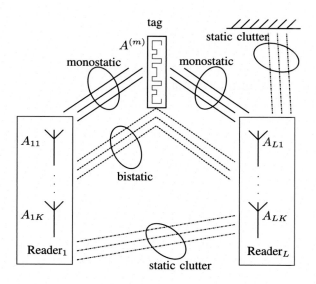

FIGURE 9.13

System Model of the MIMO RFID setup including monostatic, bistatic, and multistatic scenarios [HLMW16].

A fundamentally different approach is presented in [DSW17], where the authors propose to exploit the magnetic near-field for indoor localization, which can be a promising alternative due to its insensitivity to the environment and strong spatial gradients. The use of flat/planar coils allows for convenient integration of the agent coil into a mobile device (e.g., a smartphone or wristband) and flush mounting of the anchor coils to walls. From the analysis of the CRB on the position error for active near-field 3D localization with unknown agent orientation, and its dependence on room size and anchor count, it is shown that cm-accuracy is possible in a square room of 10 m side length. To enable such positioning accuracy in practice, an algorithm is proposed which employs a suitable scaling and alternating estimation of position and orientation, with an efficient solution for the orientation step. The resulting localization scheme performs near the CRB with high robustness and consistently low computational cost, being a potential enabler for accurate indoor 3D localization with unobtrusive infrastructure and high update rate.

9.5.2 Localization for vehicular networks

Vehicular networks are particularly challenging in terms of the fulfillment of the high-accuracy positioning requirements associated to innovative use cases such as collision avoidance, automated overtake, or high-density platooning. Vehicular networks are based on several vehicle-to-everything (V2X) communication interfaces depicted in Fig. 9.14, namely, (i) downlink and uplink transmissions between vehicle and dedicated BSs or RSUs (Road Side Units), that is, vehicle-to-infrastructure

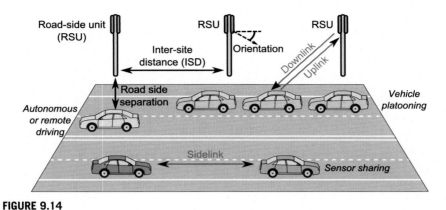

FIGURE 9.14

V2X communications interfaces and emerging vehicular applications [dPRSGKLS19].

(V2I) communications; and (ii) sidelink transmissions between vehicles or other network devices or, in other words, vehicle-to-vehicle (V2V) communications. These vehicular communications can be exploited for positioning purposes, in order to complement data from on-board sensors for precise, reliable, and secure vehicle positioning.

Uplink V2I communications can be used to compute the vehicle position with uplink TDoA (UTDoA) measurements at base stations or RSUs located along the road, as shown in Fig. 9.14. The performance limits of V2I ranging-based localization are assessed in [dPRBAZ⁺17b], considering the LTE standard and different network deployments along a highway. As it can be seen in Fig. 9.15, a high network density with an inter-site distance (ISD) of 100 m is required to achieve a UTDoA positioning accuracy around 10 m for 80% of the cases. This performance is mainly limited due to the NLOS conditions of certain UTDoA measurements. Thus, the ultimate V2I positioning performance requires dedicated network deployments and a high system bandwidth.

In [dPRSGKLS19], the authors investigate the optimization of the density of roadside deployments in order to fulfill stringent positioning requirements (vehicle positioning accuracy below 1 m for 95% of the cases). This study proposes a novel design approach that optimizes the uplink V2I positioning capabilities based on the network layout, propagation conditions and network configuration. The proposed design guidelines recommend (i) the placement of RSUs on both road sides, with a density bounded by the number of uplink measurements in LOS conditions; and (ii) the exploitation of antenna arrays for accurate vehicle localization. Considering standard propagation models, a 10-MHz signal bandwidth, and using the CRB for joint ToA and AoA, the maximum inter-site distance (ISD) to fulfill high-accuracy positioning requirements is found to be around 40 m and 230 m in urban and rural environments, respectively. As Fig. 9.16 illustrates, the combined use of ToA and AoA measurements is needed to maximize the ISD nearly one and a half times, with re-

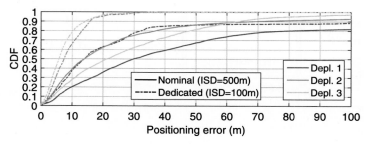

FIGURE 9.15

CDF of UTDoA position accuracy for nominal and dedicated network deployments along the highway, over 3D-UMa model with distance-dependent LOS conditions and 20-MHz system bandwidth [dPRBAZ+17b].

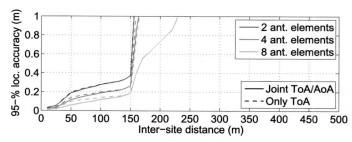

FIGURE 9.16

95% location accuracy based on CRB of joint ToA and AoA for vehicle locations close to the serving RSU in a highway scenario, using a 10-MHz bandwidth in AWGN conditions with SNR levels and LOS probabilities computed with the 3D-RMa model over 50 000 simulations [dPRSGKLS19].

spect to the use of only ToA measurements, thanks to the relaxation of the positioning problem and the improved signal measurements with antenna arrays.

Cooperative passive coherent location (CPCL)

Cooperative positioning is also a promising approach to exploit V2X communications in current and future LTE and 5G standards. Nowadays, V2X networks are mainly used to exchange vehicle status information, such as precise position and speed, which has been obtained from the vehicle's on-board navigation system. But, V2X transmissions can also be used for cooperative positioning, thanks to the inherent 5G radio features that can enable high-accuracy positioning. In this sense, Cooperative Passive Coherent Location (CPCL) is proposed in [TADG+19] as a distributed MIMO radar service that takes advantage of the 5G radio transmissions for positioning. This approach is based on the bi-static radar principle with cooperative transmitters and receivers. A preliminary field test of this concept is shown in Fig. 9.17. A vehicle transmits OFDM signals that reflect in the target moving vehicle, and these reflections are received by two sensor vehicles. Thanks to the cooperation

FIGURE 9.17

Measurement setup of a bi-static V2X radar scenario for cooperative positioning of a target moving vehicle [TADG$^+$19].

between transmitter and receiver, each sensor vehicle is able to estimate the time delay and Doppler shift of the reflected signals on the target vehicle for its localization, by using the known transmitted signals. Further research is necessary to tackle several CPCL challenges, such as sparse frequency-time resource allocation and advanced distributed detection and estimation schemes.

In [SdPMSR18], the use of CPCL is investigated for detection and tracking of various road users, employing signals with carrier frequencies in the range of 4 to 8 GHz (C-band). With transmitter and receiver spatially separated, the targets appear as time-variant signal components with a specific delay and Doppler shift. Radio measurements were performed with a Medav RUSK channel sounder at $f_c = 5.2$ GHz with BW $= 120$ MHz and distance resolution of 240 m at an old airstrip. Comparing the measurements with ground truth trajectories obtained with GNSS signals, one observes the direct relationship between delay and Doppler shift with the actual position (see Fig. 9.18).

Reference [SHRT17] investigates the synthesis of *bistatic* range profiles of small objects which also obtained a growing interest in radar in recent years, especially due to the growing use of unmanned aircraft systems (UAS). The presented framework combines LOS propagation, measured UAS characteristics in terms of measured radar cross sections and radar system models by means of the underlying geometric parameters in a fully polarimetric model.

9.5.3 Multi-system hybrid localization strategies

Future communications applications are expected to demand high-accuracy positioning requirements, such as sub-meter positioning above 95% of cases. To cope with such stringent accuracy and availability requirements, hybrid localization strategies

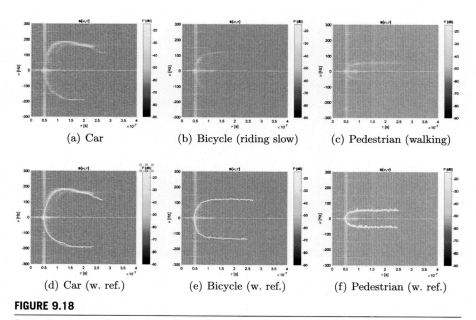

FIGURE 9.18

Delay-Doppler spectra for different passive targets.

are envisaged to overcome the (more) limited positioning capabilities of standalone systems. The combination of multi-constellation Global Navigation Satellite Systems (GNSS) and 5G technologies looks particularly promising in harsh urban environments, where both satellite and terrestrial systems complement each other in terms of accuracy and availability of observables. The performance limits of such hybrid positioning solutions must be evaluated with system-level simulations. In [dPRRR+19], for instance, a novel hybrid methodology is proposed to characterize both GNSS and 5G ranging observables, in order to avoid the high computational burden of physical-layer simulations. The GNSS observables are simulated based on exiting User Equivalent Range Error (UERE) models, while the abstraction of 5G ranging observables is based on the generation of ranging error models following 3GPP standard physical-layer simulations. Using this methodology, system-level simulations indicate that hybrid GNSS and 5G approaches are needed in harsh urban environments, in order to achieve a positioning accuracy below 10 m on the 80% of the cases. Moreover, further enhancements are required in both technologies, in terms of accuracy and availability, to obtain sub-meter localization in deep urban canyons.

In fact, deep urban canyons are specific scenarios of interest to evaluate the achievable positioning capabilities of hybrid GNSS and 5G solutions, due to the reduced number of satellite and terrestrial ranging observables. Current simulation methodologies are based on independent simulation models for satellite and terrestrial links, which does not necessarily represent specific real-world environments. In order to assess more realistic scenarios, a 3D city model is exploited in [dPRGD+19] to assess the achievable positioning capabilities of hybrid 5G round-trip time (RTT) and multi-constellation GNSS positioning. This study evaluates the deterministic LOS conditions between mobile and transmitter positions, where the transmitters can be either GNSS satellites or 5G base stations. Considering a deep urban canyon, computer simulation results indicate that the addition of only one RTT LOS measurement to the GNSS observables is sufficient to significantly improve the GNSS standalone positioning, due the relaxation of the positioning problem, improved geometry and enhanced observable accuracy.

9.5.4 Location awareness-based network optimization

Radiofrequency spectrum is a scarce resource for network operators. One of the main approaches to optimize spectrum resources consists in reducing the coverage of each mobile network cell and increasing their number. This has led to the deployment of small cell networks. However, in order to cope with the increasing complexity of the network infrastructure, automatic systems are required to perform the proper identification of failures in small cell scenarios. In addition, high network densities make classical approaches (only based on network performance metrics) unable to properly tackle the characteristics of these small cell scenarios, as compared with macrocell ones in Table 9.2. Thus, the use of non-cellular information, and especially UE location, is necessary to support the identification of failures in the network.

Table 9.2 Characteristics of macrocell outdoor and small cell indoor deployments [FGFL+16].

Characteristic	Macrocell outdoor	Small cell indoor
Robustness	Reliable elements	Vulnerable elements
Infrastructure	Dedicated	Shared
Cell size	Kilometers	10s of meters
Deployment	Planned	Typically unplanned
Coverage areas	Relatively defined	Irregular, overlapped
Monitoring	Large number of KPIs and alarms	Reduced: simplified KPIs and alarms
Performance variations	"Slow" (hours), large number of users	"Fast" (min/secs): very variable user distributions

A novel approach for context and location-awareness applied to self-healing in mobile cellular networks is presented in [FGFL+16]. The challenges for small cell networks self-healing are identified and assessed and, then, the use of UE context and direct reporting are proposed as a means to overcome the main issues in those scenarios. From this analysis, the self-healing framework shown in Fig. 9.19 is proposed to integrate context information (e.g. time of the day and terminal position) in small cell networks. Supported by this framework, a specific context-based detection and diagnosis algorithm is presented in [FGFL+16, Figure 4]. The proposed system is implemented in a femtocell deployment. Here, the use of UE context supports the automatic analysis of radio measurements, distinguishing between normal

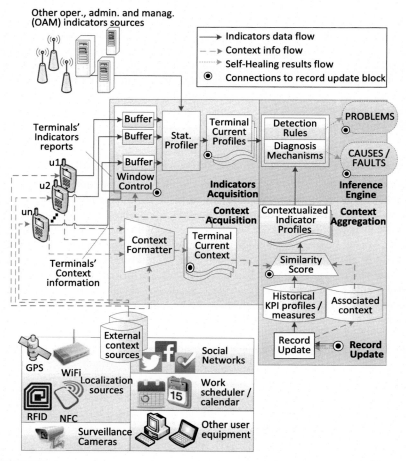

FIGURE 9.19

Context-aware self-healing framework [FGFL⁺16].

receive power degradation and failure-related ones. It also supports the fusion of the information gathered by different terminals and the joint categorization of the status of the cells in different levels of failure and their causes. Further related works can be found on the general context-aware network management [FAGB⁺15,FPSB18a], optimization [AGFDB16,AGFG⁺16], and troubleshooting [FBAGM15,FBAG16].

Another relevant approach is to consider location-aware communications. For instance, [AHM19a] investigates how angle-of-arrival (AoA) information can be exploited by machine-/deep-learning (ML/DL) approaches to perform beam selection in the uplink of a multi-user MIMO mmWave communication system. More specifically, [AHM19a] considers a hybrid beamforming setup comprising an analog beamforming (ABF) network followed by a zero-forcing (ZF) baseband processing

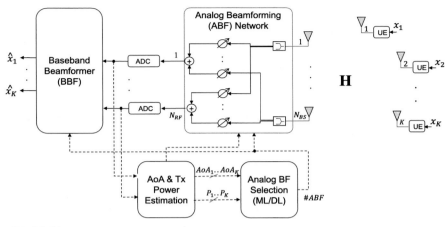

FIGURE 9.20

System model [AHM19a].

block. Unlike other contributions, AoA (and received power) information is not assumed to be known or externally supplied. Instead they are estimated by the same system via a Capon (minimum variance) spectral estimator. The underlying goal is to select the optimal configuration for the ABF network based on the estimated AoAs of the various user equipments. To that aim, authors compare two supervised machine-learning approaches, i.e., k-nearest neighbors (kNN) and support vector classifiers (SVC), and a feed-forward multilayer perceptron (MLP) deep neural network, while considering a codebook of analog beamformers with adjustable beamwidth (rather than phased-arrays or DFT-based codebooks) as a means to enhance/guarantee system performance. The system model is depicted in Fig. 9.20. Computer simulation results in [AHM19a] reveal that classification accuracy exhibits some degradation in the presence of received power fluctuations which is larger for ML approaches (actual channel state information (CSI)). With estimated CSI, the degradation is more severe, both for ML and DL schemes. This is mostly due to the limited spatial resolution of the Capon-based AoA estimation block operating on a filtered received signal, and the impact of additive noise on CSI estimates (yet the impact of the latter is moderate). In general, MLP outperforms both kNN and SVC methods in terms of classification accuracy. However, ML and DL approaches with moderate classification performance still exhibit a sum-rate performance which is close to the optimal one. This follows from the fact that most classification errors occur among adjacent regions. Only when classification accuracy severely degrades due to CSI estimation and/or fluctuations in the received power, the sum-rate degradation becomes noticeable, in particular, for the high-SNR regime.

Similar to the previous approach, drones can exploit AoA estimates to perform beamforming, in order to reduce uplink interference levels with neighboring cells, as it is proposed in [STKK19]. Given the LOS conditions of drones with cellular towers, the uplink transmissions of drones increase interference at neighboring cells. The introduction of antenna arrays in professional UAV can enable AoA estimation of the neighboring cell transmission. Assuming channel reciprocity, this angle information can then be used by drones for beamforming, which is expected to significantly reduce the uplink interference levels.

Besides, the work in [HRL+16] studies the statistics of the channel capacity for measured wideband indoor channels and compares it with those of the decomposed channel CIR provided by the multipath-assisted indoor navigation and tracking (MINT) algorithm (cf. Section 9.4). The MINT algorithm uses the signal model (9.7) to decompose the received signal into individual MPCs that can be associated with the local geometry. The authors demonstrate the contributions of individual MPCs to the channel capacity with real UWB measurements. A few observations can be made from the results: 1) The robustness of a wireless link is thus strongly related to the power of some "dominant" MPCs and the probability those components get obstructed. 2) The capacity varies significantly more at lower signal bandwidth, because of stronger interference by DM when the time resolution (i.e., the resolvability of MPCs) is lower. Overall, the statistics of the channel capacity can be closely predicted by the proposed model. Furthermore, the influence of certain MPCs can be quantified, which is seen as an indication of the link robustness considering path blockages.

9.6 Testbed and prototyping activities

Testbeds and prototypes are needed in order to conduct experimental research. In this section, we succinctly describe several experimental platforms developed and currently used by a number of institutions all over Europe.

9.6.1 Testbeds for GNSS-based localization activities

The swiftly evolving landscape of GNSS signals and systems demands rapid prototyping tools in order to explore receivers' full capability, including radically new uses of those signals. That flexibility is hard to find in today's GNSS receiver technology, realized by application-specific integrated circuits (ASIC) and system-on-chip (SoC) implementations with high development costs and very limited degree of reconfigurability, thus hampering experimentation and fair trials of new approaches.

An example of testbed aimed to the research on Global Navigation Satellite Systems signals is GESTALT (see Fig. 9.21), a loose acronym for GNSS Signal Testbed, a facility located at CTTC's headquarters in Castelldefels, Spain. The facility is equipped with an assortment of GNSS antennas; GNSS signal generators for controlled experiments; state-of-the-art radio-frequency front-ends able to work concurrently in three GNSS frequency bands, with configurable bandwidth, frequency downshifting and filtering; digitation working at sample rates as high as 80 MSps with 8-bit, coherent I/Q samples; high-speed interfaces to a host computer; and an

(a) Antenna platform.   (b) Lab rack.

FIGURE 9.21

GESTALT Testbed for experimentation with GNSS signals.

open source GNSS software receiver in charge of signal processing and generation of suitable outputs in standard formats [AFPC15].

A key GESTALT feature is its openness. In addition to the fact that it can be fully operated remotely, the core software DSP receiver engine is a free and open source project with a lively community of users and developers (http://gnss-sdr.org). Accordingly, a partial testbed replication can be done on a limited budget with commodity computers and low cost, over-the-counter antennas and radio-frequency front-ends. This allows for the practical implementation of new concepts, reproducible research and short assessment and validation times.

Besides, a set of software-defined GNSS receivers run in the cloud can be used to overcome the limited computational resources of IoT nodes needing localization functionalities. In this solution, the GNSS receiver is no longer a physical device but a virtualized function which is provided as a service. In [FPA+17], authors proposed a system architecture based on optical networks and automated orchestration tools to deliver continuous service with high-accuracy performance to users with high-bandwidth connectivity, which could be provided by 5G networks.

9.6.2 Massive MIMO testbeds for localization

Massive MIMO provides an opportunity for superior positioning schemes based on the highly accurate AOA/AOD information that can be obtained via the use of massive antenna arrays (see Section 9.2). In particular, it presents an opportunity for a simple, single BS localization by utilizing the LOS AOA and mobile station range information.

An overview of the University of Bristol's Massive MIMO testbed can be found in [HHB+16a]. Fig. 9.22 shows the distributed system, each cabinet with radios driving

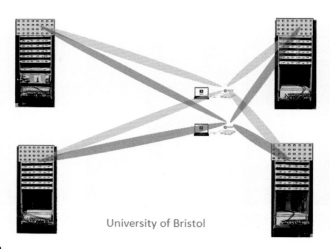

University of Bristol

FIGURE 9.22

Bristol Massive MIMO testbed.

a 32-element subarray, to act as a base station. The two USRPs and a laptop shown on the figure, are the user equipment.

Localization schemes that employs DOA estimation can be approached in two different ways when massive MIMO arrays are considered. One way is to perform DOA on individual, separate base stations, and then use triangulation to get the location of the target or user equipment (UE). Another way is to take advantage of the large geometries of linear or rectangular arrays to perform DOA using different subsets of antennas on the array. This allows DOA estimation and subsequently triangulation, to be performed using just a single BS. This approach however, depends on array size and geometry. It is suitable for LOS scenarios with smaller coverage areas, like indoor environments. Linear or rectangular arrays will suffer from the Geometric Dilution of Precision (GDOP) problem, but use of subarrays for DOA estimation allows localization algorithms to have a choice over the selection of best DOA estimates, based on predetermined confidence criterion.

9.6.3 Testbeds for localization activities based on battery-less tags

In [CDA+17] and references therein, most of the aspects related to the design of UWB-RFID systems based on backscatter modulation (e.g., design of dual-band UWB/UHF rectennas, energy harvesting circuits, and signal processing techniques) are discussed. The experimental campaign carried out in a real-world environment indicates that an accuracy of about 4 cm in a 5x5 area can be achieved. In [GDD+16], the same technology was applied to successfully sort pieces of luggage in airports. The items were placed as close as 20 cm to each other.

Complementarily, in the LOST project (funded by the European Space Agency) researchers from the University of Bologna and the Université Catholique de Lou-

FIGURE 9.23

Test of the UWB-RFID battery-less localization system on the Mars Rover prototype in the Automation and Robotics Laboratories at ESA.

vain, investigated radio technologies able to localize with centimeter-level accuracy battery-less tags inside the International Space Station (ISS), which is a quite challenging "indoor" scenario. Differently from the previous test bed, here the tag accumulates the energy from the UHF link and, once it is charged, it sends out a train of UWB pulses through a small energy efficient UWB pulse generator. In this way longer ranges can be obtained with the same accuracy: more than 10 meters both in wireless power transfer and UWB localization. The functionality and performance of the LOST system have been successfully tested directly on the Mars Rover prototype in the Automation and Robotics Laboratories at ESA as shown in Fig. 9.23.

Finally, the authors in [SKP19] deployed an experimental indoor positioning system based on Bluetooth Low Power (BLE) tags. The localization system is 1D since it is intended to be used in narrow corridors. Each of the anchors comprises four antennas (i.e., SIMO localization system) and, for this particular scenario, four anchors were deployed. The anchors communicate with one another via Zig-Bee and with a Raspberry Pi which acts as a gateway to a Local Area Network relaying all reports to the PC application for further processing, calculation of the location of the tags and showing them on the hosted website. Whereas arithmetic average performs better in combining of RSS from antennas, a weighted average outperforms the arithmetic average in combining distances from anchors. The average error of the presented system is 0.92 m, where in 90% of measurements the error is less than 2.4 m.

Perspectives

Chiara Buratti[a,b]**, Laurent Clavier**[c,b]**, and Andreas F. Molisch**[d,b]

[a]*University of Bologna, Bologna, Italy*
[c]*Université de Lille, Lille, France*
[d]*University of Southern California, Los Angeles, CA, United States*

While IRACON is ending, 5G is being deployed all around the world. The total number of mobile subscriptions[1] at the end of 2019 was around 7.9 billion. While 3G and 2G mobile subscriptions tend to decrease, the number of 4G subscriptions increased, reaching a total of 4.3 billion, or 55 percent of all mobile subscriptions. At the same time, 5G started its deployment. The world's first commercial 5G services were launched in 2018 and 5G is expected to penetrate the market much faster than 4G, the very early adoptions in China being one of the reasons. The number of 5G subscriptions reached around 13 million at the end of 2019 and are expected to represent nearly one third of all subscriptions by 2025, so 2.6 billion.

Looking at these impressive numbers, one may think that 5G is not only a done deal, but that all the goals of wireless communications for the foreseeable future are met. But, as it has been discussed all along in the IRACON project, 5G is not fulfilling all its original promises and telecommunication networks based on it do not reach the ultimate performance limits. The road to those goals will be long and difficult, with many alternative routes, external threats, and dead ends. Yet it will be exciting, requiring the understanding of increasingly complex phenomena.

In this chapter we provide some discussions about which topics are left for future research. We intend to present these perspectives in two stages. The first is to identify certain promises of 5G that are not yet possible and where research and engineering must work hand in hand to come up with effective solutions. The second will look beyond 5G, what could be called 6G, namely, what applications can be envisioned and which could be the key technologies that will enable a seamless interaction between humans and the cyber-physical world.

---

[b] Chapter editors.
[1] Ericsson Mobility Report | November 2019 and Q4 2019 Update.

Inclusive Radio Communications for 5G and Beyond. https://doi.org/10.1016/B978-0-12-820581-5.00016-X
Copyright © 2021 Elsevier Ltd. All rights reserved.

10.1 Implementing 5G

Expectations on 5G are high, but - even though deployment has started - many of the advanced aspects are not available yet. Various 'verticals' (specifications for different applications) have been defined, whose requirements are partially conflicting with each other. While higher data rates remain an important objective (just like in previous generations), new scenarios, such as IoT characterized by a massive number of devices, critical communications for connected industries, or autonomous vehicles requiring high reliability communications, are emerging.

Along with the increased number of requirements, 5G relies on technologies that have not been previously deployed on a large scale. Massive MIMO and mmWave communications require extremely efficient hardware and software. The complexity gap with 4G is huge and the core network is highly impacted by the shift as well. Combined with the shortest ever time to market, 5G will require many adjustments and, without a doubt, improvements to move towards optimal 5G operation. The many advances in 5G take the use of dimensions, frequency, and space, to an extreme, which increases throughput and density of users and connected objects. It is the first time that such large bandwidths are introduced for intensive commercial use in spectra (mmWave) where the efficiency and the stability of the communications still leave questions. Also, the large number of beams formed by the massive number of co-located antennas enable efficient spatial/angular reuse of the resources, allowing denser networks. At the same time, the requirement to connect billions of distributed devices can only be partly addressed through network densification, but also requires some completely different approaches to find a solution.

As a consequence, in the coming years, engineers and researchers will face, at unprecedented pace and scale, the introduction of complex and new technologies in public and industrial spheres. A significant number of problems that will encounter 5G may not be solved yet. Without being exhaustive, we give here some key elements which we believe are essential to the full success of 5G.

- *Fundamentals and global network design.* Heterogeneities of future networks (devices, radio solutions, quality of service, constraints) make it very complex to assess the performance of a 5G network and to optimize its operating mode. Existing models, such as those based on stochastic geometry, either deviate from the deployment realities allowing only a macroscopic performance evaluation but not a fine adjustment of the network; or present complexities that we no longer know how to handle analytically. Fundamental solutions for quantifying network performance in a multi-dimensional space (throughput, delay, energy, reliability, etc.) and tools (probably a judicious mix between data-based approaches, derived from AI, and model-based approaches, more common in telecommunications) still need to be developed to better manage 5G. Non-Gaussian models and new ways to account for dependence (in time and space) will be necessary.
- *Better know the channel.* One of the objectives of 5G is to increase the used spectrum to allow an increase of peak and user data rates: the current versions of 3GPP

5G NR (Rel-15 and Rel-16) operate in a spectrum below 52.6 GHz. However, the use of such frequencies for outdoor / broadband applications has not been explored for large-scale deployment yet. A more comprehensive understanding of radio wave propagation is required to access efficiently the large available contiguous bandwidth above 6 GHz, especially in highly dynamic scenarios. It is important to fully capture the channel complexity and its characteristics in time, space, and polarization. Tests in real conditions (over-the-air) will also be crucial to evaluate the real impact of the environment on the wireless links. Such better understanding will also have to be reflected in future channel models.

- *Channel state information*. Many of the Physical Layer techniques rely on the knowledge of CSI. However, it becomes unrealistic to estimate the state for all the channels involved when considering massive MIMO and/or IoT, especially in dynamic environments. To help the system estimate the CSI, deterministic channel models or location-aware channel-database can be used. However, their real-time use in localization, signal processing, and resource allocation algorithms still remains in its infancy. Machine Learning approaches could also be a significant help.

- *Reliable links in mobility*. Another advanced mechanism deployed in 5G is massive multiple antenna systems and beam forming. This can become very important in the mmWave bands to combat the high isotropic free-space path loss we face when reducing the wavelength. It is crucial to develop efficient beam steering and tracking. Enabling radio links in the mmWave band between vehicles and infrastructure or among vehicles, would allow to transfer at a high data rate in a smart mobility context. While the current 5G standard foresees appropriate mechanisms, they have not been fully implemented in the existing deployments, and may furthermore be too slow for use in some highly dynamic environments.

- *Ultra reliable communication*: beyond the traditional requirements of high rate transmission for, e.g., video streaming, some verticals like industry automation require much more reliable communications for some critical applications. A common goal is to reach 5 nines (99.999%) for the success rate of packet delivery, even sometimes with a delay constraint. Coupled with limited resources on the nodes and harsh industrial operating environments, such a target is a real challenge that is difficult to reach in IoT conditions (low cost, low power devices). Improved or even new strategies to access the channels and carry the information are needed. On one hand, deterministic channel access, orthogonal by construction, as traditionally used in cellular networks or proposed in TSCH for Low-Power devices can ensure reliability and short delays. On the other hand NOMA can improve the network capacity and increase the number of connected devices. The best compromise between orthogonal multiple access, requiring a precise scheduling, and NOMA with a reduced scheduling is an important topic.

- *Density*. Recent works forecast more than 40 billion connected devices by 2025, and more than half of them related to the IoT, from sensors to connected cars and wearables. It is not sure that the current status of 5G or LPWAN technologies can support these deployment densities. This will impact the whole network from the physical layer to the data management.

- *Positioning accuracy.* While no target is given today in 5G for positioning accuracy, 3GPP tries to achieve an accuracy of less than 3 m to improve 5G NR location awareness. Several challenges remain when dealing with indoor localization, especially in the virtual reality contexts, where very high accuracy is requested, or in the framework of logistics-type applications.
- *Latency.* Today in 5G NR, the URLLC target for the user plane of 1 ms of latency was abolished because it could not be achieved, and now 5 ms is the goal. Realizing the original 1 ms has to take into account the whole network, from the Physical link to the IP core. Edge computing could play an important role in solving this challenge, by avoiding entering the core of the network and by being very close to the source for a faster reaction.
- *Network and terminal energy efficiency.* Energy is becoming one of the main concerns of society. Telecommunication networks play a particular role in this context, being both a tool for reducing consumption, whether in the home or by enabling remote meetings, as well as being a large consumer. At the same time, in applications such as IoT, energy is the resource that sometimes defines the life of the network, when the batteries cannot be replaced. In all cases, saving energy is a crucial factor posing challenges both at the hardware level and in the organization of the network. Fully autonomous nodes, with harvested energy, can also be developed and near sensor computing becomes an important challenge. New hardware design can significantly reduce the energy consumption of devices, for instance implementing artificial neural networks or spiking neural networks.
- *Co-existence issues.* Connectivity will be ensured through multiple radio access technologies in licensed and unlicensed spectrum. How these technologies will live together remains an unsolved question, both for coexistence between different 5G services, and between 5G and other applications, such as WiFi. Especially in the ISM bands, inter network interference could become a significant problem. Duty-cycle or carrier sensing approaches are not sufficient or sometimes not adapted to mitigate the generated interference. Better modeling and understanding are needed to define adapted co-existence rules. Exploiting more unlicensed bands such as 60 GHz, to mitigate the traffic in existing bands, may also be helpful. Such possibilities need to be further incorporated to offer wider solutions for networks.
- *Network orchestration.* Providing increased peak data rates to a few users or low rates to many, with a reduced latency requires a new approach to signal processing and physical layer communication techniques. But handling the wide varieties of applications, requesting significantly different performance, also requires an improved resource management and orchestration. SDN and NFV are strong current trends, where each step of the transmission is controlled in software, offering significant gains in cost, flexibility, ease of deployment and efficiency. However, orchestration of networks characterized by high dynamicity, due to user mobility and changing of traffic load distribution, is still an open issue. In addition, effective real-time network resource orchestration in the presence of multi-dimensional resources, many service types, and unknown traffic models, is another challenge. Dynamic network slicing, according to which a network operator may generate,

in a dynamic way, dedicated virtual networks to support the optimized delivery of any service toward a wide range of users, will be an important element of future networks.

- *Security and privacy.* One cannot list important topics that still need research effort without mentioning the security and privacy aspects. Security threats, ranging from access to the physical network structure to pure software attacks, are getting more and more numerous and it is essential to protect the integrity and the verifiability of the received data. IoT also brings new risks with many low cost devices transmitting information. The low cost makes this information difficult to protect and increases the many threats to users' privacy, whether they come from organizations with a legal right to access data and data brokers or criminals seeking to invade privacy.

It is difficult to cover the full list of areas that still needs research and engineering. The new network technologies, for 5G or IoT, definitely need time to achieve what we expect from them. The multiplicity of verticals, the new frequency bands, the intense spatial usage (massive MIMO and ultra dense networks) are requiring a wide range of different optimizations, that is a network able to understand the end user needs and to adapt to them.

10.2 Preparing 6G

10.2.1 Applications

Nowadays we are moving toward a society of fully automated and remote management systems. Autonomous systems are becoming popular in every sector of society, and require to embed cities, vehicles, homes, and industries with millions of sensors. Hence, a high data-rate with reliable connectivity will be required to support these applications. Although 5G will provide better QoS as compared with 4G, it will not have the capacity to deliver a completely automated and intelligent network, that provides everything as a service, and a completely immersive experience. In contrast, it is expected that 6G will be able to jointly meet all the stringent network demands (e.g., ultra-high reliability, capacity, efficiency, and low latency) in a holistic fashion, in view of the foreseen economic, social, technological, and environmental context of the 2030 era. Some key prospects and applications of 6G wireless communication are briefly described below.

Full and immersive experience. This application aims at providing to humans immersive experience with machine/things, making the human-to-machine communication seamless. To this aim, a digital real-time experience that mimics the full resolution of human perception is needed. Extended reality, five-sense communication and haptic communication will be essential for the realization of this full experience. Extended Reality (XR) services, including Augmented Reality (AR), Mixed Reality (MR), and Virtual Reality (ViR), use 3D objects and artificial intelligence as key driving elements. ViR uses headsets to generate realistic sensations and replicate a real environment or create an imaginary world. AR is a live view of a physical real world, whose elements are augmented by various computer-generated sensor inputs; it uses the existing reality and adds to it by using a device of some sort. MR merges the real and the virtual worlds to create new atmospheres and visualizations to interact in real-time. With MR the artificial and real world contents can respond to one another in real-time. XR refers to all combined real and virtual environments and human-machine interactions generated by computer technology and wearables. It brings together AR, ViR, and MR. XR will require a data rate above 1 Tbps, as opposed to the 20 Gbps target defined for 5G. Additionally, to meet the latency requirements that enable real-time user interaction in the immersive environment the per-user data rate needs to touch the Gbps, in contrast to the more relaxed 100 Mbps 5G target. A true XR environment engages all five senses, requesting the *communication and transfer of information related to the five senses* of hearing, sight, taste, smell, and touch. This technology uses the neurological process through sensory integration. It detects the sensations from the human body and the environment and uses the body effectively within the environment and local circumstances. The research in this field is at a very early stage. Finally, a full experience involves the *haptic communication*, that is a branch of nonverbal communication that uses the sense of touch: remote users will be able to enjoy haptic experiences through real-time interactive systems. The implementation of haptic systems and applications will be facilitated by the superior features of 6G communication networks.

Industry 4.0. The industry today undergoes a major transformation due to the increasing role of the new technologies - robotics, the IoT, and AI. This revolution, which is often referred to as Industry 4.0, will make it possible to increase productivity and safety, ensure the quality of products, and contribute to reducing material and energy wastage. Automation comes with its own set of requirements in terms of reliable and real-time communication. For example, high-precision manufacturing requires very high reliability - up to the order of 5 nines or more - and extremely low latency - in the order of 0.1 to 1 ms of round trip time. Furthermore, industrial control networks require real-time data transfer and strong determinism, which translates into a very low delay jitter, in the order of 1 μs. While existing technologies, such as LoRa, NB-IoT, and even 5G, could be useful for monitoring purposes, they have issues when it comes to controlling loops. This calls for the design of novel transmission techniques, for example moving to the use of sub-THz and THz bands (allowing extreme high-data rates and low-latency), and the need a precise knowledge of the radio channel in such harsh environments.

eHealth. 6G will revolutionize the health-care sector, eliminating time and space barriers through remote surgery and guaranteeing health-care workflow optimizations. Besides the high cost, the current major limitation is the lack of real-time tactile feedback, together with the challenge to meet the stringent requirements, that are continuous connection availability, ultra low latency (sub-ms), and mobility support. In addition, absorption of radio signals by living tissues, including human and animals, from radio devices operating in the proximity of or inside a body is another concern that hinders wide-spread deployment of these applications.

Unmanned mobility. The evolution towards fully autonomous transportation systems offers safer traveling, improved traffic management, and support for infotainment, with a market of 7 trillion USD. Future transportation scenarios will be characterized by high mobility and involve cars, trains, and unmanned aerial vehicles flying at low altitudes. Connecting autonomous vehicles demands unprecedented levels of reliability and low latency (i.e., above 7 nines and below 1 ms, respectively), even in ultra-high mobility scenarios (up to 1000 km/h), to guarantee passenger safety, a requirement that is hard to satisfy with existing technologies. In addition, the increasing number of sensors per vehicle will demand very-high data rates (e.g., Terabytes generated per driving hour), beyond current network capacity. The 6G system will promote the real deployment of self-driving cars (autonomous cars or driverless cars). A self-driving car perceives its surroundings by combining a variety of sensors, such as light detection and ranging, radar, Global Positioning System (GPS), sonar, odometry, and inertial measurement units. The 6G system will support reliable vehicle-to-everything and vehicle-to-server connectivity. Also UAVs, a.k.a. drones, represent a huge potential for various scenarios. Swarms of drones may be used to provide network connectivity and capacity when and where needed. The ground-based controller and the system communications between the UAVs and the ground will be supported by 6G networks. Full 3D vision of the networks will have to be handled, increasing complexity in comparison to the mainly 2D networks considered in 5G.

10.2.2 Technologies

The large range of new applications will motivate research in a wide range of new technologies, both to improve the efficiency of existing concepts, and to arrive at completely new structures. We suggest the following in a (necessarily incomplete) list of interesting topics that is arranged to progress from the physical layer to the higher layers:

- *New spectral bands.* While mmWave systems were a hallmark of 5G, it can be expected that 6G will exploit even larger swaths of hitherto unused spectrum. The spectrum bands between 140 and 1000 GHz offer the bandwidths required for applications requiring hundreds of Gbit/s. Related research will range from channel measurement, to hardware technology that can efficiently generate such frequencies, to algorithms for multi-antenna technology in those ranges. Going even higher in frequency, free-space optics based on either Light Emitting Diodes (LEDs) or lasers is promising for extremely high data rate applications.
- *New hardware for information processing.* With the end of Moore's law and the impossibility to evacuate more heat from the electronic components, new chip designs have been proposed (e.g., neuro inspired like spiking neural networks or quantum computing). Such components will modify the way to process information and could significantly impact the communication networks.
- *Spectrum adaptivity and aggregation.* The usage of spectrum will be made more efficient by adaptive use of the available spectrum. For example, each UE might be

able to transmit in different bands, but will select the most suitable spectrum based on the propagation conditions, existing interference, and interference to existing primary users. This will differ from standard cognitive radio in the ability to exploit multiple bands with different propagation characteristics and beamforming capabilities.

- *New modulation formats and coding approaches.* While research for 5G in this area has mostly concentrated on variants of OFDM, the requirements of new applications for extreme broadband (sampling and processing with hundreds of Gsamples/s) and extremely low energy consumption merits the investigation of completely new modulation formats and detection methods. The transceiver structures need to be considered holistically, involving investigations into clock-free architectures and backscatter communications. On the coding side, codes with short block lengths deserve further attention, as the polar codes used in 5G might not be sufficient. Since in delay-sensitive applications retransmissions might not be permissible, residual error rate and latency also need to be traded off with each other.

- *New spatial multiplexing methods.* Massive MIMO, usually realized with hybrid beamforming, was a major factor in 5G. For 6G, a number of new topics arise for the spatial multiplexing of data streams. For the field of massive MIMO, promising avenues for use of low-resolution Analog-to-Digital Converters (ADCs) need to be further pursued. For the reduction of complexity, index modulation, such as spatial modulation, is also worth further exploration. For high-speed LOS connections over relatively short distances, Orbital Angular Moment technology is well suited for multiplexing of data streams. While proof-of-principle implementations exist for mmWave and optical frequencies, various implementation aspects and practical optimization require more work. Another rapidly emerging area of interest is *large intelligent surfaces*, which could, e.g., occupy a whole wall.

- *Reduction of channel estimation effort.* The acquisition, and possible feedback, of CSI and other control information may consume a considerable percentage of the radio resources for a user. This is especially true for massive MIMO, high-speed vehicles, and trains, but also for IoT systems where the transmitted packets are short and therefore the CSI acquisition overhead becomes *relatively* important. A variety of overhead reduction methods have been proposed, ranging from exploiting low-rank signal space, to combined CSI acquisition and demodulation, to the use of noncoherent or differential modulation that largely obviates the need for CSI; further refinement of these ideas, and derivations of other suitable ones should be part of 6G research.

- *Physical-layer security and quantum communications.* Security has emerged as one of the most important topics for IoT. In addition to traditional cryptographic methods, *physical-layer security* has become an important topic for keeping information safe from adversaries. While the information-theoretic basis of physical-layer security has been well established, the practical implementation is currently lacking. Authentication is also important and can exploit the hardware imperfections of RF devices. These imperfections can be extracted from transient or steady state received signals. Many different fingerprints can be obtained thanks to clock shifts, digital-to-analog converters, power amplifiers, or RF oscillators... Furthermore, a key variable, namely the channel strength between transmitter and snooper, is generally unknown, and schemes that can improve security under those circumstances, are needed. On the other hand, quantum communications are provably secure, and research into their realization at optical frequencies, as well as adaptation to lower frequencies, will be important for 6G.

- *Integrated sensing and communications.* Industrial, vehicular, and IoT applications often need to sense the environment and then communicate the results to other nodes. While such sensor networks have been explored since the early 2000s, a new emphasis lies in the integration of the communications and the sensing functionality. This is especially true for radar-based sensing, since channel estimation based on pilot tones is a form of bi-static radar. Similarly, Lidar (which is often used in self-driving vehicles) can be naturally combined with optical communications.

- *Massive multiple access.* Traditional multiple access schemes rely on orthogonality of users in time, frequency, code, and/or space. NOMA provides a new paradigm for multiple access, relying on different strength of signals to separate users (note that NOMA is really based on the pioneering work of Poor and Verdu on multi-user detection in the 1980s). Another important question is how to provide access to thousands or more devices, each of which might have only a very low data rate, to a single base station. Traditional orthogonalization of the users is not feasible, since contention-free access is inefficient when the duty cycle of the UEs is very low, while Carrier Sense Multiple Access (CSMA)-like schemes are not efficient when the SNR between devices is low, as can be anticipated for many IoT applications. New schemes like SigFox and LoRa have been proposed, but a search for more efficient methods will be a cornerstone of 6G.

- *Cell-free massive MIMO.* The continuous densification of networks has made inter-cell interference the dominant factor limiting the throughput per user. The idea of having a UE connected to several of the infrastructure nodes that are closest to it, without the formation of explicit cell boundaries within which, and between which, the UE moves, promises to reduce the interference and at the same time make the SINR and thus rate more uniform. While similar concepts have been explored in the past, under the name of base station cooperation, network MIMO, and cloud-RAN, the implementation that minimizes overhead and avoids creating "cluster boundaries" at which SINR would be low promises greater gains.

- *Vehicular and aerial networks.* One of the most important applications for 6G will be vehicular networks. While V2I was already considered a goal for 5G, progress has been slow, and is mostly concentrated on creating low-data-rate links that can carry some fundamental control information. The transmission of raw sensor data, in particular camera and Lidar images, between cars, will be a topic for 6G. The networking between many, highly mobile, nodes will present significant challenges from physical to network layer. Similar issues will arrive in aerial networks, which encompass both the communication between ground and UAVs for the purpose of communicating control and sensor information, and for setups in which

UAVs could serve as flying base stations or relay stations. The joint optimization of flight patterns and resource allocation will require significant research.

- *Integrated communication, computing, and caching.* Many of the services for mobile devices require extensive computations, and the computational capabilities of those devices are often not sufficient - cases in point are video games, as well as control for factory automation. This has given rise to Mobile Edge Computing (MEC), where a device offloads computations to edge servers. A more general version is known as Augmented Information services, where the source and destination of the information need not be the same, and the necessary computations can be performed on multiple devices that lie along a multi-hop route over which the information is forwarded. The load of computation and communication can be reduced by caching of information close to the destination - current edge caching investigations mostly concentrate on wireless video files, but in a more general context, the "3C" - computation, communication, and caching - should be optimized in a joint manner, and take into account a variety of constraints such as energy consumption, latency, and cost for hardware.

Some final conclusions: Communication is an essential means to make things happen. It is worth remembering how fundamentally wireless communications have changed our lives- from interactions between people to availability of content. These concluding words are written in the middle of a pandemic, where communications via cellphones and WiFi are a lifeline - both figuratively and literally - for hundreds of millions of people. In the future, wireless communications will have an even broader purview, enabling new human-machine and machine-to-machine interactions that will make our lives more convenient and contribute to better well-being.

Even though so much has been done in the past 5 years, including significant contributions from IRACON members, much more remains to be done so that networks can deliver a completely new experience for the relationship between humans or between humans and the physical world. Due to constantly evolving, and always more challenging, requirements and applications, wireless is far from "mature". The COST actions, ranging from IRACON, to IC 1004, COST 2100, COST 273, COST 259, COST 231, all the way back to COST 207, have played an essential role in fostering this research, crossing national boundaries and opening research and technology for the benefit of all participants, as well as companies and citizens. The role that COST has played in creating collaborations and synergies across national borders cannot be overemphasized. The research and cooperation between European countries, and between Europe and other countries, are key elements of these crucial technologies at the heart of the profound transformations experienced by our society, ranging from digital content to industrial, urban, environmental, mobility, energy, and many more applications.

Bibliography

[36.15] TR 36.873, Study on 3D channel model for LTE, Technical report, June 2015, Technical Report, V12.2.0, 3GPP.

[3GP08a] 3GPP, User equipment (UE) / mobile station (MS) over the air (OTA) antenna performance; conformance testing, Technical Report 3GPP TS 34.144, 3GPP, 2008.

[3GP08b] 3GPP, User equipment (UE) and mobile station (MS) over the air performance requirements, Technical Report 3GPP TS 25.144, 3GPP, 2008.

[3GP09] 3GPP, R4-092042, Simulation assumptions and parameters for FDD HeNB RF requirements, 3GPP TSG RAN WG4 Meeting 51. TR R4-092042, 3rd Generation Partnership Project (3GPP), 2009.

[3GP10] 3GPP, Technical Specification Group Radio Access Network, Technical Report 3GPP TR 36.814, Version 9.0.0, 3GPP, Valbonne, France, March 2010, http://www.3gpp.org/dynareport/36814.htm. (Accessed 5 May 2016).

[3GP16] 3GPP, 3GPP TR 22.891: technical specification group services and system aspects; feasibility study on new services and markets technology enablers; Stage 1; (Release 14), Technical report, ETSI, March 2016.

[3GP17a] 3GPP, 3GPP TR 38.913: Study on Scenarios and Requirements for Next Generation Access Technologies; (Release 14), Technical report, ETSI, June 2017.

[3GP17b] 3GPP, Study on channel model for frequencies from 0.5 to 100 GHz, Technical Report 3GPP TR 38.901, 3GPP RAN, 2017.

[3GP17c] 3GPP, Universal terrestrial radio access (e-UTRA); user equipment (UE) over the air (OTA) antenna performance; conformance testing, Technical Report TS 37.544 V14.1.0, 3GPP, April 2017.

[3GP18] 3GPP, 5G New Radio - study on test methods, Technical Report 3GPP TR 38.810, 3GPP RAN4, 2018.

[3GP19] 3GPP, User Equipment (UE) radio transmission and reception; Part 2: Range 2 Standalone (Release 16), Technical Report 3GPP TS 38.101-2 V16.1.0, 3GPP RAN4, October 2019.

[5GP15] 5GPP, 5G automotive vision, 5G Public Private Partnership, Oct. 2015.

[80211] IEEE Standard for Local and metropolitan area networks–Part 15.4: Low-Rate Wireless Personal Area Networks (LR-WPANs), IEEE Std 802.15.4-2011 (Revision of IEEE Std 802.15.4-2006), September 2011, pp. 1–314.

[AAC+17] I. Al-Samman, M. Artuso, H. Christiansen, A. Doufexi, M. Beach, Envisioning spectrum management in virtualised C-RAN, in: 2017 IEEE Wireless Communications and Networking Conference (WCNC), March 2017, pp. 1–6. Also presented as TD(16)02040.

[AB17a] Agota Antal, Vasile Bota, Performance analysis of a generic cooperative HARQ algorithm over Rayleigh faded channels, in: COST-IRACON, 2017, TD05032.

[AB17b] Agota Antal, Vasile Bota, Performance analysis of a repetition redundancy HARQ algorithm with decode-forward two-hop relaying over Rayleigh channels, in: COST-IRACON, 2017, TD04015.

[AB18] S.L. Ambroziak, C. Buratti, IRACON Experimental Facilities Report, Technical report, March 2018.

[ABC+14] Jeffrey G. Andrews, Stefano Buzzi, Wan Choi, Stephen V. Hanly, Angel Lozano, Anthony C.K. Soong, Jianzhong Charlie Zhang, What will 5G be?, IEEE Journal on Selected Areas in Communications 32 (6) (2014) 1065–1082.

[ABK+17] J.G. Andrews, T. Bai, M.N. Kulkarni, A. Alkhateeb, A.K. Gupta, R.W. Heath, Modeling and analyzing millimeter wave cellular systems, IEEE Transactions on Communications 65 (1) (January 2017) 403–430.

[ABLH19] K.A. Al Mallak, M. Beach, T. Loh, G. Hilton, Influence of antenna beamwidth on the observed characteristics of an outdoor vehicular, Technical Report TD(19)11056, Gdańsk, Poland, September 2019.

[ACBL18a] M.Z. Aslam, Y. Corre, E. Björnson, E.G. Larsson, Large-scale massive MIMO network evaluation using ray-based deterministic simulations, in: 2018 IEEE 29th Annual International Symposium on Personal, Indoor and Mobile Radio Communications (PIMRC), Bologna, Italy, Sep. 2018, pp. 1–5. Also available as TD(18)08045.

[ACBL18b] Mohammed Zahid Aslam, Yoann Corre, Emil Bjornson, Erik G. Larsson, Using Ray-based Deterministic Simulations for Large-scale System Level Evaluation of a Massive MIMO Network, Technical Report TD-18-08045, Podgorica, Montenegro, October 2018.

[ACK+16] Slawomir J. Ambroziak, Luis M. Correia, Ryszard J. Katulski, Michal Mackowiak, Carla Oliveira, Jaroslaw Sadowski, Kenan Turbic, An off-body channel model for body area networks in indoor environments, IEEE Transactions on Antennas and Propagation 64 (9) (September 2016) 4022–4035.

[ACL17] M.Z. Aslam, Y. Corre, Y. Lostanlen, Effect of human crowd obstruction on the performance of an urban small-cell millimeter-wave access network, in: 2017 IEEE 86th Vehicular Technology Conference (VTC-Fall), Sep. 2017, pp. 1–5.

[ACS18] R. Aleksiejunas, A. Cesiul, K. Svirskas, Spatially consistent LOS/NLOS model for time-varying MIMO channels, in: 2018 Baltic URSI Symposium (URSI), May 2018, pp. 61–64. Also available as TD(18)06027.

[ACT16a] Slawomir J. Ambroziak, Luis M. Correia, Kenan Turbic, Radio channel measurements in body-to-body communications in different scenarios, in: Proc. URSI AP-RASC 2016 - URSI Asia-Pacific Radio Science Conference, Seoul, Korea, August 2016. Also available as TD(16)01003.

[ACT16b] Slawomir J. Ambroziak, Luis M. Correia, Kenan Turbic, Radio Channel Measurements in Body-to-Body Communications in Different Scenarios, Technical Report TD(16)01003, Lille, France, May 2016.

[ADMO+16] Ehsan Ahvar, Nafiseh Daneshgar-Moghaddam, Antonio M. Ortiz, Gyu Myoung Lee, Noel Crespi, On analyzing user location discovery methods in smart homes: a taxonomy and survey, Journal of Network and Computer Applications 76 (2016) 75–86.

[AFPC15] Javier Arribas, Carles Fernandez-Prades, Pau Closas, GESTALT: a testbed for experimentation and validation of GNSS software receivers, in: Proceedings of the ION GNSS+ 2015, Tampa (FL), USA, Sept. 2015.

[AGBFM15] Alejandro Aguilar-Garcia, Raquel Barco, Sergio Fortes, Pablo Muñoz, Load balancing mechanisms for indoor temporarily overloaded heterogeneous femtocell networks, EURASIP Journal on Wireless Communications and Networking 2015 (1) (Feb. 2015) 29.

[AGC+18] Carlos Andreu, Concepcion Garcia-Pardo, Sergio Castello-Palacios, Ana Valles-Lluch, Narcis Cardona, Frequency dependence of UWB in-body radio channel characteristics, IEEE Microwave and Wireless Components Letters 28 (4) (April 2018) 359–361. Also available as TD(18)07014.

[AGE19] Jens Abraham, Golsa Ghiaasi, Torbjorn Ekman, Characterisation of Channel Hardening Using the Diversity Order of the Effective Channel, Technical Report TD-19-09072, Dublin, Ireland, January 2019.

[AGF+16] Carlos Andreu, Concepcion Garcia-Pardo, Alejandro Fornes-Leal, Narcis Cardona, Sergio Castello-Palacios, Ana Valles-Lluch, Spatial in-body channel characterization using an accurate UWB phantom, IEEE Transactions on Microwave Theory and Techniques 64 (11) (November 2016) 3995–4002. Also available as TD(17)03025.

[AGFCB15] A. Aguilar-Garcia, S. Fortes, E. Colin, R. Barco, Enhancing RFID indoor localization with cellular technologies, EURASIP Journal on Wireless Communications and Networking 2015 (1) (2015).

[AGFDB16] A. Aguilar-Garcia, S. Fortes, A. Fernández Durán, R. Barco, Context-aware self-optimization: evolution based on the use case of load balancing in small-cell networks, IEEE Vehicular Technology Magazine 11 (1) (March 2016) 86–95.

[AGFG+16] A. Aguilar-Garcia, S. Fortes, A. Garrido, A. Fernández Durán, R. Barco, Improving load balancing techniques by location awareness at indoor femtocell networks, EURASIP Journal on Wireless Communications and Networking 2016 (2016).

[AGFMG+15] A. Aguilar-Garcia, S. Fortes, Mariano Molina-García, Jaime Calle-Sánchez, José I. Alonso, Aaron Garrido, Alfonso Fernández-Durán, R. Barco, Location-aware self-organizing methods in femtocell networks, Computer Networks 93 (Part 1) (2015) 125–140. Also available as TD(17)05052.

[AHM19a] C. Antón-Haro, X. Mestre, Learning and data-driven beam selection for mmWave communications: an angle of arrival-based approach, IEEE Access 7 (2019) 20404–20415. Also available as TD(18)08042.

[AHM19b] Carles Anton-Haro, Xavier Mestre, Learning approaches to beam selection for hybrid analog beamforming, in: COST-IRACON, 2019, TD09005.

[AHZ+12] Bo Ai, Ruisi He, Zhangdui Zhong, Ke Guan, Binghao Chen, Pengyu Liu, Yuanxuan Li, Radio wave propagation scene partitioning for high-speed rails, International Journal of Antennas and Propagation 2012 (2012) 815232, 7 pp. Also available as TD(16)01010.

[AKGH] G. Artner, W. Kotterman, G. Galdo, M.A. Hein, Conformal Automotive Antennas in the Roof for Vehicle-to-X Communications, Technical Report TD(18)08020, Podgorica, Montenegro, October.

[AKMZ17a] G. Artner, J. Kowalewski, C.F. Mecklenbräuker, T. Zwick, Automotive pattern reconfigurable antennas concealed in a chassis cavity, in: Proc. International Symposium on Wireless Personal Multimedia Communications (WPMC 2017), Yogyakarta, Indonesia, December 2017, pp. 265–270. Also available as TD(18)06002.

[AKMZ17b] G. Artner, J. Kowalewski, C.F. Mecklenbräuker, T. Zwick, Pattern reconfigurable antenna with four directions hidden in the vehicle roof, in: Proc.

International Workshop on Antenna Technology (iWAT 2017), Athens, Greece, March 2017, pp. 82–85. Also available as TD(17)03002.

[AKMZ19] G. Artner, J. Kowalewski, C.F. Mecklenbräuker, T. Zwick, Electronically steerable parasitic array radiator flush-mounted for automotive LTE, in: Proc. European Conference on Antennas and Propagation (EuCAP 2019), Krakow, Poland, March 2019. Also available as TD(19)09009.

[AKS17] H. Ahmadi, K. Katzis, M. Zeeshan Shakir, A novel airborne self-organising architecture for 5G+ networks, in: 2017 IEEE 86th Vehicular Technology Conference (VTC-Fall), 2017, pp. 1–5. Also available as TD(17)05041.

[Ale17a] R. Aleksiejunas, MIMO channel reconstruction from lower dimensional multiple antenna measurements, Wireless Personal Communications 96 (1) (Sep 2017) 543–562. Also available as TD(16)01034.

[Ale17b] Rimvydas Aleksiejunas, Spatially consistent LOS/NLOS state model for statistical Monte Carlo simulations, Technical report, Lille, France, May 2017, TD(17)04039.

[ALU16] Nicolas Amiot, Mohamed Laaraiedh, Bernard Uguen, An advanced graph-based ray tracing method for radio propagation modelling and localization, Technical report, Lille, France, May 2016, TD(16)01049.

[AMG⁺16] H. Ahmadi, I. Macaluso, I. Gomez, L. DaSilva, L. Doyle, Virtualization of spatial streams for enhanced spectrum sharing, in: 2016 IEEE Global Communications Conference (GLOBECOM), Dec 2016, pp. 1–6. Also presented as TD(16)02008.

[And96] Makoto Ando, Physical optics, in: E. Yamashita (Ed.), Analysis Methods for Electromagnetic Wave Problems, vol. 2, 1996.

[AO18] Omar Ahmadien, Mehmet Kemal Ozdemir, FM Band Channel Modelling and Measurements, Technical Report TD(18)07021, Cartagena, Spain, May 2018.

[Aoy16] Takahiro Aoyagi, Consideration of Directivity of Antennas for High Frequency Wireless Body Area Networks During Human Movements, Technical Document TD(16)02019, COST Action CA15104 (IRACON), Durham, UK, October 2016.

[AP18a] Ramoni Adeogun, Troels Pedersen, Modelling polarimetric power delay spectrum for indoor wireless channels via propagation graph formalism, in: 2018 2nd URSI Atlantic Radio Science Meeting (AT-RASC), May 2018. Also available as TD(18)07046.

[AP18b] Ramoni Adeogun, Troels Pedersen, Propagation graph based model for polarized multiantenna wireless channels, in: 2018 IEEE Wireless Communications and Networking Conference (WCNC), Apr. 2018. Also available as TD(18)06020.

[APB18] Ramoni Adeogun, Troels Pedersen, Ayush Bharti, Transfer function computation for complex indoor channels using propagation graphs, in: 2018 IEEE 29th Annual International Symposium on Personal, Indoor and Mobile Radio Communications (PIMRC), Bologna, Italy, Sep. 2018, pp. 566–567. Also available as TD(18)08038.

[ARG17] M.L.M. Altozano, S.L. Ramirez, M.T. Genovés, Load balance performance analysis with a quality of experience perspective in LTE networks, Technical Report TD(17)05003, IRACON Cost Action CA15104, Lisbon, Portugal, February 2017.

[Art19a]   G. Artner, Channel static antennas, Technical Report TD(19)10014, Oulu, Finland, May 2019.

[Art19b]   G. Artner, Channel static antennas for compensating the movements of a partner antenna, Technical Report TD(19)10026, Oulu, Finland, May 2019.

[Art19c]   G. Artner, Channel static antennas for mobile devices, Technical Report TD(19)10061, Oulu, Finland, May 2019.

[AS17]   Alaa Alameer, Aydin Sezgin, Joint beamforming and network topology optimization of green cloud radio access networks, in: COST-IRACON, 2017, TD03010.

[ASAAA17]   S. Al-Sarawi, M. Anbar, K. Alieyan, M. Alzubaidi, Internet of things (IoT) communication protocols: review, in: 2017 8th International Conference on Information Technology (ICIT), May 2017, pp. 685–690.

[AT18]   Jesus Alonso-Zarate, Carles Anton-Haro, Achilleas Tsitsimelis, Charalampos Kalalas, Suitability of LTE Random Access Schemes for State Estimation in Smart Electricity Grids, Technical Report TD(18)06003, Nicosia, Cyprus, 2018.

[ATA$^+$18]   Reza Aminzadeh, Arno Thielens, Sam Agneessens, Patrick Van Torre, Matthias Van den Bossche, Stefan Dongus, Marloes Eeftens, Anke Huss, Roel Vermeulen, René de Seze, Paul Mazet, Elisabeth Cardis, Hendrik Rogier, Martin Röösli, Luc Martens, Wout Joseph, A multi-band body-worn distributed radio-frequency exposure meter: design, on-body calibration and study of body morphology, Sensors 18 (1) (jan 2018) 272.

[ATC17]   Slawomir J. Ambroziak, Kenan Turbic, Luis M. Correia, An approach to mean path loss model estimation for off-body channels, in: Proc. IS-MICT'17 - 11th International Symposium on Medical Information and Communication Technology, Lisbon, Portugal, February 2017. Also available as TD(17)03006.

[ATC19]   Slawomir J. Ambroziak, Kenan Turbic, Luis M. Correia, Empirical Validation of the Polarised Off-Body Channel Model with Dynamic Users, Technical Document TD(19)10018, COST Action CA15104 (IRACON), Oulu, Finland, May 2019.

[AVB$^+$19]   M.J. Arpaio, E.M. Vitucci, M. Barbiroli, V. Degli-Esposti, D. Masotti, F. Fuschini, Ray-launching narrowband analysis of UAV-to-ground propagation in urban environment, in: 2019 International Symposium on Antennas and Propagation (ISAP), Oct. 2019. Also available as TD(19)10055.

[AWM14]   A. Asadi, Q. Wang, V. Mancuso, A survey on device-to-device communication in cellular networks, IEEE Communications Surveys and Tutorials 16 (4) (April 2014) 1801–1819.

[BA02]   Kenneth P. Burnham, David R. Anderson, Model Selection and Multimodel Inference, 2nd edition, Springer, New York, NY, USA, 2002.

[BAF17]   Estefania Crespo Bardera, Ana Garcia Armada, Matilde Sanchez Fernandez, Achievable rates and applications of a textile massive MIMO hub, in: COST-IRACON, 2017, TD03050.

[BB18]   Manijeh Bashar, Alister G. Burr, Cell-free massive MIMO with limited backhaul, in: COST-IRACON, 2018, TD07071.

[BBB14]   Paolo Baracca, Federico Boccardi, Nevio Benvenuto, A dynamic clustering algorithm for downlink CoMP systems with multiple antenna UEs, EURASIP Journal on Wireless Communications and Networking 2014 (1) (Aug 2014) 125.

[BBB17]   P.D. Bojović, Ž. Bojović, D. Bajić, V. Šenk, IP session continuity in het-
          erogeneous mobile networks using software defined networking, Journal of
          Communications and Networks 19 (6) (Dec 2017) 563–568. Also available
          as TD(18)06044.

[BBC17]   Manijeh Bashar, Alister G. Burr, Kanapathippillai Cumanan, Low complex-
          ity scheduling and beamforming in massive MIMO systems with COST
          2100 channel model, in: COST-IRACON, 2017, TD04030.

[BBG+18] T. Blazek, T. Berisha, E. Gashi, B. Krasniqi, C.F. Mecklenbrauker,
          A Stochastic Performance Model for Dense Vehicular Ad-Hoc Networks,
          Technical Report TD-08-027, Podgorica, Montenegro, October 2018.

[BBG+19] T. Blazek, T. Berisha, E. Gashi, B. Krasniqi, C.F. Mecklenbräuker,
          A stochastic performance model for dense vehicular Ad-Hoc networks, in:
          13th European Conference on Antennas and Propagation (EuCAP 2019),
          Krakow, Poland, March 2019. Also available as TD(18)08027.

[BBM16]   Henry Brice, Mark Beach, Evangelos Mellios, Massive MIMO Propaga-
          tion Models, Technical Report TD-16-02015, Durham, United Kingdom,
          October 2016.

[BBM18]   Alister Burr, Manijeh Bashar, Dick Maryopi, Cooperative access networks:
          optimum fronthaul quantization in distributed massive MIMO and cloud
          RAN, in: COST-IRACON, 2018, TD06038.

[BBS19]   T. Balan, A. Balan, F. Sandu, SDR implementation of a D2D security cryp-
          tographic mechanism, IEEE Access 7 (2019) 38847–38855. Also available
          as TD(19)10031.

[BC17]   Alister Burr, Cheng Chen, BER performance of convolutionally coded
          BPSK layered hierarchical decode and forward relaying, in: COST-
          IRACON, 2017, TD03062.

[BCF18a] M. Barahman, L.M. Correia, L.S. Ferreira, A fair computational resource
          management strategy in C-RAN, in: 2018 International Conference on
          Broadband Communications for Next Generation Networks and Multime-
          dia Applications, Graz, Austria, July 2018. Also available as TD(18)06040.

[BCF18b] M. Barahman, L.M. Correia, L.S. Ferreira, A Real-time Computational
          Resource Management in C-RAN, Technical Report TD(18)07064, Carta-
          gena, Spain, May 2018.

[BCF19a] M. Barahman, L.M. Correia, L.S. Ferreira, Optimizing Computational Re-
          sources Usage in C-RAN, Technical Report TD(19)10037, Oulu, Finland,
          May 2019.

[BCF19b] M. Barahman, L.M. Correia, L.S. Ferreira, A Real-time Computational
          Resource Usage Optimization in C-RAN, Technical Report TD(19)11018,
          Gdańsk, Poland, September 2019.

[BCX17]   T. Brown, P. Chambers, R. Xie, Indicative measurements of mmWave build-
          ing penetration and exploitation of gaps, Technical Report TD(17)04036,
          Lund, Sweden, May 2017.

[Ber95]   J.E. Berg, A recursive method for street microcell path loss calculations, in:
          Proc. Sixth IEEE International Symposium on Personal, Indoor and Mobile
          Radio Communications (PIMRC'95), vol. 1, Sep. 1995, pp. 140–143.

[Ber17]   Taulant Berisha, Measurement and analysis of LTE coverage for vehicular
          use cases in live networks, Technical Report TD(17)04057, Lund, Sweden,
          May 2017.

[BGB+19] T. Blazek, G. Ghiaasi, C. Backfrieder, G. Ostermayer, C.F. Mecklenbräuker, Performance modeling and analysis for vehicle-to-anything connectivity in representative high-interference channels, IET Microwaves, Antennas and Propagation (2019) 1–9. Also available as TD(19)09006.

[BGLF+16] T.H. Barratt, A.A. Goulianos, A. Loaiza Freire, T.M. Stone, E. Mellios, P. Cain, A.R. Nix, M.A. Beach, Non-specular diffuse scattering measurements and estimation of material permittivity at millimetre wave frequencies, Technical Report TD-16-01081, Lille, France, May 2016.

[BGN+18] M. Barbi, C. Garcia-Pardo, A. Nevarez, V. Ponds, N. Cardona, UWB RSS-based localization for capsule endoscopy using a multilayer phantom and in-vivo measurements, in: IRACON 7th MC and Technical Meeting - Cartagena 2018, May 2018, pp. 1–9. Also available as TD(18)05024.

[BHB18] M. Bashar, K. Haneda, A.G. Burr, Spatial consistency of clusters in mm-Wave ray-tracing results for 5G communications, Technical Report TD(18)07056, Cartagena, Spain, June 2018.

[BHK18] A. Burr, K. Haneda, A. Karttunen, On clustering in multipath channel models, Technical Report TD(18)07047, Cartagena, Spain, June 2018.

[BHL+14] Federico Boccardi, Robert W. Heath, Angel Lozano, Thomas L. Marzetta, Petar Popovski, Five disruptive technology directions for 5G, IEEE Communications Magazine 52 (2) (2014) 74–80.

[BHM16] K. Bregar, A. Hrovat, M. Mohorčič, NLOS channel detection with multilayer perceptron in low-rate personal area networks for indoor localization accuracy improvement, in: Proceedings of the 8th Jožef Stefan International Postgraduate School Students' Conference, Ljubljana, Slovenia, 2016. Also available as TD(17)04050.

[Bit19] J. Bito, Moving interference source prediction for millimetre wave mesh network, Technical Report TD-09-076, Dublin, Ireland, January 2019.

[BK16] J. Baumgarten, T. Kurner, Device-To-Device Graph-Oriented Resource Allocation in LTE Uplink using SC-FDMA, Technical Report TD(16)02058, IRACON Cost Action CA15104, Durham, United Kingdom, October 2016.

[BK19] T. Brown, M. Khalily, The importance of far field effect in diffraction at mmWave, Technical Report TD(19)09047, Dublin, Ireland, jan 2019.

[BLA+ed] T. Blazek, M. Lindorfer, M. Ashury, G. Ostermayer, C.F. Mecklenbräuker, Scenario performance classification for highway vehicle-to-vehicle communications, Technical Report TD(19)11023, IRACON COST Action CA15104, 2019.

[BLM+14] Naga Bhushan, Junyi Li, Durga Malladi, Rob Gilmore, Dean Brenner, Aleksandar Damnjanovic, Ravi Sukhavasi, Chirag Patel, Stefan Geirhofer, Network densification: the dominant theme for wireless evolution into 5G, IEEE Communications Magazine 52 (2) (2014) 82–89.

[BM18] T. Blazek, C.F. Mecklenbräuker, Measurement-based burst-error performance modeling for cooperative intelligent transport systems, IEEE Transactions on Intelligent Transportation Systems (January 2018) 1–10. Also available as TD(17)03011.

[BMB17] Henry Brice, Evangelos Mellios, Mark Beach, Models for massive MIMO mobility scenarios, in: COST-IRACON, 2017, TD05045.

[BMD16] A. Bamba, F. Mani, R. D'Errico, E-band millimeter wave indoor channel characterization, in: 2016 IEEE 27th Annual International Symposium on

Personal, Indoor, and Mobile Radio Communications (PIMRC), Sep. 2016, pp. 1–6.

[BMG+17] J. Blumenstein, R. Marsalek, T. Gotthans, R. Nissel, M. Rupp, On mutual information of measured 60 GHz wideband indoor MIMO channels: time domain singular values, in: 2017 IEEE 28th Annual International Symposium on Personal, Indoor, and Mobile Radio Communications (PIMRC), October 2017, pp. 1–5. Also available as TD(17)05019.

[BNMJ19] Klemen Bregar, Roman Novak, Mihael Mohorcic, Tomaž Javornik, Combining measurements and simulations for evaluation of tracking algorithms, in: COST IRACON (CA15104) 9th Scientific Meeting, TD(19)09032, Dublin, Ireland, Jan 16-19 2019, pp. 1–6. Also available as TD(19)09032.

[BPG+18] M. Barbi, S. Perez-Simbor, C. Garcia-Pardo, C. Andreu, N. Cardona, Localization for capsule endoscopy at UWB frequencies using an experimental multilayer phantom, in: 2018 IEEE Wireless Communications and Networking Conference Workshops (WCNCW), April 2018, pp. 390–395. Also available as TD(17)05023.

[BPMS17] Lazar Berbakov, Bogdan Pavkovic, Vladana Markovic, Marina Svetel, Architecture and partial implementation of the remote monitoring platform for patients with movement disorders, in: Zinc Conference, 2017. Also available as TD(17)03023.

[BPN+19] Said Benaissa, David Plets, Denys Nikolayev, Margot Deruyck, Leen Verloock, Günter Vermeeren, Luc Martens, Eli De Poorter, Frank André Maurice Tuyttens, Bart Sonck, Wout Joseph, Experimental Characterization of the In-To-Out-Body Path Loss at 433 MHz for Dairy Cows, Technical Document TD(19)09020, COST Action CA15104 (IRACON), Dublin, Ireland, January 2019.

[BPT+16a] S. Benaissa, D. Plets, E. Tanghe, G. Vermeeren, L. Martens, B. Sonck, F.A.M. Tuyttens, L. Vandaele, J. Hoebeke, N. Stevens, W. Joseph, Characterization of the on-body path loss at 2.45 GHz and energy efficient WBAN design for dairy cows, IEEE Transactions on Antennas and Propagation 64 (11) (Nov 2016) 4848–4858.

[BPT+16b] Said Benaissa, David Plets, Emmeric Tanghe, Leen Verloock, Luc Martens, Jeroen Hoebeke, Bart Sonck, Frank Tuyttens, Leen Vandaele, Nobby Stevens, Wout Joseph, Experimental characterisation of the off-body wireless channel at 2.4 GHz for dairy cows in barns and pastures, Computers and Electronics in Agriculture 127 (2016) 593–605.

[Bre01] L. Breiman, Random forests, Machine Learning 45 (2) (2001) 5–32, PMID: 1573-0565.

[BRS16] T. Balan, D. Robu, F. Sandu, LISP optimisation of mobile data streaming in connected societies, in: Mobile Information System, 2016.

[BRS17] T. Balan, D. Robu, F. Sandu, Multihoming for mobile internet of multimedia things, Mobile Information Systems 2017 (2017). Also available as TD(17)03020.

[BS19] Tuncer Baykas, Mohamed Hashir Syed, Measurements and Modeling for Indoor to Outdoor Channels in TVWS Bands, Technical Report TD(19)09065, Dublin, Ireland, 2019.

[BSA15] M. Barbi, K. Sayrafian, M. Alasti, Impact of the energy detection threshold on performance of the IEEE 802.15.6 CSMA/CA, in: 2015 IEEE Conference on Standards for Communications and Networking (CSCN), Oct 2015, pp. 224–228. Also available as TD(16)02029.

[BSA16] Martina Barbi, Kamran Sayrafian, Mehdi Alasti, Low complexity adaptive schemes for energy detection threshold in the IEEE 802.15.6 CSMA/CA, in: 2016 IEEE Conference on Standards for Communications and Networking, CSCN 2016, Berlin, Germany, October 31 - November 2, 2016, 2016, pp. 237–242. Also available as TD(17)04023.

[BSL95] Yaakov Bar-Shalom, Xiao-Rong Li, Multitarget-multisensor tracking: principles and techniques, vol. 19, YBs Storrs, CT, 1995.

[BSOM17] T. Berisha, P. Svoboda, S. Ojak, C.F. Mecklenbräuker, SegHyPer: segmentation- and hypothesis based network performance evaluation for high speed train users, in: 2017 IEEE International Conference on Communications (ICC), May 2017, pp. 1–6. Also available as TD(17)03008.

[BSV18] Emanuel Teixeira, Bruno Silva, Sofia Sousa, Fernando J. Velez, Insights on spectrum sharing in heterogeneous networks with small cells, Technical Report TD:(18)07020, Cartagena, Spain, Sept. 2018.

[BTB18] L. Berbakov, N. Tomašević, M. Batić, Specification of the Internet of Things platform for energy efficient living, Technical Report TD(18)07044, Cartagena, Spain, May 2018.

[BTY+ed] B.D. Beelde, E. Tanghe, M. Yusuf, D. Plets, E.D. Poorter, W. Joseph, 60 GHz path loss modelling inside ships, in: 14th European Conference on Antennas and Propagation (EuCAP 2020), March 2020, pp. 1–5. Also available as TD(19)11035.

[Bul47] K. Bullington, Radio propagation at frequencies above 30 megacycles, Proceedings of the IRE 35 (10) (Oct. 1947) 1122–1136.

[Bur16] Alister Burr, Non-linear PNC mapping for hierarchical wireless network, in: COST-IRACON, 2016, TD02033.

[BV18a] Bota Vasile, Mihaly Varga, A minimum average-delay and maximum spectral efficiency configuration algorithm for relay-assisted HARQ transmissions under reliability and maximum-delay constraints, in: COST-IRACON, 2018, TD08021.

[BV18b] C. Buratti, R. Verdone, Joint scheduling and routing with power control for centralized wireless sensor networks, Wireless Networks 24 (5) (July 2018) 1699–1714. Also available as TD(16)01035.

[CACL17] R. Charbonnier, M.Z. Aslam, Y. Corre, Y. Lostanlen, Mixing deterministic and stochastic propagation for assessing mmWave small-cell networks, in: 2017 11th European Conference on Antennas and Propagation (EuCAP), Paris, France, Mar. 2017, pp. 136–140. Also available as TD(17)03035.

[CAMR18] M. Calvo-Fullana, C. Antón-Haro, J. Matamoros, A. Ribeiro, Stochastic routing and scheduling policies for energy harvesting communication networks, IEEE Transactions on Signal Processing 66 (13) (July 2018) 3363–3376. Also available as TD(17)03053.

[CAN18] Yor Castaño, Juan Arango, Andrés Navarro, Spatiotemporal gait variables using wavelets for an objective analysis of Parkinson disease, in: Studies in Health Technology and Informatics, vol. 249, 2018, pp. 173–178. Also available as TD(18)07061.

[Car16] N. Cardona, Cooperative Radio Communications for Green Smart Environments, River Publishers, Aalborg, Denmark, 2016.

[CAR18a] Krzysztof K. Cwalina, Slawomir J. Ambroziak, Piotr Rajchowski, An off-body narrowband and ultra-wide band channel model for body area networks in a ferry environment, Applied Sciences 8 (6) (June 2018) 1–16. Also available as TD(18)07002.

[CAR$^+$18b] Krzysztof K. Cwalina, Slawomir J. Ambroziak, Piotr Rajchowski, Jaroslaw Sadowski, Jacek Stefanski, A novel bitrate adaptation method for heterogeneous wireless body area networks, Applied Sciences 8 (7) (2018). Also available as TD(19)09017.

[CARC17] Krzysztof K. Cwalina, Slawomir J. Ambroziak, Piotr Rajchowski, Luis M. Correia, Radio channel measurements in 868 MHz off-body communications in a ferry environment, in: Proc. URSI'17 - General Assembly and Scientific Symposium of the International Union of Radio Science, Montreal, QC, Canada, August 2017. Also available as TD(17)04008.

[CB16] Cheng Chen, Alister Burr, SER analysis of QPSK modulated physical layer network coding for system-level simulation, in: COST-IRACON, 2016, TD02030.

[CB18] Krzysztof Cichon, Hanna Bogucka, Efficient Detection and Clustering Mechanisms for Spectral-Activity Information Exchange, Technical Report TD(18)08014, Podgorica, Montenegro, October 2018.

[CBC$^+$17] W. Cerroni, C. Buratti, S. Cerboni, G. Davoli, C. Contoli, F. Foresta, F. Callegati, R. Verdone, Intent-based management and orchestration of heterogeneous openflow/IoT SDN domains, in: 2017 IEEE Conference on Network Softwarization (NetSoft), July 2017, pp. 1–9. Also available as TD(17)05047.

[CBS17] A. Czylwik, S. Bieder, M. Sichma, Extreme wideband arbitrary waveform generator based on frequency multiplexing, Technical Report TD-17-04011, Universität Duisburg-Essen, Duisburg, Germany, May 2017.

[CCAL16] Y. Corre, R. Charbonnier, M.Z. Aslam, Y. Lostanlen, Assessing the performance of a 60-GHz dense small-cell network deployment from ray-based simulations, in: 2016 IEEE 21st International Workshop on Computer Aided Modelling and Design of Communication Links and Networks (CAMAD), Oct 2016, pp. 213–218.

[CCLG16] S. Chessa, A. Corradi, L. Foschini, M.E. Girolami, Mobile crowdsensing through social and ad hoc networking, IEEE Communications Magazine 54 (7) (2016) 108–114, PMID: 16141383.

[CDA$^+$17] A. Costanzo, D. Dardari, J. Aleksandravicius, N. Decarli, M. Del Prete, D. Fabbri, M. Fantuzzi, A. Guerra, D. Masotti, M. Pizzotti, A. Romani, Energy autonomous UWB localization, IEEE Journal of Radio Frequency Identification 1 (3) (Sept 2017) 228–244.

[CDG$^+$18] Denis Couillard, Ghassan Dahman, Marie-Eve Grandmaison, Gwenael Poitau, Francois Gagnon, Robust Broadband Maritime Communications: Theoretical and Experimental Validation, Technical Report TD(18)06034, Nicosia, Cyprus, 2018.

[CDMJ19] G. Castellanos, M. Deruyck, L. Martens, W. Joseph, Performance evaluation of direct-link backhaul for UAV-aided emergency networks, Sensors 19 (15) (2019). Also available as TD(19)11021.

[CFMAHR17] M. Calvo-Fullana, J. Matamoros, C. Antón-Haro, A. Ribeiro, Control-aware scheduling policies for energy harvesting sensors, September 2017. Also available as TD(17)05021.

[CGF$^+$16]   S. Castelló-Palacios, C. Garcia-Pardo, A. Fornes-Leal, N. Cardona, A. Vallés-Lluch, Tailor-made tissue phantoms based on acetonitrile solutions for microwave applications up to 18 GHz, Technical Report TD-16-02009, Universitat Politécnica de Valéncia, Valencia, Spain, October 2016.

[CGMP18]   Claudia Carciofi, Paolo Grazioso, Francesco Matera, Valeria Petrini, Study of coexistence between different services in novel 5G Frequency Bands, Technical Report TD(18)07012, Cartagena, Spain, May, June 2018.

[CGP$^+$17]   Claudia Carciofi, Doriana Guiducci, Valeria Petrini, Eva Spina, Giuseppe De Sipio, Domenico Massimi, Enrico Scognamiglio, Vincenzo Sorrentino, Alessandro Casagni, Li Guoyue, Zhenbo Lai, Richard Rudd, Paolo Grazioso, Spectrum sharing between LTE-TDD and VSAT DVB-S in C-band: experimental campaign on consumer VSAT receivers, Technical Report TD(17)05036, Graz, Austria, September 2017.

[CGV$^+$19]   Claudia Carciofi, Andrea Garzia, Simona Valbonesi, Alessandro Gandolfo, Roberto Franchelli, Electromagnetic field levels massive monitoring in 5G scenarios: dynamic and standard measurements comparison, Technical Report TD(19)09040, Dublin, Ireland, January 2019.

[CHA18]   Chen-Hu Kun, Ana Garcia Armada, Comparative of single-carrier and multi-carrier waveforms combined with massive-MIMO, in: COST-IRACON, 2018, TD06006.

[CKB18]   Krzysztof Cichon, Adrian Kliks, Hanna Bogucka, Correlation-based clustering procedure towards energy-efficient cooperative spectrum sensing, in: COST-IRACON, 2018, TD07045.

[CKS$^+$19]   D. Czaniera, M. Käske, G. Sommerkorn, C. Schneider, R.S. Thomä, G. Del Galdo, M. Boban, J. Luo, Investigation on stationarity of V2V channels in a highway scenario, in: 2019 13th European Conference on Antennas and Propagation (EuCAP), March 2019, pp. 1–5. Also available as TD(19)09059.

[CL09]   Y. Corre, Y. Lostanlen, Three-dimensional urban EM wave propagation model for radio network planning and optimization over large areas, IEEE Transactions on Vehicular Technology 58 (7) (Sep. 2009) 3112–3123.

[CLC$^+$19]   G. Callebaut, G. Leenders, S. Crul, C. Buyle, L. Van der Perre, Lora physical layer evaluation for point-to-point links and coverage measurements in diverse environments, Technical Report TD(19)09091, Dublin, Ireland, January 2019.

[CLL$^+$19]   F. Challita, P. Laly, M. Lienard, E. Tanghe, W. Joseph, Davy P. Gaillot, Massive MIMO for Industrial Scenarios: Measurement-Based Polarimetric Analysis, Technical Report TD-19-09024, Dublin, Ireland, January 2019.

[CLMG18]   F. Challita, P. Laly, M. Lienard, D.P. Gaillot, Hybrid Virtual polarimetric Massive MIMO measurements at 1.35 GHz, Technical Report TD-18-07043, Cartagena, Spain, May 2018.

[CMA16]   M. Calvo-Fullana, J. Matamoros, C. Antón-Haro, Sensor selection and power allocation strategies for energy harvesting wireless sensor networks, IEEE Journal on Selected Areas in Communications 34 (12) (Dec 2016) 3685–3695. Also available as TD(16)01077.

[CMC16]   J.M. Conrat, I. Maaz, J.C. Cousin, Impact of antenna position on performances in relay-assisted wireless network, Number TD(16)01002, Lille, France, May 2016.

[CMG⁺18] F. Challita, M. Lienard, D.P. Gaillot, M-T. Martinez-Ingles, J-M. Molina-Garcia-Pardo, Evaluation of an antenna selection strategy for reduced massive MIMO complexity, in: COST-IRACON, 2018, TD07010.

[CML⁺18] F. Challita, M. Martinez-Ingles, M. Lienard, J. Molina-Garcia-Pardo, D.P. Gaillot, Line-of-sight massive MIMO channel characteristics in an indoor scenario at 94 GHz, IEEE Access 6 (2018) 62361–62370. Also available as TD(17)04004.

[Cor16] Intel Corporation, New WID: Radiated performance requirements for the verification of multi-antenna reception of UEs, Technical Report 3GPP RP-160603, 3GPP RAN, March 2016.

[COS18] CATR, OPPO, Samsung, Study on radiated test methodology for the verification of multi-antenna reception performance of NR UEs, Technical Report 3GPP RP-181402, 3GPP, jun 2018.

[Cox72] D.C. Cox, Delay Doppler characteristics of multipath propagation at 910 MHz in a suburban mobile radio environment, IEEE Transactions on Antennas and Propagation 20 (5) (Sep. 1972) 625–635.

[CPB⁺19] Claudia Carciofi, Samuela Persia, Marina Barbiroli, Daniele Bontempelli, Giuseppe Anania, Radio Frequency Electromagnetic Field Exposure Assessment for future 5G networks, Technical Report TD(19)09041, Dublin, Ireland, January 2019.

[CPGPC⁺18] Sergio Castelló-Palacios, Concepcion Garcia-Pardo, Narcís Cardona, Ana Vallés-Lluch, Reza Aminzadeh, Günter Vermeeren, Wout Joseph, Elaboration of simple gel phantoms for 5G/mmWave communications, in: IEEE 29th Annual International Symposium on Personal, Indoor, and Mobile Radio Communications (PIMRC), Bologna, Italy, 2018, pp. 1215–1219.

[CPGPFL⁺16] Sergio Castello-Palacios, Concepcion Garcia-Pardo, Alejandro Fornes-Leal, Narcis Cardona, Ana Valles-Lluch, Tailor-made tissue phantoms based on acetonitrile solutions for microwave applications up to 18 GHz, IEEE Transactions on Microwave Theory and Techniques 64 (11) (nov 2016) 3987–3994.

[CPGPFL⁺18] S. Castello-Palacios, C. Garcia-Pardo, A. Fornes-Leal, N. Cardona, A. Vallés-Lluch, Full-spectrum phantoms for cm-Wave and medical wireless communications, in: 12th European Conference on Antennas and Propagation (EuCAP 2018), Institution of Engineering and Technology, 2018, pp. 1–3.

[CPYA17] X. Cai, B. Peng, X. Yin, Y. Antonio Pérez, Hough-transform-based cluster identification and modelling for V2V channels based on measurements, IEEE Transactions on Vehicular Technology 67 (5) (December 2017) 3838–3852. Also available as TD(19)10067.

[CPYY18] Xuesong Cai, Bile Peng, Xuefeng Yin, Antonio Perez Yuste, Hough-transform-based cluster identification and modeling for V2V channels based on measurements, IEEE Transactions on Vehicular Technology 67 (5) (2018) 3838–3852. Also available as TD(19)10067.

[CSM⁺19] G. Cuozzo, M. Skocaj, S. Mignardi, M. Cavaletti, C. Buratti, R. Verdone, A 2.4 GHz LoRa Technology Solution for Industrial IoT, Technical Report TD(19)10053, Oulu, Finland, May 2019.

[CSTD19] D. Czaniera, S. Schieler, R.S. Thomä, G. Del Galdo, Utilization of the V2X radio channels temporal stationarity for communication and radar sensing, Technical Report TD(19)11024, Gdańsk, Poland, September 2019.

[CTI01] CTIA, Test plan for wireless device over-the-air performance, Technical report, CTIA, 2001.

[CTI16] CTIA, Test plan for 2 × 2 downlink MIMO and transmit diversity over-the-air performance (v1.1), Technical report, CTIA, 2016.

[CTJ⁺17] L. Chen, S. Thombre, K. Jarvinen, E.S. Lohan, A.K. Alen-Savikko, H. Leppakoski, M.Z.H. Bhuiyan, S. Bu-Pasha, G.N. Ferrara, S. Honkala, J. Lindqvist, L. Ruotsalainen, P. Korpisaari, H. Kuusniemi, Robustness, security and privacy in location-based services for future IoT: a survey, IEEE Access 5 (April 2017) 2169–3536.

[CTS⁺16] Y. Corre, T. Tenoux, J. Stéphan, F. Letourneux, Y. Lostanlen, Analysis of outdoor propagation and multi-cell coverage from ray-based simulations in sub-6 GHz and mmWave bands, in: 2016 10th European Conference on Antennas and Propagation (EuCAP), Apr. 2016, pp. 1–5. Also available as TD(16)01045.

[CVP18] A. Colpaert, E. Vinogradov, S. Pollin, Aerial coverage analysis of cellular systems at LTE and mmWave frequencies using 3D city models, Sensors 18 (12) (2018). Also available as TD(19)09010.

[CW95] D.J. Cichon, W. Wiesbeck, The Heinrich Hertz wireless experiments at Karlsruhe in the view of modern communication, in: Proc. 1995 International Conference on 100 Years of Radio, vol. 2, Sep. 1995, pp. 1–6.

[CYCP16] X. Cai, X. Yin, X. Cheng, A. Pérez Yuste, An empirical random-cluster model for subway channels based on passive measurements in UMTS, IEEE Transactions on Communications 64 (8) (Aug 2016) 3563–3575. Also available as TD(16)01080.

[CYCW17] Jiajing Chen, Xuefeng Yin, Xuesong Cai, Stephen Wang, Measurement-based Massive MIMO Channel Modeling for Outdoor LoS and NLoS Environments, Technical Report TD-17-03034, Lisbon, Portugal, February 2017.

[CYTK17] Jiajing Chen, Xuefeng Yin, Li Tian, Myung-Don Kim, Millimeter-wave channel modeling based on a unified propagation theory, IEEE Communications Letters 21 (2) (Feb. 2017) 246–249.

[CZT19] J. Chen, M. Zhu, F. Tufvesson, LTE multipath component delay based simultaneous localization and mapping, Technical Report TD(19)10013, Oulu, Finland, May 2019.

[DA17] L. Diez, R. Agüero, Generic system level simulation for advanced resource management solutions: a holistic approach for complex network deployments, Technical Report TD(17)04062, IRACON Cost Action CA15104, Lund, Sweden, May 2017.

[DAB⁺19] N. Dreyer, G. Artner, F. Backwinkel, M. Hein, T. Kürner, Evaluating automotive antennas for cellular radio communications, in: Proc. Int. Conference on Connected Vehicles and Expo (ICCVE), Graz, Austria, November 2019. Also available as TD(19)09053.

[dAVCA18] Rooderson Martines de Andrade, Fernando J. Velez, Kun Chen, Ana Garcia Armada, Implementation of low-complexity hybrid analogue-digital solutions in CAP-MIMO, in: COST-IRACON, 2018, TD07019.

[DBKU⁺17] M. Dialounke-Balde, A. Karttunen, B. Uguen, K. Haneda, H.L.S. Nguyen, Time varying multi-path components at 28 GHz from mobile channel sounding, Technical Report TD(17)04067, Lund, Sweden, may 2017.

[DC99] E. Damosso, L.M. Correia, Digital mobile radio towards future generation systems, Number COST Action 231 Final report, 1999.

[DCB⁺18] J. Doré, Y. Corre, S. Bicais, J. Palicot, E. Faussurier, D. Ktenas, F. Bader, Above-90 GHz spectrum and single-carrier waveform as enablers for efficient Tbit/s wireless communications, in: 2018 25th International Conference on Telecommunications (ICT), June 2018, pp. 274–278. Also available as TD(18)08044.

[DCCS17a] C.A.L. Diakhate, J. Conrat, J. Cousin, A. Sibille, Frequency-dependence of channel delay spread in an outdoor environment, in: 2017 IEEE 28th Annual International Symposium on Personal, Indoor, and Mobile Radio Communications (PIMRC), Oct 2017, pp. 1–5.

[DCCS17b] C.A.L. Diakhate, J. Conrat, J. Cousin, A. Sibille, Millimeter-wave outdoor-to-indoor channel measurements at 3, 10, 17 and 60 GHz, in: 2017 11th European Conference on Antennas and Propagation (EUCAP), March 2017, pp. 1798–1802.

[DCCS18] C.A.L. Diakhate, J. Conrat, J. Cousin, A. Sibille, Antenna aperture impact on channel delay spread in an urban outdoor scenario at 17 and 60 GHz, in: 12th European Conference on Antennas and Propagation (EuCAP 2018), April 2018, pp. 1–5. Also available as TD(18)07016.

[DCF⁺09] Davide Dardari, Andrea Conti, Ulric Ferner, Andrea Giorgetti, Moe Z. Win, Ranging with ultrawide bandwidth signals in multipath environments, Proceedings of the IEEE 97 (2) (Feb 2009) 404–426, Special Issue on UWB Technology & Emerging Applications.

[DCT⁺18] G. Davoli, W. Cerroni, S. Tomovic, C. Condoli, C. Buratti, F. Callegati, Intent-based service management for heterogeneous software-defined infrastructure domains, International Journal of Network Management (2018). Also available as TD(16)02050.

[DDGG16] Davide Dardari, Nicoló Decarli, Anna Guerra, Francesco Guidi, The future of ultra-wideband localization in RFID, in: 2016 IEEE International Conference on RFID (RFID) (IEEE RFID 2016), Orlando, USA, May 2016.

[DDR19] C. Schneider, S. Skoblikov, J. Luo, M. Boban, G. Del Galdo, D. Dupleich, R. Muller, R. Thoma, Multi-band vehicle to vehicle channel measurements from 6 GHz to 60 GHz at "t" intersection, Technical Report TD(19)11012, Gdansk, Poland, sep 2019.

[D'E19a] R. D'Errico, Channel characterization of indoor scenarios in sub-THz bands, Technical Report TD(19)11038, Gdansk, Poland, September 2019.

[D'E19b] R. D'Errico, Cluster characteristics and dense components in industrial environments, Technical report, Gdansk, Poland, Sep. 2019, TD(19)11026.

[Dey66] J. Deygout, Multiple knife-edge diffraction of microwaves, IEEE Transactions on Antennas and Propagation 16 (4) (July 1966) 825–873.

[DEZV⁺17] V. Degli-Esposti, M. Zoli, E.M. Vitucci, F. Fuschini, M. Barbiroli, J. Chen, A method for the electromagnetic characterization of construction materials based on Fabry-Pérot resonance, IEEE Access 5 (October 2017) 24938–24943. Also available as TD(17)05040.

[dFEC⁺17] M. de Freitas, M. Egan, L. Clavier, A. Goupil, G.W. Peters, N. Azzaoui, Capacity bounds for additive symmetric α-stable noise channels, IEEE Transactions on Information Theory 63 (8) (August 2017) 5115–5123.

[DFGTV17] F. Derogarian, J.C. Ferreira, V.M. Grade Tavares, F.J. Velez, An Energy-Efficient Routing Protocol for WSNs Combining Source Routing and Minimum Cost Forwarding, Technical Report TD(17)03071, Lisbon, Portugal, February 2017.

[DGGV17] F. Derogarian, G. Gardašević, F.J. Velez, Development of minimum cost forwarding (MCF) and source routing MCF routing protocols over 6TiSCH in OpenWSN, Technical Report TD(17)04077, Lund, Sweden, May 2017.

[DGR⁺19] L. Duarte, R. Gomes, C. Ribeiro, A. Hammoudeh, M. Sanchez, R.F.S. Caldeirinha, An agile full-duplex 5G testbed for OTA testing at 28 GHz: from GbE-based to OTA live streaming of 8 HD videos, Technical Report TD-19-09042, Polytechnic Institute of Leiria, Leiria, Portugal, January 2019.

[dHBGV19] Miguel Martínez del Horno, Luis Orozco Barbosa, Ismael García-Varea, In-motion calibration of Wi-Fi-based indoor tracking systems for smartphones, in: COST IRACON (CA15104) 9th Scientific Meeting, TD(19)090007, Dublin, Ireland, Jan 16-19 2019, pp. 1–8. Also available as TD(19)090007.

[DJKed] N. Dreyer, E.A. Janas, T. Kürner, Evaluation of veins and ns-3 for 802.11p based vehicular communication, Technical Report TD(19)10028, IRACON COST Action CA15104, 2019.

[DK19] N. Dreyer, T. Kürner, An analytical raytracer for efficient D2D path loss predictions, in: 2019 13th European Conference on Antennas and Propagation (EuCAP), Krakow, Poland, Mar. 2019, pp. 1–5. Also available as TD(18)08028.

[DKS⁺19] M. Dobereiner, M. Kaske, A. Schwind, C. Andrich, M.A. Hein, R.S. Thoma, G.D. Galdo, Joint high-resolution delay-Doppler estimation for bi-static radar measurements, in: 2019 16th European Radar Conference (EuRAD), Oct. 2019, pp. 145–148. Also available as TD(19)09046.

[DLV⁺18] V. Degli-Esposti, J.S. Lu, E.M. Vitucci, F. Fuschini, M. Barbiroli, J. Blaha, H.L. Bertoni, Efficient RF coverage prediction through a fully discrete, GPU-parallelized ray-launching model, in: 12th European Conference on Antennas and Propagation (EuCAP 2018), Apr. 2018, pp. 1–5. Also available as TD(18)06046.

[DM18] S. Dwivedi, J. Medbo, Window loss measurements and model validation, Technical Report TD(18)07058, Cartagena, Spain, May 2018.

[DMM⁺16] N. Dreyer, A. Moller, Z.H. Mir, F. Filali, T. Kurner, A Data Traffic Steering Algorithm for IEEE 802.11p/LTE Hybrid Vehicular Networks, September 2016, pp. 1–6. Also available as TD(16)01025.

[DMM⁺17] M. Deruyck, A. Marri, S. Mignardi, L. Martens, W. Joseph, R. Verdone, Performance evaluation of the dynamic trajectory design for an unmanned aerial base station in a single frequency network, in: IEEE 28th Annual International Symposium on Personal, Indoor, and Mobile Radio Communications (PIMRC), October 2017. Also available as TD(17)04018.

[DMS18] S. Dwivedi, J. Medbo, D. Sundman, Elevation angle dependence of building entry loss, in: 12th European Conference on Antennas and Propagation (EuCAP 2018), April 2018, pp. 1–5.

[dPRBAZ⁺17a] J.A. del Peral-Rosado, M.A. Barreto-Arboleda, F. Zanier, M. Crisci, G. Seco-Granados, J.A. López-Salcedo, Pilot placement for power-efficient uplink positioning in 5G vehicular networks, in: 2017 IEEE 18th International Workshop on Signal Processing Advances in Wireless Communications (SPAWC), Sapporo, Japan, IEEE, Jul 2017, pp. 1–5. Also available as TD(17)04056.

[dPRBAZ⁺17b] J.A. del Peral-Rosado, M.A. Barreto-Arboleda, F. Zanier, G. Seco-Granados, J.A. López-Salcedo, Performance limits of V2I ranging localization with LTE networks, in: 2017 14th Workshop on Positioning, Navigation and Communications (WPNC), Bremen, Germany, October 2017, pp. 1–5. Also available as TD(17)05049.

[dPRGD⁺19] J.A. del Peral-Rosado, F. Gunnarsson, S. Dwivedi, S.M. Razavi, O. Renaudin, J.A. López-Salcedo, G. Seco-Granados, Exploitation of 3D city maps for hybrid 5G RTT and GNSS positioning simulations, Technical Report TD(19)11045, Gdańsk, Poland, September 2019.

[dPRLSSG17] J.A. del Peral-Rosado, J.A. López-Salcedo, Go. Seco-Granados, Impact of frequency-hopping NB-IoT positioning in 4G and future 5G networks, in: 2017 IEEE International Conference on Communications Workshops (ICC Workshops), IEEE, May 2017, pp. 815–820. Also available as TD(17)03029.

[dPRRLSSG17] José A. del Peral-Rosado, Ronald Raulefs, José A. López-Salcedo, Gonzalo Seco-Granados, Survey of cellular mobile radio localization methods: from 1G to 5G, IEEE Communications Surveys and Tutorials 20 (2) (December 2017) 1124–1148.

[dPRRR⁺19] J.A. del Peral-Rosado, O. Renaudin, R. Raulefs, E. Domínguez, A. Fernandez-Cabezas, F. Blázquez-Luengo, G. Cueto-Felgueroso, A. Chassaigne, D. Bartlett, F. Grec, L. Ries, R. Prieto-Cerdeira, J.A. López-Salcedo, G. Seco-Granados, Physical-layer abstraction for hybrid GNSS and 5G positioning evaluations, in: 2019 IEEE 90th Vehicular Technology Conference (VTC2019-Fall), Honolulu, USA, September 2019, pp. 1–6. Also available as TD(19)11044.

[dPRSGKLS19] J.A. del Peral-Rosado, G. Seco-Granados, S. Kim, J.A. López-Salcedo, Network design for accurate vehicle localization, IEEE Transactions on Vehicular Technology 68 (5) (May 2019) 4316–4327. Also available as TD(18)07030.

[DRM⁺] Margot Deruyck, Daniela Renga, Michela Meo, Luc Martens, Wout Joseph, Reducing the impact of solar energy shortages on the wireless access network powered by a PV panel system and the power grid.

[DSW17] Gregor Dumphart, Eric Slottke, Armin Wittneben, Robust near-field 3D localization of an unaligned single-coil agent using unobtrusive anchors, in: IEEE 28th Annual International Symposium on Personal, Indoor, and Mobile Radio Communications (PIMRC), Oct. 2017. Also available as TD(17)05022.

[DT18] B. Dordevic, V. Timcenko, The hypervisor-based and container-based virtualization in IoT environment, Technical Report TD(18)07039, Cartagena, Spain, May 2018.

[DTC⁺19] Stavros Domouchtsidis, Christos G. Tsinos, Symeon Chatzinotas, Bjorn Ottersten, RF domain symbol level precoding for large-scale antenna array systems, in: COST-IRACON, 2019, TD09073.

[DTRM15]   A. Duran, M. Toril, F. Ruiz, A. Mendo, Self-optimization algorithm for outer loop link adaptation in LTE, IEEE Communications Letters 19 (11) (September 2015) 2005–2008.

[DUM⁺16]   B. Denis, B. Uguen, F. Mani, R. D'errico, N. Amiot, Joint orientation and position estimation from differential RSS measurements at on-body nodes, in: 2016 IEEE 27th Annual International Symposium on Personal, Indoor, and Mobile Radio Communications (PIMRC), IEEE, Sep 2016, pp. 1–6. Also available as TD(16)01061.

[Duped]   D. Dupleich, Characterization of the propagation channel in conference room scenario at 190 GHz, in: 14th European Conference on Antennas and Propagation (EuCAP 2020), March 2020, pp. 1–5, Technical Report TD(19)10024, IRACON COST Action CA15104, 2019.

[DWMJ16]   M. Deruyck, J. Wyckmans, L. Martens, W. Joseph, Emergency ad-hoc networks by using drone mounted base stations for a disaster scenario, in: IEEE 12th International Conference on Wireless and Mobile Computing, Networking and Communications (WiMOB), October 2016. Also available as TD(16)02001.

[DWP⁺17]   M. Deruyck, J. Wyckmans, D. Plets, L. Martens, W. Joseph, Spreading the traffic load in emergency ad-hoc networks deployed by drone mounted base stations, in: URSI France 2017 Workshop Radio Science for Humanity, 2017. Also available as TD(17)03024.

[D19]   Raffaele D'Errico, Cluster Characteristics and Dense Components in Industrial Environments, Technical Report TD-19-11026, Gdansk, Poland, September 2019.

[EBDI14]   H. Elshaer, F. Boccardi, M. Dohler, R. Irmer, Downlink and uplink decoupling: a disruptive architectural design for 5G networks, in: 2014 IEEE Global Communications Conference, IEEE, December 2014, pp. 1798–1803.

[EC16]   RADIO SPECTRUM POLICY GROUP European Commission. STRATEGIC ROADMAP TOWARDS 5G FOR EUROPE, Opinion on spectrum related aspects for next-generation wireless systems (5G), Technical Report RSPG16-032 FINAL, 2016.

[ECC13]   ECC, Least Restrictive Technical Conditions suitable for Mobile/Fixed Communication Networks (MFCN), including IMT, in the frequency bands 3400-3600 MHz and 3600-3800 MHz, Technical report, 2013.

[ECdF⁺17]   M. Egan, L. Clavier, M. de Freitas, L. Dorville, J.M. Gorce, A. Savard, Wireless communication in dynamic interference, in: GLOBECOM 2017 - 2017 IEEE Global Communications Conference, December 2017, pp. 1–6. Also available as TD(16)01067 and TD(17)04073.

[ECZ⁺18]   M. Egan, L. Clavier, C. Zheng, M. de Freitas, J.M. Gorce, Dynamic interference for uplink SCMA in large-scale wireless networks without coordination, EURASIP Journal on Wireless Communications and Networking (August 2018). Also available as TD(16)02041.

[EdFC⁺16]   M. Egan, M. de Freitas, L. Clavier, A. Goupil, G.W. Peters, N. Azzaoui, Achievable rates for additive isotropic α-stable noise channels, in: 2016 IEEE International Symposium on Information Theory (ISIT), July 2016, pp. 1874–1878.

[EFVG07] Vittorio Degli Esposti, Franco Fuschini, Enrico Vitucci, Daniele Graziani, Measurement and modeling of scattering from buildings, IEEE Transactions on Antennas and Propagation 55 (1) (Jan. 2007) 143–153.

[EHD17] P. Ezran, Y. Haddad, M. Debbah, Availability optimization in a ring-based network topology, Computer Networks 124 (September 2017) 27–32. Also available as TD(17)05053.

[ELLPGMGP19] E. Egea-Lopez, F. Losilla, J. Pascual-Garcia, J.M. Molina-Garcia-Pardo, Vehicular networks simulation with realistic physics, IEEE Access 7 (2019) 44021–44036. Also available as TD(19)10001.

[Eme97] D.T. Emerson, The work of Jagadis Chandra Bose: 100 years of millimeter-wave research, IEEE Transactions on Microwave Theory and Techniques 45 (12) (Dec. 1997) 2267–2273.

[EMSO11] Akihiro Enomoto, Michael Moore, Tatsuya Suda, Kazuhiro Oiwa, Design of self-organizing microtubule networks for molecular communication, Nano Communication Networks 2 (March 2011) 16.

[Eva15] Dave Evans, The Internet of Things - How the Next Evolution of the Internet is Changing Everything, CISCO white paper, 2015.

[FAC+17] Manuel M. Ferreira, Slawomir J. Ambroziak, Filipe D. Cardoso, Jaroslaw Sadowski, Luis M. Correia, Fading Modelling in Maritime Container Terminal Environments, Technical Report TD(17)03032, Lisbon, Portugal, 2017.

[FAGB+15] S. Fortes, A. Aguilar-García, R. Barco, F. Barba, J.A. Fernández-Luque, A. Fernández-Durán, Management architecture for location-aware self-organizing LTE/LTE-A small cell networks, IEEE Communications Magazine 53 (1) (January 2015) 294–302.

[FAGB+17] S. Fortes, A. Aguilar-Garcia, Eduardo Baena, Mariano Molina-García, Jaime Calle-Sánchez, José I. Alonso, Aaron Garrido, Alfonso Fernández-Durán, R. Barco, Location-aware enhanced self-healing in femtocell networks, Technical report TD(17)05052, Graz, Austria, September 2017.

[FAZ19] G. Fontanesi, H. Ahmadi, A. Zhu, Over the sea UAV based communication, arXiv preprint, arXiv:1905.04954, 2019. Also available as TD(19)10066.

[FBAG16] S. Fortes, R. Barco, A. Aguilar-Garcia, Location-based distributed sleeping cell detection and root cause analysis for 5G ultra-dense networks, EURASIP Journal on Wireless Communications and Networking 2016 (1) (June 2016) 1–18.

[FBAGM15] S. Fortes, R. Barco, A. Aguilar-García, P. Muñoz, Contextualized indicators for online failure diagnosis in cellular networks, Computer Networks 82 (April 2015) 96–113.

[FBV+18] L. Feltrin, C. Buratti, E. Vinciarelli, R. De Bonis, R. Verdone, LoRaWAN: evaluation of link- and system-level performance, IEEE Internet of Things Journal 5 (3) (June 2018) 2249–2258. Also available as TD(17)04075.

[FCM+18] L. Feltrin, M. Condoluci, T. Mahmoodi, M. Dohler, R. Verdone, NB-IoT: performance estimation and optimal configuration, in: European Wireless 2018; 24th European Wireless Conference, May 2018. Also available as TD(18)08003.

[FFCM17] A.F.G. Ferreira, D.M.A. Fernandes, A.P. Catarino, J.L. Monteiro, Localization and positioning systems for emergency responders: a survey, IEEE Communications Surveys and Tutorials 19 (4) (2017) 2836–2870.

[FGFL⁺16] S. Fortes, A. Aguilar Garcia, J.A. Fernandez-Luque, A. Garrido, R. Barco, Context-aware self-healing: user equipment as the main source of information for small-cell indoor networks, IEEE Vehicular Technology Magazine 11 (1) (March 2016) 76–85. Also available as TD(16)01075.

[FHZ⁺17] F. Fuschini, S. Häfner, M. Zoli, R. Müller, E.M. Vitucci, D. Dupleich, M. Barbiroli, J. Luo, E. Schulz, V. Degli-Esposti, R.S. Thomä, Analysis of in-room mm-wave propagation: ray tracing simulations and directional channel measurements, vol. 38, Jun 2017. Also available as TD(16)01038.

[FJZP] Wei Fan, Yilin Ji, Fengchun Zhang, Gert F. Pedersen, Channel estimation algorithms and their impact on wideband millimeter wave channel characteristics, in: 2018 2nd URSI Atlantic Radio Science Meeting (AT-RASC), pp. 1–4. Also available as TD(17)04066.

[FLET20] J. Flordelis, X. Li, O. Edfors, F. Tufvesson, Massive MIMO extensions to the COST 2100 channel model: modeling and validation, IEEE Transactions on Wireless Communications 19 (1) (Jan. 2020) 380–394.

[FLGPC⁺17] Alejandro Fornes-Leal, Concepcion Garcia-Pardo, Narcís Cardona, Sergio Castello-Palacios, Accurate broadband measurement of electromagnetic tissue phantoms using open-ended coaxial systems, in: 2017 11th International Symposium on Medical Information and Communication Technology (ISMICT), IEEE, feb 2017, pp. 32–36.

[FLGPF⁺18] Alejandro Fornes-Leal, Concepcion García-Pardo, Matteo Frasson, Sergio Castelló-Palacios, Andrea Nevárez, Vicente Pons Beltrán, Narcís Cardona, Variability of the dielectric properties due to tissue heterogeneity and its influence on the development of EM phantoms, in: IEEE 29th Annual International Symposium on Personal, Indoor, and Mobile Radio Communications (PIMRC), 2018, pp. 365–369.

[FMB⁺16] L. Feltrin, S. Mijovic, C. Buratti, A. Stajkic, E. Vinciarelli, R. Verdone, R. De Bonis, Performance evaluation of LoRa technology: experimentation and simulation, Technical Report TD(16)01036, Lille, France, May 2016.

[FMPV17] L. Feltrin, A. Marri, M. Paffetti, R. Verdone, Preliminary evaluation of NB-IOT technology and its capacity, Technical Report TD(17)04078, Lund, Sweden, May 2017.

[For15] 5G Forum, 5G new wave - Towards future societies in the 2020s, https://www.itfind.or.kr/admin/getFile.htm?identifier=02-003-160415-000003, accessed Jan 2018, March 2015.

[FP17] B.C. Fargas, M.N. Petersen, GPS-free geolocation using LoRa in low-power WANs, in: 2017 Global Internet of Things Summit (GIoTS), IEEE, jun 2017, pp. 1–6.

[FPA⁺17] C. Fernández-Prades, C. Pomar, J. Arribas, J.M. Fàbrega, J. Vilà-Valls, M. Svaluto Moreolo, R. Casellas, R. Martinez, M. Navarro, F.J. Vilchez, R. Muñoz, R. Vilalta, L. Nadal, A. Mayoral, A cloud optical access network for virtualized GNSS receivers, in: Proc. 30th Int. Tech. Meeting Sat. Div. Inst. Navig., Portland, OR, Sep. 2017, pp. 3796–3815.

[FPK⁺18] A.L. Freire, T. Pelham, D. Kong, L. Sayer, V. Sgardoni, F. Tila, E. Mellios, M. Beach, A. Nix, G. Steinböck, Polarimetric diffuse scattering channel measurements at 26 GHz and 60 GHz, in: 2018 IEEE 29th Annual International Symposium on Personal, Indoor and Mobile Radio Communications (PIMRC), Sep. 2018, pp. 210–214. Also available as TD(18)07068.

[FPSB18a] S. Fortes, D. Palacios, I. Serrano, R. Barco, Applying social event data for the management of cellular networks, IEEE Communications Magazine 56 (11) (November 2018) 36–43. Also available as TD(18)07067.

[FPSB18b] S. Fortes, D. Palacios, I. Serrano, R. Barco, Data from Social Sources in Cellular OAM, Technical Report TD-07-067, Cartagena, Spain, May 2018.

[FRT+17] Jose Flordelis, Fredrik Rusek, Fredrik Tufvesson, Erik G. Larsson, Ove Edfors, Massive MIMO performance—TDD versus FDD: what do measurements say?, in: COST-IRACON, 2017, TD04017.

[FSRP+19] S. Fortes, J.A. Santoyo-Ramón, D. Palacios, E. Baena, R. Mora-García, M. Medina, P. Mora, R. Barco, The campus as a smart city: University of Málaga environmental, learning, and research approaches, Sensors 19 (06) (march 2019) 1349, pp. 1–23. Also available as TD(19)11025.

[FTC+19] L. Feltrin, G. Tsoukaneri, M. Condoluci, C. Buratti, T. Mahmoodi, M. Dohler, R. Verdone, Narrowband-IoT: a survey on downlink and uplink perspectives, IEEE Wireless Communications 26 (1) (February 2019) 78–86. Also available as TD(18)06016.

[FTH+99] B.H. Fleury, M. Tschudin, R. Heddergott, D. Dahlhaus, K.I. Pedersen, Channel parameter estimation in mobile radio environments using the SAGE algorithm, IEEE Journal on Selected Areas in Communications 17 (3) (1999) 434–450.

[FVV+14] A.J. Fehske, I. Viering, J. Voigt, C. Sartori, S. Redana, G.P. Fettweis, Small-cell self-organizing wireless networks, Proceedings of the IEEE 102 (3) (March 2014) 334–350.

[FXVJ18] T. Feilen, C. Xu, L. Vandendorpe, L. Jacques, 1-bit localization scheme for radar using dithered quantized compressed sensing, Technical Report TD(18)08022, Podgorica, Montenegro, Oct 2018.

[FZJ+16] W. Fan, F. Zhang, T. Jämsä, M. Gustafsson, P. Kyösti, G.F. Pedersen, Reproducing standard SCME channel models for massive MIMO base station radiated testing, in: Proc. 11th European Conference on Antennas and Propagation EUCAP 2016, Davos, CH, April 2016, pp. 3658–3662. Also available as TD(16)02056.

[FZW19] W. Fan, F. Zhang, Z. Wang, Over-the-air testing of 5G communication systems: validation of the test environment in simple-sectored multiprobe anechoic chamber setups, IEEE Antennas & Propagation Magazine (October 2019). Also available as TD(19)11006.

[GA10] M. Gregori, I.F. Akyildiz, A new nanonetwork architecture using flagellated bacteria and catalytic nanomotors, IEEE Journal on Selected Areas in Communications 28 (4) (May 2010) 612–619.

[Gab96] Camelia Gabriel, Compilation of the Dielectric Properties of Body Tissues at RF and Microwave Frequencies, Technical report, US Dept of the Air Force, jan 1996.

[GAEE18] G. Ghiaasi, J. Abraham, E. Eide, T. Ekman, Measured Channel Hardening in an Indoor Multiband Scenario, Technical Report TD-18-07023, Cartagena, Spain, May 2018.

[GAP+18a] K. Guan, B. Ai, B. Peng, D. He, G. Li, J. Yang, Z. Zhong, T. Kürner, Towards realistic high-speed train channels at 5G millimeter-wave band-part I: paradigm, significance analysis, and scenario reconstruction, IEEE Transactions on Vehicular Technology 67 (10) (Oct 2018) 9112–9128. Also available as TD(18)06008.

[GAP+18b] K. Guan, B. Ai, B. Peng, D. He, G. Li, J. Yang, Z. Zhong, T. Kürner, Towards realistic high-speed train channels at 5G millimeter-wave band-part II: case study for paradigm implementation, IEEE Transactions on Vehicular Technology 67 (10) (Oct 2018) 9129–9144. Also available as TD(19)09050.

[GBLO19a] R.G. Gomez, S.R. Boque, M.G. Lozano, J. Olmos, Using COST IC1004 Vienna scenario to test C-RAN optimisation algorithms, Technical Report TD(19)09083, Dublin, Ireland, January 2019.

[GBLO19b] R.G. Gomez, S.R. Boque, M.G. Lozano, J. Olmos, A weighted-sum Multi-Objective optimization for Dynamic Resource Allocation with QoS constraints in realistic C-RAN, Technical Report TD(19)10068, Oulu, Finland, May 2019.

[GBLO19c] R.G. Gomez, S.R. Boque, M.G. Lozano, J. Olmos, Predicting Required Computational Capacity in C-RAN networks by the use of different Machine Learning strategies, Technical Report TD(19)11010, Gdańsk, Poland, September 2019.

[GBM19] David Gregoratti, Carlos Buelga, Xavier Mestre, Detection of row-sparse matrices with row-structure constraints, in: COST-IRACON, 2019, TD09019.

[GCC+17] Doriana Guiducci, Claudia Carciofi, Claudio Cecchetti, Elisa Ricci, Manuela Vaser, Elio Restuccia, Gianmarco Fusco, Analysis of Experimental Results related to the introduction of LTE in 2300-2400 MHz band in response to the European Commission, Technical Report TD(17)03045, Lisbon, Portugal, May, June 2017.

[GCO17] Quentin Gueuning, Christophe Craeye, Claude Oestges, NFFT-based inhomogeneous plane-wave physical optics dedicated to radio-channel estimation in urban environment, Technical report, Lisbon, Portugal, Feb. 2017, TD(17)03059.

[GCO19] Q. Gueuning, C. Craeye, C. Oestges, Sampling rules for an inhomogeneous plane-wave representation of scattered fields, IEEE Transactions on Antennas and Propagation 67 (3) (Mar. 2019) 1697–1708.

[GCP+16] Doriana Guiducci, Claudia Carciofi, Valeria Petrini, Eva Spina, Pravir Chawdhry, Spectrum sharing in 5G networks: the Italian first world's Licensed Shared Access pilot in the 2.3-2.4 GHz band, Technical Report TD(16)01026, Lille, France, May, June 2016.

[GCP+17] Doriana Guiducci, Claudia Carciofi, Valeria Petrini, Sergio Pompei, Eva Spina, Giuseppe De Sipio, Domenico Massimi, Domenico Spoto, Fabrizio Amerighi, Tommaso Magliocca, Pravir Chawdhry, Seppo Yrjola, Vesa Hartikainen, Lucia Tudose, Jesus L. Santos, Vicent F. Guasch, Jose Costa-Requena, Heikki Kokkinen, Luigi Ardito, Fausto Grazioli, Donatella Caggiati, Massimiliano Gianesin, Pierre-Jean Muller, Sharing analysis in a live LTE network in the 2.3-2.4 GHz band: compliance regulatory and technical results, Technical Report TD(17)03039, Lisbon, Portugal, May, June 2017.

[GCS17] J. Guita, L.M. Correia, M. Serrazina, Balancing the load in LTE urban networks via inter-frequency handover, Technical Report TD(17)03075, IRACON Cost Action CA15104, Lisbon, Portugal, February 2017.

[GCV+16] Doriana Guiducci, Claudia Carciofi, Simona Valbonesi, Marina Barbiroli, Valeria Petrini, Eva Spina, Pravir Chawdhry, EMF exposure assessment in a real femtocell environment under 5G paradigm, Technical Report TD(16)01027, Lille, France, May, June 2016.

[GDD⁺16] F. Guidi, N. Decarli, D. Dardari, F. Natali, E. Savioli, M. Bottazzi, A low complexity scheme for passive UWB-RFID: proof of concept, IEEE Communications Letters 20 (4) (April 2016) 676–679.

[GERT15] X. Gao, O. Edfors, F. Rusek, F. Tufvesson, Massive MIMO performance evaluation based on measured propagation data, IEEE Transactions on Wireless Communications 14 (7) (July 2015) 3899–3911.

[GFA⁺16] Concepcion Garcia-Pardo, Alejandro Fornes-Leal, Carlos Andreu, Narcis Cardona, Sergio Castello-Palacios, Ana Valles-Lluch, Raul Chavez-Santiago, Ilangko Balasingham, Experimental ultra wideband path loss models for implant communications, in: Proc. PIMRC'16 - 27th Annual International Symposium on Personal, Indoor, and Mobile Radio Communications, Valencia, Spain, September 2016. Also available as TD(16)02036.

[GFD⁺15] X. Gao, J. Flordelis, G. Dahman, F. Tufvesson, O. Edfors, Massive MIMO channel modeling extension of the COST 2100 model, in: Joint NEWCOM/COST Workshop on Wireless Communications, JNCW, Oct. 2015.

[GFdPT18] Sara Gunnarsson, Jose Flordelis, Liesbet Van der Perre, Fredrik Tufvesson, Channel Hardening in Massive MIMO: Experimental Validation and Model Parameters, Technical Report TD-18-08048, Podgorica, Montenegro, October 2018.

[GFT⁺18] Gordana Gardašević, Hossein Fotouhi, Ivan Tomasic, Maryam Vahabi, Mats Björkman, Maria Lindén, A heterogeneous IoT-based architecture for remote monitoring of physiological and environmental parameters, in: Mobyen Uddin Ahmed, Shahina Begum, Jean-Baptiste Fasquel (Eds.), Internet of Things (IoT) Technologies for HealthCare, Springer International Publishing, Cham, 2018, pp. 48–53. Also available as TD(17)05027.

[GJV16] L. Gupta, R. Jain, G. Vaszkun, Survey of important issues in UAV communication networks, IEEE Communications Surveys and Tutorials 18 (2) (2016) 1123–1152.

[GKAP18a] C. Gavrilă, C. Kertesz, M. Alexandru, V. Popescu, Reconfigurable IoT gateway based on a SDR platform, in: 2018 International Conference on Communications (COMM), June 2018, pp. 345–348. Also available as TD(19)09045.

[GKAP18b] C. Gavrilă, C. Kertesz, M. Alexandru, V. Popescu, SDR-based gateway for IoT and M2M applications, in: 2018 Baltic URSI Symposium (URSI), May 2018, pp. 71–74. Also available as TD(19)09044.

[GKDGH18] G. Artner, W. Kotterman, G. Del Galdo, M.A. Hein, Conformal automotive roof-top antenna cavity with increased coverage to vulnerable road users, IEEE Antennas and Wireless Propagation Letters 17 (12) (December 2018) 2399–2403. Also available as TD(18)07048.

[GKG⁺17] S. Grebien, J. Kulmer, F. Galler, M. Goller, E. Leitinger, H. Arthaber, K. Witrisal, Range estimation and performance limits for UHF-RFID backscatter channels, IEEE Journal of Radio Frequency Identification 1 (1) (March 2017) 39–50.

[Gla19] A.A. Glazunov, Impact of deficient array antenna elements on downlink massive MIMO performance, Technical Report TD-19-10023, Universiteit Twente, Enschede, the Netherlands, May 2019.

[GLFW18]   Stefan Grebien, Erik Leitinger, Bernard H. Fleury, Klaus Witrisal, Super-resolution channel estimation including the dense multipath component – a sparse variational Bayesian approach, Technical report, Podgorica, Montenegro, Oct. 2018, TD(18)08030.

[GLGP13]   G. Gómez, J. Lorca, R. Garcí, Q. Pérez, Towards a QoE-driven resource control in LTE and LTE-A networks, Journal of Computer Networks and Communications 2013 (December 2013).

[GLH⁺17]   K. Guan, X. Lin, D. He, B. Ai, Z. Zhong, Z. Zhao, D. Miao, H. Guan, T. Kürner, Scenario modules and ray-tracing simulations of millimeter wave and terahertz channels for smart rail mobility, in: 2017 11th European Conference on Antennas and Propagation (EUCAP), March 2017, pp. 113–117. Also available as TD(17)03070.

[GLRT17]   C. Gijón, S. Luna-Ramírez, M. Toril, A new QoE-based approach for HO parameter tuning in LTE networks, Technical Report TD(17)05020, IRACON Cost Action CA15104, Graz, Austria, September 2017.

[GMBT20]   C. Gustafson, K. Mahler, D. Bolin, F. Tufvesson, The COST IRACON geometry-based stochastic channel model for vehicle-to-vehicle communication in intersections, IEEE Transactions on Vehicular Technology 69 (3) (March 2020) 2365–2375. Also available as TD(19)09087.

[GMMLLH11] Javier Gutierrez-Meana, Jose A. Martinez-Lorenzo, Fernando Las-Heras, High frequency techniques: the physical optics approximation and the modified equivalent current approximation (MECA), in: Electromagnetic Waves - Propagation in Complex Matters, INTECH Open Access Publisher, 2011.

[GMMP15]   L. Galluccio, S. Milardo, G. Morabito, S. Palazzo, SDN-WISE: design, prototyping and experimentation of a stateful SDN solution for WIreless SEnsor networks, in: 2015 IEEE Conference on Computer Communications (INFOCOM), April 2015, pp. 513–521.

[GO19]   Julian David Villegas Gutierrez, Claude Oestges, On spatial loss analysis through physical clusters, Technical Report TD-19-11042, Gdansk, Poland, September 2019.

[GPdP17]   Carl Gustafson, Sofie Pollin, Liesbet Van der Perre, Initial Results on the Performance of Fixed mm-Wave Wireless Links with Relaying, Technical Report TD(17)04082, Lund, Sweden, May 2017.

[GPV19]   G. Gardašević, P. Plavsic, D. Vasiljevic, Experimental IoT testbed for testing the 6TiSCH and RPL coexistence, in: IEEE Conference on Computer Communications (IEEE INFOCOM 2019), April 2019. Also available as TD(19)10021.

[GQM⁺16]   Kexin Guo, Zhirong Qiu, Cunxiao Miao, Abdul Hanif Zaini, Chun-Lin Chen, Wei Meng, Lihua Xie, Ultra-wideband-based localization for quadcopter navigation, Unmanned Systems 4 (1) (Feb. 2016) 23–34, World Scientific Publishing Company.

[GTB⁺17]   S.K. Goudos, A. Tsiflikiotis, D. Babas, K. Siakavara, C. Kalialakis, G.K. Karagiannidis, Evolutionary design of a dual band e-shaped patch antenna for 5G mobile communications, in: 2017 6th International Conference on Modern Circuits and Systems Technologies (MOCAST), May 2017, pp. 1–4. Also available as TD(17)04020.

[GTE13]   X. Gao, F. Tufvesson, O. Edfors, Massive MIMO channels – measurements and models, in: 2013 Asilomar Conference on Signals, Systems and Computers, Nov. 2013, pp. 280–284.

[GTJ⁺15] Davy P. Gaillot, Emmeric Tanghe, Wout Joseph, Pierre Laly, Viet-Chi Tran, Martine Lienard, Luc Martens, Polarization properties of specular and dense multipath components in a large industrial hall, IEEE Transactions on Antennas and Propagation 63 (7) (July 2015) 3219–3228.

[Gud91] M. Gudmundson, Correlation model for shadow fading in mobile radio systems, Electronics Letters 27 (23) (Nov 1991) 2145–2146.

[GV18] G. Gardašević, D. Vasiljević, Distributed Scheduling in 6TiSCH Networks based on 6top Protocol, Technical Report TD(18)06026, Nicosia, Cyprus, January 2018.

[GVBV18] G. Gardašević, D. Vasiljević, C. Buratti, R. Verdone, Experimental characterization of joint scheduling and routing algorithm over 6TiSCH, in: 2018 European Conference on Networks and Communications (EuCNC), June 2018. Also available as TD(17)03022.

[GVV16] G. Gardašević, D. Vasiljević, M. Veletić, Testing OpenWSN and Contiki OSs Performances on OpenMote Platform, Technical Report TD(16)01014, Lille, France, May 2016.

[GYL11] R.K. Ganti, F. Ye, H. Lei, Mobile crowdsensing: current state and future challenges, IEEE Communications Magazine 49 (11) (November 2011) 32–39.

[GZH⁺19] H. Groll, E. Zöchmann, M. Hofer, H. Hammoud, S. Sangodoyin, G. Ghiaasi, T. Zemen, J. Blumenstein, A. Prokes, A.F. Molisch, C.F. Mecklenbräuker, 60 GHz V2I channel variability for different elevation angle switching strategies, Technical Report TD(19)11051, Gdańsk, Poland, September 2019.

[GZP⁺19] H. Groll, E. Zöchmann, S. Pratschner, M. Lerch, D. Schützenhöfer, M. Hofer, J. Blumenstein, S. Sangodoyin, T. Zemen, A. Prokeš, A.F. Molisch, S. Caban, Sparsity in the delay-Doppler domain for measured 60 GHz vehicle-to-infrastructure communication channels, in: 2019 IEEE International Conference on Communications Workshops (ICC Workshops), May 2019, pp. 1–6. Also available as TD(19)09054.

[HAG⁺18] D. He, B. Ai, K. Guan, Z. Zhong, B. Hui, J. Kim, H. Chung, I. Kim, Channel measurement, simulation, and analysis for high-speed railway communications in 5G millimeter-Wave band, IEEE Transactions on Intelligent Transportation Systems 19 (10) (Oct 2018) 3144–3158. Also available as TD(17)03004.

[HAG⁺19] D. He, B. Ai, K. Guan, L. Wang, Z. Zhong, T. Kürner, The design and applications of high-performance ray-tracing simulation platform for 5G and beyond wireless communications: a tutorial, IEEE Communications Surveys and Tutorials 21 (1) (2019) 10–27. Also available as TD(19)09049.

[HAM⁺18] Ruisi He, Bo Ai, Andreas F. Molisch, Gordon L. Stüber, Qingyong Li, Zhangdui Zhong, Jian Yu, Clustering enabled wireless channel modeling using big data algorithms, IEEE Communications Magazine 56 (5) (2018) 177–183.

[Han16] K. Haneda, et al., 5G 3GPP-like channel models for outdoor urban microcellular and macrocellular environments, in: Proc. 2016 IEEE 83rd Vehicular Technology Conference, VTC Spring, Nanjing, China, May 2016.

[Hat80] M. Hata, Empirical formula for propagation loss in land mobile radio services, IEEE Transactions on Vehicular Technology 29 (12) (Dec. 1980) 317–325.

[HB16] S. Hussain, C. Brennan, An image visibility based pre-processing method for fast ray tracing in urban environments, in: 2016 10th European Conf. Ant. Prop. (EuCAP 2016), Apr. 2016, pp. 1–5. Also available as TD(16)01070.

[HB17] Qinhui Huang, Alister Burr, Compute-and-forward in cell-free massive MIMO: great performance with low backhaul load, in: COST-IRACON, 2017, TD03061.

[HBFR14] Thomas L. Hansen, Mihai A. Badiu, Bernard H. Fleury, Bhaskar D. Rao, A sparse Bayesian learning algorithm with dictionary parameter estimation, in: 2014 IEEE 8th Sensor Array and Multichannel Signal Processing Workshop (SAM), IEEE, 2014, pp. 385–388.

[HBJ18] A. Hrovat, K. Bregar, T. Javornik, Measurement based ultra-wideband channel model for mobile communications in tunnels, in: 12th European Conference on Antennas and Propagation (EuCAP 2018), April 2018, pp. 1–5. Also available as TD(18)07054.

[HBPH17] Balint Horvath, Zsolt Badics, Jozsef Pavo, Peter Horvath, Validation of numerical models of portable wireless devices for near-field simulation, IEEE Transactions on Magnetics 53 (6) (jun 2017) 1–4.

[HBPH19] Balint Horvath, Zsolt Badics, Jozsef Pavo, Peter Horvath, TD(19)11047 - Validation of Numerical Models of Portable Wireless Devices for Specific Absorption Rate Evaluation, Technical report, Gdansk (Poland), 2019.

[HCA+16] Ruisi He, Wei Chen, Bo Ai, Andreas F. Molisch, Wei Wang, Zhangdui Zhong, Jian Yu, Seun Sangodoyin, On the clustering of radio channel impulse responses using sparsity-based methods, IEEE Transactions on Antennas and Propagation 64 (6) (2016) 2465–2474. Also available as TD(19)10011.

[HH15] E. Hossain, M. Hasan, 5G cellular: key enabling technologies and research challenges, IEEE Instrumentation & Measurement Magazine 18 (3) (June 2015) 11–21.

[HHB+16a] P. Harris, W.B. Hasan, H. Brice, B. Chitambira, M. Beach, E. Mellios, A. Nix, S. Armour, A. Doufexi, An overview of massive MIMO research at the University of Bristol, in: Radio Propagation and Technologies for 5G (2016), 2016.

[HHB+16b] Paul Harris, Wael Boukley Hasan, Henry Brice, Mark Beach, Evangelos Mellios, Andrew Nix, Simon Armour, Angela Doufexi, Massive MIMO mobility measurements in LOS with power control, in: COST-IRACON, 2016, TD02031.

[HHDB16] Wael Boukley Hasan, Paul Harris, Angela Doufexi, Mark Beach, Preliminary investigation of uplink power control for massive MIMO, in: COST-IRACON, 2016, TD01084.

[HHDB17] Wael Boukley Hasan, Paul Harris, Angela Doufexi, Mark Beach, User grouping for massive MIMO in TDD, in: COST-IRACON, 2017, TD03069.

[HHJ17] K. Haneda, M. Heino, J. Jarvelainen, Dominant eigenmode gains of millimeter-wave antenna arrays on a mobile phone, Technical Report TD-17-03067, Aalto University, Espoo, Finland, February 2017.

[HHPB17] B. Horváth, P. Horváth, J. Pávó, Z. Badics, Validation of numerical models of portable wireless devices for specific absorption rate evaluation, Technical Report TD-17-04072, Budapest University of Technology and Economics, Budapest, Hungary, May 2017.

[HHS18] Marjo Heikkilä, Aki Hekkala, Ossi Saukko, Use of digital maps in context of radio propagation simulations, Technical report, Cartagena, Spain, May 2018, TD(18)07053.

[HK16] Aki Hekkala, Pekka Kyösti, Generality of map-based model, Technical report, Lille, France, May 2016, TD(16)01033.

[HKC17] Hyunwoo Hwangbo, Yang Sok Kim, Kyung Jin Cha, Use of the smart store for persuasive marketing and immersive customer experiences: a case study of Korean apparel enterprise, Mobile Information Systems 2017 (2017).

[HKS+18] N. Haeean, M. Käske, G. Sommerkorn, C. Schneider, R. Thomä, Extended TDL modeling for V2X channels, Technical Report TD(18)07036, Cartagena, Spain, May 2018.

[HKVG16] F. Hutu, D. Kibloff, G. Villemaud, J. Gorce, Experimental validation of a wake-up radio architecture, January 2016, pp. 155–158. Also available as TD(16)01050.

[HLA+17] Ruisi He, Qingyong Li, Bo Ai, Yang Li-Ao Geng, Andreas F. Molisch, Vinod Kristem, Zhangdui Zhong, Jian Yu, A kernel-power-density-based algorithm for channel multipath components clustering, IEEE Transactions on Wireless Communications 16 (11) (2017) 7138–7151. Also available as TD(19)09029.

[HLHI18] Rreze Halili, Ahma Luan, Enver Hamiti, Mimoza Ibrani, TD(18)06019 - Comparative analysis of downlink signal levels emitted by GSM 900, GSM 1800, UMTS, and LTE base stations and evaluation of daily personal-exposure induced by wireless operating networks, Technical report, Nicosia (Cyprus), 2018.

[HLM+16] S. Hinteregger, E. Leitinger, P. Meissner, J. Kulmer, K. Witrisal, Bandwidth dependence of the ranging error variance in dense multipath, in: 2016 24th European Signal Processing Conference (EUSIPCO), IEEE, Aug 2016, pp. 733–737. Also available as TD(16)02017.

[HLMW16] Stefan Hinteregger, Erik Leitinger, Paul Meissner, Klaus Witrisal, MIMO gain and bandwidth scaling for RFID positioning in dense multipath channels, in: 2016 IEEE International Conference on RFID (RFID), May 2016. Also available as TD(16)02018.

[HLS93] Simon Haykin, John Litva, Terence J. Shepherd, Radar Array Processing, Springer, 1993.

[HMDP18] Dávid Hrabčák, Martin Matis, L'ubomír Doboš, Ján Papaj, Tools for evaluation of social relations in mobility models, Telecommunication Systems 68 (3) (Jul 2018) 409–424. Also available as TD(17)04010.

[HMJV+17] Paul Harris, Steffen Malkowsky, Joao Vieira, Fredrik Tufvesson, Wael Boukley Hasan, Liang Liu, Mark Beach, Simon Armour, Ove Edfors, Temporal analysis of measured LOS massive MIMO channels with mobility, in: COST-IRACON, 2017, TD03043.

[HNK16] K. Haneda, S.L.H. Nguyen, A. Khatun, Attainable capacity of spatial radio channels: a multiple-frequency analysis, in: 2016 IEEE Globecom Workshops (GC Wkshps), Dec 2016, pp. 1–5. Also available as TD(16)01053.

[HPC+16] Tom Haute, Eli Poorter, Pieter Crombez, Filip Lemic, Vlado Handziski, Niklas Wirström, Adam Wolisz, Thiemo Voigt, Ingrid Moerman, Performance analysis of multiple indoor positioning systems in a healthcare environment, International Journal of Health Geographics 15 (1) (2016) 7.

[HRK⁺15] R. He, O. Renaudin, V. Kolmonen, K. Haneda, Z. Zhong, B. Ai, C. Oestges, A dynamic wideband directional channel model for vehicle-to-vehicle communications, IEEE Transactions on Industrial Electronics 62 (12) (Dec 2015) 7870–7882.

[HRL⁺16] A.N. Hong, M. Rath, E. Leitinger, S. Hinteregger, K.N. Van, K. Witrisal, Channel capacity analysis of indoor environments for location-aware communications, in: 2016 IEEE Globecom Workshops (GC Wkshps), Washington, DC, USA, December 2016. Also available as TD(17)05018.

[HRT⁺17] R. Hadani, S. Rakib, M. Tsatsanis, A. Monk, A.J. Goldsmith, A.F. Molisch, R. Calderbank, Orthogonal time frequency space modulation, in: IEEE Wireless Communications and Networking Conference (WCNC), March 2017.

[HST⁺17] P. Hanpinitsak, K. Saito, J. Takada, M. Kim, L. Materum, Multipath clustering and cluster tracking for geometry-based stochastic channel modeling, IEEE Transactions on Antennas and Propagation 65 (11) (Nov 2017) 6015–6028. Also available as TD(16)02013.

[HST⁺18] B. Hanssens, K. Saito, E. Tanghe, L. Martens, W. Joseph, J. Takada, Modeling the power angular profile of dense multipath components using multiple clusters, IEEE Access 6 (2018) 56084–56098. Also available as TD(17)03056.

[HSZ⁺16] Yanjun Han, Yuan Shen, Xiao-Ping Zhang, Moe Z. Win, Huadong Meng, Performance limits and geometric properties of array localization, IEEE Transactions on Information Theory 62 (2) (2016) 1054–1075.

[HT17] Magnus M. Halldorsson, Tigran Tonoyan, Optimal aggregation throughput is nearly constant, in: COST-IRACON, 2017, TD03007.

[HTG⁺18] Brecht Hanssens, Emmeric Tanghe, Davy P. Gaillot, Martine Liénard, Claude Oestges, David Plets, Luc Martens, Wout Joseph, An extension of the RiMAX multipath estimation algorithm for ultra-wideband channel modeling, EURASIP Journal on Wireless Communications and Networking 2018 (1) (2018) 164. Also available as TD(16)01019.

[htt18] Horizon 2020 Clear5G project, Horizon2020, website, UoS, https://cordis.europa.eu/project/id/761745, November 2018.

[HWZ⁺19] D. He, X. Wang, C. Zheng, K. Guan, B. Ai, Z. Zhong, Millimeter wave channel characterization and modeling for intra-wagon communication, Technical Report TD(19)10025, Oulu, Finland, May 2019.

[icn] International Commission on Non-Ionizing Radiation Protection.

[IEE12] IEEE standard for local and metropolitan area networks - part 15.6: wireless body area networks, IEEE Std 802.15.6-2012, Feb 2012, pp. 1–271.

[IEE17] IEEE Standard for Information technology–Telecommunications and information exchange between systems - Local and metropolitan area networks–Specific requirements - Part 11: Wireless LAN Medium Access Control (MAC) and Physical Layer (PHY) Specifications Amendment 2: Sub 1 GHz License Exempt Operation, IEEE Std 802.11ah-2016 (Amendment to IEEE Std 802.11-2016, as amended by IEEE Std 802.11ai-2016), May 2017, pp. 1–594.

[IOS⁺19] Anastasia Yu. Ivchenko, Yu.N. Orlov, Andrey K. Samuylov, Dmitri A. Moltchanov, Yuliya V. Gaidamaka, Konstantin K. Samuylov, Luis M. Correira, Technical report, 2019. Also available as TD(19)09048.

[IR09]   ITU-R, Report ITU-R M.2135-1 (12/2009) Guidelines for evaluation of radio interface technologies for IMT-Advanced, Technical Report M Mobile, radiodetermination, amateur and related satellites services, 2009.

[IR13]   ITU-R, Recommendation ITU-R P.1546-5 (09/2013) Method for point-to-area predictions for terrestrial services in the frequency range 30 MHz to 3000 MHz, Technical Report P Series Radiowave propagation, 2013.

[IR15]   ITU-R, Recommendation ITU-R P.452-16 (07/2015) Prediction procedure for the evaluation of interference between stations on the surface of the Earth at frequencies above about 0.1 GHz, Technical Report P Series Radiowave propagation, 2015.

[IR16]   ITU-R, Recommendation ITU-R P.525-3 (09/2016) Calculation of free-space attenuation, Technical Report P Series Radiowave propagation, 2016.

[IR17]   ITU-R, Recommendation ITU-R P.2108-0 (06/2017) Prediction of clutter loss, Technical Report P Series Radiowave propagation, 2017.

[IRA18]   COST IRACON, COST 2100 channel model implementation, https://github.com/cost2100, Oct. 2018.

[IT12]   D. Inserra, A.M. Tonello, Performance analysis of a novel antenna array calibration approach for direction finding systems, Transactions on Emerging Telecommunications Technologies 23 (8) (Sep. 2012) 777–788.

[ITU15]   ITUR, ITU-R recommendation p.2040-1, effects of building materials and structures on radiowave propagation above about 100 MHz, Technical Report P.2040-1, Geneva, Switzerland, July 2015.

[ITU16]   ITU-R, The Concept of Transmission Loss for Radio Links, September 2016, Recommendation P.341-6.

[JBGS18] Tommi Jämsä, Guo Bolun, Mingming Gan, Gerhard Steinböck, A unified channel model for terrestrial and non-terrestrial networks, Technical report, Cartagena, Spain, May 2018, TD(18)07072.

[JFK+19] Yilin Ji, Wei Fan, Pekka Kyosti, Jinxing Li, Gert Frolund Pedersen, Antenna Correlation Under Geometry-Based Stochastic Channel Models, Technical Report TD-19-11008, Gdansk, Poland, September 2019.

[JHFP18] Y. Ji, J. Hejselbæk, W. Fan, G.F. Pedersen, A map-free indoor localization method using ultrawideband large-scale array systems, IEEE Antennas and Wireless Propagation Letters (August 2018). Also available as TD(18)07013.

[JHK16]   Jan Järveläinen, Katsuyuki Haneda, Aki Karttunen, Indoor propagation channel simulations at 60 GHz using point cloud data, IEEE Transactions on Antennas and Propagation 64 (10) (Oct 2016) 4457–4467.

[JJHK18] Dushantha N.K. Jayakody, M.N. Jamal, S.A. Hassan, Rabia Khan, A new approach to cooperative NOMA using distributed space time block coding, in: COST-IRACON, 2018, TD06010.

[JJS+18] S. Jovanovic, M. Jovanovic, T. Skoric, B. Milovanovic, S. Jokic, K. Katzis, D. Bajic, A mobile crowd sensing application for hypertensive patients, Sensors 19 (2) (2018) 400–415. Also available as TD(18)08015.

[JK16]   Xiwen Jiang, Florian Kaltenberger, OpenAirInterface massive MIMO testbed: a 5G innovation platform, in: COST-IRACON, 2016, TD02044.

[JK17]   Xiwen Jiang, Florian Kaltenberger, TDD channel reciprocity calibration in hybrid beamforming massive MIMO systems, in: COST-IRACON, 2017, TD04085.

[JS16]   Tommi Jämsä, Gerhard Steinböck, 3GPP 3D channel model for 5G, Technical report, Lille, France, May 2016, TD(16)01066.

[JS17]   L.G. Jaimes, R. Steele, Incentivization for health crowdsensing, in: IEEE 15th Intl Conf on Dependable, Autonomic and Secure Computing, 15th Intl Conf on Pervasive Intelligence and Computing, 3rd Intl Conf on Big Data Intelligence and Computing and Cyber Science and Technology Congress, IEEE, 2017, pp. 139–146.

[JY19]   Tommi Jämsä, Wu Young, Radar multi-path channel modeling for autonomous driving, Technical report, Oulu, Finland, May 2019, TD(19)10039.

[KAC17]   Pawel T. Kosz, Slawomir J. Ambroziak, Luis M. Correia, Radio channel measurements in off-body communications in a ferry passenger cabin, in: Proc. URSI'17 - General Assembly and Scientific Symposium of the International Union of Radio Science, Montreal, QC, Canada, August 2017. Also available as TD(17)05013.

[KACS19]   Pawel T. Kosz, Slawomir J. Ambroziak, Luis M. Correia, Jacek Stefanski, Fading Analysis in Off-Body Channels in a Straight Metallic Corridor in a Passenger Ferry Environment, Technical Document TD(19)09036, COST Action CA15104 (IRACON), Dublin, Ireland, January 2019.

[KAS$^+$18]   Pawel T. Kosz, Slawomir J. Ambroziak, Jacek Stefanski, Krzysztof K. Cwalina, Luis M. Correia, Kenan Turbic, An empirical system loss model for body area networks in a passenger ferry environment, in: Proc. URSI'18 - Baltic URSI Symposium, Poznan, Poland, May 2018. Also available as TD(18)07001.

[Kay93]   Steven Kay, Fundamentals of Statistical Signal Processing: Estimation Theory, Prentice Hall Signal Processing Series, 1993.

[KB19]   I. Kavanagh, C. Brennan, Validation of a volume integral equation method for indoor propagation modelling, IET Microwaves, Antennas and Propagation 13 (6) (June 2019) 705–713. Also available as TD(17)03036, TD(17)05010 and TD(18)07070.

[KFK17]   P. Kyösti, W. Fan, J. Kyröläinen, Assessing measurement distances for OTA testing of massive MIMO base station at 28 GHz, in: Proc. 11th European Conference on Antennas and Propagation EUCAP 2017, Paris, France, March 2017, pp. 3679–3683. Also available as TD(17)03015.

[KFPL16]   P. Kyösti, W. Fan, G. Pedersen, M. Latva-Aho, On dimensions of OTA set-up for massive MIMO base stations radiated testing, IEEE Access 4 (September 2016) 5971–5981. Also available as TD(16)01043.

[KH19a]   A. Karttunen, K. Haneda, Large-scale parameter estimation in channel sounding with limited dynamic range, in: 2019 13th European Conference on Antennas and Propagation (EuCAP), Mar. 2019, pp. 1–5. Also available as TD(18)06022.

[KH19b]   P. Kyösti, P. Heino, Fading channel emulation for massive MIMO testing using a conductive phase matrix setup, Technical Report TD-19-09058, Keysight Technologies Finland oy, Oulu, Finland, January 2019.

[KHF$^+$18]   P. Kyösti, L. Hentilä, W. Fan, A. Hekkala, M. Rumney, On radiated performance evaluation of massive MIMO devices in multiprobe anechoic chamber OTA setups, IEEE Transactions on Antennas and Propagation 66 (10) (May 2018) 5485–5497. Also available as TD(17)04060 and TD(17)04061.

[KHG+17] J. Kulmer, S. Hinteregger, B. Großwindhager, M. Rath, M.S. Bakr, E. Leitinger, K. Witrisal, Using DecaWave UWB transceivers for high-accuracy multipath-assisted indoor positioning, in: 2017 IEE Int. Conf. on Commun. Workshops (ICC Workshops), Paris, France, May 2017. Also available as TD(17)03066.

[KHH+19] Pasi Koivumaki, Mikko Heino, Katsuyuki Haneda, Mikko Puranen, Johanna Pesola, Dynamic Polarimetric Wideband Channel Sounding in an Elevator Shaft, Technical Report TD(19)09022, Dublin, Ireland, 2019.

[KHK+18] P. Kyösti, L. Hentilä, J. Kyröläinen, F. Zhang, W. Fan, M. Latva-aho, Emulating dynamic radio channels for radiated testing of massive MIMO devices, in: Proc. 12th European Conference on Antennas and Propagation (EuCAP), London, UK, January 2018. Also available as TD(18)06035.

[Kil13] P.-S. Kildal, Rethinking the wireless channel for OTA testing and network optimization by including user statistics: RIMP, pure-LOS, throughput and detection probability, in: Intern. Symposium on Antennas and Propagation (ISAP), Nanjing, PRC, October 2013.

[KIU+16] Minseok Kim, Tatsuki Iwata, Kento Umeki, Karma Wangchuk, Jun-ichi Takada, Shigenobu Sasaki, Simulation based mm-wave channel model for outdoor open area access scenarios, in: 2016 URSI Asia-Pacific Radio Science Conference, August 2016, pp. 1292–1295. Also available as TD(16)01044.

[KIU+17] M. Kim, T. Iwata, K. Umeki, J-I. Takada, S. Sasaki, Mm-wave indoor channel cluster analysis by multipath component extraction from directional antenna scanning, Technical Report TD(17)03055, Lisbon, Portugal, February 2017.

[KIU+20] M. Kim, T. Iwata, K. Umeki, S. Sasaki, J. Takada, Millimeter-wave radio channel characterization using multi-dimensional sub-grid CLEAN algorithm, IEICE Transactions on Communications E103-B (7) (July 2020). Also available as TD(17)03055 and TD(19)09030.

[KJD17] Konstantinos Katzis, Richard Jones, George Despotou, Totally connected healthcare with TV white spaces, in: Studies in Health Technology and Informatics, vol. 238, July 2017, p. 68. Also available as TD(17)04029.

[KJNH17] A. Karttunen, J. Järveläinen, S.L.H. Nguyen, K. Haneda, Above 6 GHz multipath cross-polarization ratio in indoor and outdoor scenarios, Technical Report TD(17)03052, Lisbon, Portugal, February 2017.

[KKH18] Pekka Kyösti, Jukka Kyröläinen, Lassi Hentilä, Fixing randomness of 3GPP clustered delay line models, Technical report, Podgorica, Montenegro, Oct. 2018, TD(18)08047.

[KKWJ17] Jakub Kmiecik, Pawel Kulakowski, Krzysztof Wojcik, Andrzej Jajszczyk, In-body communication between mobile proteins using FRET, Technical Report TD(17)04058, COST Action CA15104 (IRACON), Lund, Sweden, May 2017.

[KKWJ18] Jakub Kmiecik, Pawel Kulakowski, Krzysztof Wojcik, Andrzej Jajszczyk, Information transfer from FRET-based nanonetworks to nerve cells, Technical Report TD(18)07041, COST Action CA15104 (IRACON), Cartagena, Spain, May 2018.

[KLDG16] W. Kotterman, M.H. Landmann, G. Del Galdo, On stochastically emulating continuous scattering structures by discrete sources for OTA testing of DuT with highly directive antennas, Technical report, Technische Universität Ilmenau, Ilmenau, Germany, May 2016. Also available as TD(16)01022.

[KLHC18] Kevin Kolpatzeck, Xuan Liu, Lars Haring, Andreas Czylwik, System-theoretical modeling and analysis of phase control in a photonically steered terahertz phased array transmitter, in: COST-IRACON, 2018, TD07066.

[KLHNK+18] A. Karttunen, S. Le Hong Nguyen, P. Koivumäki, K. Haneda, T. Hentilä, A. Asp, A. Hujanen, I. Huhtinen, M. Somersalo, S. Horsmanheimo, J. Aurinsalo, Window and wall penetration loss on-site measurements with three methods, in: Proc. of the 18th IEEE European Conference on Antennas and Propagation (EuCAP 2018), London, UK, April 2018. Also available as TD(18)06022.

[KLMLa17] P. Kyösti, J. Lehtomäki, J. Medbo, M. Latva-aho, Map-based channel model for evaluation of 5G wireless communication systems, IEEE Transactions on Antennas and Propagation 65 (12) (Dec. 2017) 6491–6504.

[KMMZ19] Jerzy Kowalewski, Jonathan Mayer, Tobias Mahler, Thomas Zwick, Evaluation of a Wideband Pattern-Reconfigurable Multiple Antenna System for Vehicular Applications, Technical Report TD-19-11002, Gdansk, Poland, September 2019.

[KMZ16] Jerzy Kowalewski, Tobias Mahler 1, Jonathan Mayer, Thomas Zwick, A miniaturized pattern reconfigurable antenna for automotive applications, Technical Report TD-16-02045, Durham, United Kingdom, October 2016.

[KNHS18] P. Koivumäki, S.L.H. Nguyen, K. Haneda, G. Steinböck, A study of polarimetric diffuse scattering at 28 GHz for a shopping center facade, Technical Report TD(18)07022, Cartagena, Spain, June 2018.

[KO17] Amit Kachroo, M. Kemal Ozdemir, Reduced rank based joint DoA estimation with mutual coupling in massive MIMO networks, in: COST-IRACON, 2017, TD04044.

[KPB+17] V. Kalokidou, T. Pelham, M. Beach, P. Legg, A. Lunness, Link performance evaluation and channel propagation for mm-Wave systems, in: Proc. 2017 European Conference on Networks and Communications (EuCNC), Oulu, Finland, June 2017. Also available as TD(17)03047.

[KPVC16] D. Kibloff, S.M. Perlaza, G. Villemaud, L.S. Cardoso, On the duality between state-dependent channels and wiretap channels, in: 2016 IEEE Global Conference on Signal and Information Processing (GlobalSIP), December 2016, pp. 953–958. Also available as TD(16)01073.

[KRCJS17] Pawel T. Kosz, Piotr Rajchowski, Krzysztof K. Cwalina, Jaroslaw Sadowski, Jacek Stefanski, Measurements of High-Speed Data Transmission Radio Link for Special Applications, Technical Report TD-05-012, Graz, Austria, 2017.

[KSH19] Pasi Koivumäki, Gerhard Steinböck, Katsuyuki Haneda, Impacts of point cloud modeling on accuracy of ray-based wave propagation simulations, Technical report, Oulu, Finland, May 2019, TD(19)10044.

[KSW17] Pawel Kulakowski, Kamil Solarczyk, Krzysztof Wojcik, Routing based on FRET for in-body nanonetworks, Technical Report TD(17)03030, COST Action CA15104 (IRACON), Lisbon, Portugal, February 2017.

[KTC+09] J. Karedal, F. Tufvesson, N. Czink, A. Paier, C. Dumard, T. Zemen, C.F. Mecklenbrauker, A.F. Molisch, A geometry-based stochastic MIMO model for vehicle-to-vehicle communications, IEEE Transactions on Wireless Communications 8 (7) (July 2009) 3646–3657.

[KTC19]   Pawel Kulakowski, Kenan Turbic, Luis M. Correia, From nanocommuni-cations to body area networks: a perspective on truly personal communica-tions, Technical Report TD(19)09075, COST Action CA15104 (IRACON), Dublin, Ireland, January 2019.

[KV96]   H. Krim, M. Viberg, Two decades of array signal processing research: the parametric approach, IEEE Signal Processing Magazine 13 (4) (July 1996) 67–94.

[KWA+07]   Johan Karedal, Shurjeel Wyne, Peter Almers, Fredrik Tufvesson, Andreas F. Molisch, A measurement-based statistical model for industrial ultra-wideband channels, IEEE Transactions on Wireless Communications 6 (8) (2007).

[KWAC+19]   Pawel Kulakowski, Krzysztof Wojcik, Rafael Asorey-Cacheda, Sebastian Canovas-Carrasco, Garcia-Sanchez Antonio-Javier, Joan Garcia-Haro, An architecture model of an in-body nanonetwork for disease detection, Tech-nical Report TD(19)10056, COST Action CA15104 (IRACON), Oulu, Fin-land, May 2019.

[KWG+18]   Josef Kulmer, Fuxi Wen, Nil Garcia, Henk Wymeersch, Klaus Witrisal, Im-pact of rough surface scattering on stochastic multipath component models, in: 2018 IEEE 29th Annual International Symposium on Personal, Indoor and Mobile Radio Communications (PIMRC), Sep. 2018. Also available as TD(18)07073.

[KWKJ18]   Jakub Kmiecik, Krzysztof Wojcik, Pawel Kulakowski, Andrzej Jajszczyk, Nanocommunication using FRET signals for in-body medical systems, Technical Report TD(18)06025, COST Action CA15104 (IRACON), Nicosia, Cyprus, January 2018.

[KWTC19]   Pawel Kulakowski, Krzysztof Wojcik, Kenan Turbic, Luis M. Correia, In-terfacing bans with nano-networks for medical applications, Technical Re-port TD(19)11037, COST Action CA15104 (IRACON), Gdansk, Poland, September 2019.

[Kyo19]   P. Kyoesti, Some considerations on correlations of hybrid beam form-ing arrays in the context of OTA testing, Technical Report TD-19-10058, Keysight Technologies Finland oy, Oulu, Finland, May 2019.

[LAG+18]   Xue Lin, Bo Ai, Ke Guan, Danping He, Ian Kavanagh, Conor Brennan, Zhangdui Zhong, Comparative analysis of ray tracing and volume electric field integral equation for indoor propagation modeling, Technical report, Nicosia, Cyprus, Jan. 2018, TD(18)06028.

[LAH+19]   J. Li, B. Ai, R. He, M. Yang, Z. Zhong, Y. Hao, A cluster-based chan-nel model for massive MIMO communications in indoor hotspot scenar-ios, IEEE Transactions on Wireless Communications 18 (8) (Aug 2019) 3856–3870. Also available as TD(19)11011.

[Lal16]   Pierre Laly, Ground to X polarimetric radio channel characterization in for-est scenarios, Technical Report TD(16)02032, Durham, UK, 2016.

[LB20]   C. Larsson, L. Bao, Availability of 7 km-long parallel 18 GHz band and e-band links for multi band solutions, in: 14th European Conference on Antennas and Propagation (EuCAP 2020), March 2020. Also available as TD(19)11043.

[LBÅ+17]   X. Li, K. Batstone, K. Åstrom, M. Oskarsson, C. Gustafson, F. Tufves-son, Robust phase-based positioning using massive MIMO with limited bandwidth, Montreal, QC, Canada, October 2017. Also available as TD(17)04032.

[LBC+16] W. Li, F. Bassi, A. Callisti, D. Dardari, L. Galluccio, M. Kieffer, G. Pasolini, Distributed defective node detection in Delay Tolerant Networks, Technical Report TD-01-083, Lille, France, May 2016.

[LBG+17] Xingqin Lin, Johan Bergman, Fredrik Gunnarsson, Olof Liberg, Sara Modarres Razavi, Hazhir Shokri Razaghi, Henrik Rydn, Yutao Sui, Positioning for the internet of things: a 3GPP perspective, IEEE Communications Magazine 55 (12) (2017) 179–185.

[LETM14] E. Larsson, O. Edfors, F. Tufvesson, T. Marzetta, Massive MIMO for next generation wireless systems, IEEE Communications Magazine 52 (2) (Feb. 2014) 186–195.

[LFBH19] A. Loaiza-Freire, M. Beach, G. Hilton, Characterisation of specular reflection of different building materials at 26 GHz, Technical Report TD(19)10033, Oulu, Finland, May 2019.

[LFGB+16] A. Loaiza-Freire, A.A. Goulianos, T.H. Barratt, T.M. Stone, D. Kong, E. Mellios, P. Cain, M. Rummey, A. Nix, M.A. Beach, Measuring and modelling of specular reflections at mm-wave frequencies, Technical Report TD(16)01016, Lille, France, May 2016.

[LFGB17] A. Loaiza-Freire, A.A. Goulianos, M.A. Beach, Analysis of RMS delay spread of outdoor channel at 60 GHz, Technical Report TD(17)04064, Lund, Sweden, May 2017.

[LG17] P.H. Lehne, A.A. Glazunovz, An Angular Model for the Spatial User Distribution inside a Cell, Technical Report TD(17)05033, IRACON Cost Action CA15104, Lisbon, Portugal, February 2017.

[LGW19] E. Leitinger, S. Grebien, K. Witrisal, Multipath-based slam exploiting AOA and amplitude information, in: Proc. IEEE ICC-19, Workshops, 2019.

[LHB+19] D. Loschenbrand, M. Hofer, L. Bernado, G. Humer, B. Schrenk, S. Zelenbaba, T. Zemen, Distributed massive MIMO channel measurements in urban vehicular scenario, in: 2019 13th European Conference on Antennas and Propagation (EuCAP), March 2019, pp. 1–5. Also available as TD(18)08031.

[LHZ17a] David Loschenbrand, Markus Hofer, Thomas Zemen, Ray-Tracer Based Channel Characteristics for Distributed Massive MIMO, Technical Report TD-17-03041, Lisbon, Portugal, February 2017.

[LHZ17b] David Loschenbrand, Markus Hofer, Thomas Zemen, Spatial Correlation for Distributed Massive MIMO in Indoor Scenarios, Technical Report TD-17-05035, Graz, Austria, September 2017.

[Li16] J. Li, LOS probability modeling for 5G indoor scenario, in: 2016 International Symposium on Antennas and Propagation (ISAP), Oct 2016, pp. 204–205.

[LJ16] Sylvie Dijkstra-Soudarissanane, Michal Golinski, Ljupco Jorguseski, Haibin Zhang, LTE Delay Assessment for Real-Time Management of Future Smart Grids, Technical Report TD(16)02003, Durham, United Kingdom, 2016.

[LKDG17] M. Lorenz, W. Kotterman, G. Del Galdo, On modeling OTA channel emulation systems, in: Proc. 12th European Conference on Antennas and Propagation (EuCAP), Ilmenau, Germany, May 2017. Also available as TD(17)04063.

[LL17] Hui Li, Buon Kiong Lau, Hybrid Mode MIMO Terminal Antenna with Low Correlation and Enhanced Bandwidth, Technical Report TD-17-04027, Lund, Sweden, May 2017.

[LLO+19] X. Li, E. Leitinger, M. Oskarsson, K. Åström, F. Tufvesson, Massive MIMO-based localization and mapping exploiting phase information of multipath components, IEEE Transactions on Wireless Communications 18 (9) (Sep. 2019) 4254–4267.

[LLY+16] T. Liu, Y. Liu, L. Yang, Y. Guo, C. Wang, BackPos: high accuracy backscatter positioning system, IEEE Transactions on Mobile Computing 15 (3) (March 2016) 586–598.

[LMTW17] E. Leitinger, F. Meyer, F. Tufvesson, K. Witrisal, Factor graph based simultaneous localization and mapping using multipath channel information, in: 2017 IEE Int. Conf. on Commun. Workshops (ICC Workshops), Paris, France, May 2017. Also available as TD(17)04012.

[LOP+12] L. Liu, C. Oestges, J. Poutanen, K. Haneda, P. Vainikainen, F. Quitin, F. Tufvesson, P.D. Doncker, The COST 2100 MIMO channel model, IEEE Wireless Communications 19 (6) (Dec. 2012) 92–99.

[LPK+18] A. Loaiza, T. Pelham, D. Kong, L. Sayer, V. Sgardoni, F. Tila, E. Mellios, M. Beach, A. Nix, G. Steinböck, Polarimetric diffuse scattering channel measurements at 26 GHz and 60 GHz, Technical Report TD1807068, Cartagena, Spain, June 2018.

[LST+07] Markus Landmann, Kriangsak Sivasondhivat, Junichi Takada, Ichirou Ida, Reiner Thomä, Polarization behavior of discrete multipath and diffuse scattering in urban environments at 4.5 GHz, EURASIP Journal on Wireless Communications and Networking 1 (Jan. 2007).

[LVD+19] J.S. Lu, E.M. Vitucci, V. Degli-Esposti, F. Fuschini, M. Barbiroli, J.A. Blaha, H.L. Bertoni, A discrete environment-driven GPU-based ray launching algorithm, IEEE Transactions on Antennas and Propagation 67 (2) (Feb. 2019) 1180–1192. Also available as TD(16)01041 and TD(17)05042.

[LWC+16] D. Liu, L. Wang, Y. Chen, M. Elkashlan, K. Wong, R. Schober, L. Hanzo, User association in 5G networks: a survey and an outlook, IEEE Communications Surveys and Tutorials 18 (2) (2016) 1018–1044.

[Lys17] Albert Lysko, CSIR achievements to the date and ongoing work in television white spaces, dynamic spectrum access and smart antenna: an overview, Technical Report TD(17)05001, Graz, Austria, September 2017.

[LYWT17] C. Ling, X. Yin, H. Wang, R.S. Thoma, Experimental characterization and multipath cluster modeling for 13-17 GHz indoor propagation channels, IEEE Transactions on Antennas and Propagation 65 (12) (Dec 2017) 6549–6561. Also available as TD(16)01076.

[LZB+16a] Leo Laughlin, Chunqing Zhang, Mark A. Beach, Kevin A. Morris, John L. Haine, Dynamic performance of electrical balance duplexing in a vehicular scenario, in: COST-IRACON, 2016, TD01082.

[LZB+16b] Leo Laughlin, Chunqing Zhang, Mark A. Beach, Kevin A. Morris, John L. Haine, Electrical balance duplexer performance in high speed rail applications, in: COST-IRACON, 2016, TD02038.

[LZB+17] Leo Laughlin, Chunqing Zhang, Mark A. Beach, Kevin A. Morris, John L. Haine, Update on electrical balance duplexer performance in high speed rail applications, in: COST-IRACON, 2017, TD03033.

[MAAV14] A. Mesodiakaki, F. Adelantado, L. Alonso, C. Verikoukis, Energy-efficient context-aware user association for outdoor small cell heterogeneous networks, in: 2014 IEEE International Conference on Communications (ICC), June 2014, pp. 1614–1619.

[MAR+18] M.A. Marotta, H. Ahmadi, J. Rochol, L. DaSilva, C.B. Both, Characterizing the relation between processing power and distance between BBU and RRH in a cloud RAN, IEEE Wireless Communications Letters 7 (3) (June 2018) 472–475. Also available as TD(19)11040.

[MBBV19] Silvia Mignardi, Chiara Buratti, Alessandro Bazzi, Roberto Verdone, Trajectories and resource management of flying base stations for C-V2X, Sensors 19 (4) (2019). Also available as TD(19)10020.

[MBC19] B. Montenegro-Villacieros, J. Bishop, J. Chareau, Clutter loss measurements and simulations at 26 GHz and 40 GHz, in: 2019 13th European Conference on Antennas and Propagation (EuCAP), March 2019, pp. 1–5.

[MBCV18] S. Mignardi, C. Buratti, V. Cacchiani, R. Verdone, Path optimization for unmanned aerial base stations with limited radio resources, in: IEEE 28th Annual International Symposium on Personal, Indoor, and Mobile Radio Communications (PIMRC), September 2018. Also available as TD(18)08023.

[MBLM13] P. Munoz, R. Barco, D. Laselva, P. Mogensen, Mobility-based strategies for traffic steering in heterogeneous networks, IEEE Communications Magazine 51 (5) (May 2013) 54–62.

[MBV18] S. Mignardi, C. Buratti, R. Verdone, On the Impact of Radio Channel over REM-Aware UAV-Aided Mobile Networks, in: 22nd International ITG Workshop on Smart Antennas (WSA 2018), March 2018. Also available as TD(18)06039.

[MCB+18] B. Montenegro-Villacieros, J. Chareau, J. Bishop, P. Viaud, T. Pinato, M. Basso, Clutter loss measurements in suburban environment at 26 GHz and 40 GHz, in: 12th European Conference on Antennas and Propagation (EuCAP 2018), April 2018, pp. 1–4.

[MCL16] L. Maviel, Y. Corre, Y. Lostanlen, Simulation tools for the design of mesh backhaul and small-cell networks in millimeter-wave band, in: 2016 IEEE 21st International Workshop on Computer Aided Modelling and Design of Communication Links and Networks (CAMAD), Oct 2016, pp. 165–169.

[MEO18] G. Makhoul, R.D. Errico, C. Oestges, Characterizing and modeling the wideband vehicle to pedestrian propagation channels, Technical Report TD(18)07029, Cartagena, Spain, May 2018.

[MFJP18] A.W. Mbugua, W. Fan, Y. Ji, G.F. Pedersen, Millimeter wave multi-user performance evaluation based on measured channels with virtual antenna array channel sounder, IEEE Access 6 (2018) 12318–12326. Also available as TD(18)07006.

[MGG+16] M. Martinez-Ingles, J.M. Garcia-Pardo, D.P. Gaillot, J. Pascual-Garcia, J. Rodriguez, L.J. Llacer, M. Lienard, Polarimetric indoor measurements at 94 GHz, in: 2016 10th European Conference on Antennas and Propagation (EuCAP), April 2016, pp. 1–3. Also available as TD(16)02034.

[MGO18] Yang Miao, Quentin Gueuning, Claude Oestges, Modeling the phase correlation of effective diffuse scattering from surfaces for radio propagation prediction with antennas at refined separation, IEEE Transactions on Antennas and Propagation 66 (3) (March 2018) 1427–1435. Also available as TD(16)01015.

[MH17]   Y. Mirsky, Y. Haddad, A Linear Downlink Power Control Algorithm for Cellular Networks, Technical Report TD(17)03057, IRACON Cost Action CA15104, Lisbon, Portugal, February 2017.

[MHD⁺17]   Martin Matis, David Hrabcak, Lubomir Dobos, Dominik Neznik, Jan Papaj, Study of channel assigning methods impact on paths selection in CR-MANET, Technical Report TD(17)05043, Graz, Austria, September 2017.

[MHRA18]   Y. Mirsky, Y. Haddad, O. Rozenblit, R. Azoulay, Predicting wireless coverage maps using radial basis networks, in: 2018 15th IEEE Annual Consumer Communications Networking Conference (CCNC), January 2018, pp. 1–4. Also available as TD(18)07017.

[MIGMGP⁺17]   M.T. Martinez-Ingles, D.P. Gaillot, J.M. Molina-Garcia-Pardo, J.V. Rodriguez, J. Pascual-Garcia, L. Juan-Liacer, M. Lienard, Experimental analysis of alignment impact in short communications at 300 GHz, in: 11th European Conference on Antennas and Propagation (EuCAP 2017), 2017, pp. 1–5. Also available as TD(17)03012.

[MIHK19]   P. Mäkikyrö, M.-M. Islam, A. Hekkala, E. Kaivanto, OTA measurement range length in 5G systems, Technical Report TD-19-10047, Keysight Technologies Finland oy, Oulu, Finland, May 2019.

[MIMGPG⁺17]   Maria-Teresa Martinez-Ingles, Jose-Maria Molina-Garcia-Pardo, Davy P. Gaillot, J. Romeu, L. Jofre, UWB Millimeter-Wave and Terahertz Monostatic Near Field Synthetic Aperture Imaging, Technical Report TD(17)05002, Graz, Austria, 2017.

[Min]   Mininet, open networking foundation - webpage, https://www.opennetworking.org/mininet/.

[MIŠ⁺11]   C. Mehlführer, J.C. Ikuno, M. Šimko, S. Schwarz, M. Wrulich, M. Rupp, The Vienna LTE simulators - enabling reproducibility in wireless communications research, EURASIP Journal on Advances in Signal Processing 2011 (1) (July 2011) 29.

[MK18]   L. Mfupe, K. Katzis, Airborne Wireless Network Infrastructure for Affordable Broadband, Emergence Services and Monitoring of Radio Astronomy, 2018. Available as TD(18)006042.

[MKK⁺16]   J. Medbo, P. Kyösti, K. Kusume, L. Raschkowski, K. Haneda, T. Jämsä, V. Nurmela, A. Roivainen, J. Meinilä, Radio propagation modeling for 5G mobile and wireless communications, IEEE Communications Magazine 54 (6) (June 2016) 144–151.

[ML17]   Zachary Miers, Buon Kiong Lau, Characteristic Mode Based MIMO Terminal Antenna Design with User Proximity, Technical Report TD-17-04028, Lund, Sweden, May 2017.

[MLV07]   Achraf Mallat, Jérôme Louveaux, Luc Vandendorpe, UWB based positioning in multipath channels: CRBs for AOA and for hybrid TOA-AOA based methods, in: Communications, 2007. ICC'07. IEEE International Conference on, IEEE, 2007, pp. 5775–5780.

[MMA⁺18]   M. Mehdi, G. Mühlmeier, K. Agrawal, R. Pryss, M. Reichert, F.J. Hauck, Referenceable mobile crowdsensing architecture: a healthcare use case, Procedia Computer Science 134 (2) (2018) 445–451, PMID: 18770509.

[MMC⁺ed]   M. Menarini, P. Marrancone, G. Cecchini, A. Bazzi, B. Masini, A. Zanella, TRUDI: Testing Environment for Vehicular Applications Running with Devices in the Loop, Technical Report TD(19)09070, IRACON COST Action CA15104, 2019.

[MMdHB18] Ismael García Varea, Miguel Martínez del Horno, Luis Orozco Barbosa, Spatial statistical analysis for the design of indoor particle filter based localization mechanisms, in: COST IRACON (CA15104) 7th Scientific Meeting, TD(18)07027, Cartagena, Spain, May 29-June 1 2018, pp. 1–10. Also available as TD(18)07027.

[MMEO16] Gloria Makhoul, Francesco Mani, Raffaele D. Errico, Claude Oestges, Time Correlation in Mobile to Mobile Indoor Channels, Technical Report TD(16)01072, Lille, France, May 2016.

[MNP+18] S. Monfared, T. Nguyen, L. Petrillo, P. De Doncker, F. Horlin, Experimental demonstration of BLE transmitter positioning based on AOA estimation, in: 2018 IEEE 29th Annual International Symposium on Personal, Indoor and Mobile Radio Communications (PIMRC), Bologna, Italy, Sep 2018, pp. 856–859. Also available as TD(18)07028.

[Mol05] Andreas F. Molisch, Wireless Communications, Wiley-IEEE Press, 2005.

[MPG+19a] M. Martinez-Ingles, J. Pascual-García, D.P. Gaillot, C.S. Borras, J. Molina-García-Pardo, Indoor 1-40 GHz channel measurements, in: 2019 13th European Conference on Antennas and Propagation (EuCAP), March 2019, pp. 1–5.

[MPG+19b] Yang Miao, Troels Pedersen, Mingming Gan, Ebgenii Vinogradov, Claude Oestges, Reverberant room-to-room radio channel prediction by using rays and graphs, IEEE Transactions on Antennas and Propagation 67 (1) (Oct. 2019) 484–494. Also available as TD(17)04048.

[MPGRS18] Blagoj Mitrevski, Viktor Petreski, Martin Gjoreski, Biljana Risteska Stojkoska, Framework for human activity recognition on smartphones and smartwatches, in: Proc. ICT Innovations 2018, Ohrid, Macedonia, September 2018. Also available as TD(18)08008.

[MPLC16] G. Martinez-Pinzon, K. Llamas, N. Cardona, Potential sharing between DTT and IoT services in the UHF band, in: Proc. PIMRC 2016 - IEEE 27th Int. Symp. on Pers., Indoor and Mobile Radio Commun., September 2016. Also available as TD(16)01020.

[MPSK17] J. Milos, L. Polak, I. Slanina, T. Kratochvil, Link-Level WLAN Simulator: Performance Analysis of IEEE 802.11 Technologies, Technical Report TD-04-009, Lund, Sweden, May 2017.

[MR11] A. Mahmud, R. Rahmani, Exploitation of OpenFlow in wireless sensor networks, in: Proceedings of 2011 International Conference on Computer Science and Network Technology, vol. 1, 2011, pp. 594–600.

[MSA02] A.F. Molisch, M. Steinbauer, H. Asplund, "Virtual cell deployment areas" and "cluster tracing" - new methods for directional channel modeling in microcells, in: Vehicular Technology Conference. IEEE 55th Vehicular Technology Conference (VTC Spring 2002), vol. 3, May 2002, pp. 1279–1283.

[MSA16] J. Medbo, N. Seifi, H. Asplund, Frequency dependency of measured highly resolved directional propagation channel characteristics, in: European Wireless 2016; 22nd European Wireless Conference, May 2016, pp. 1–6. Also available as TD(16)02035.

[MSA+17] J. Medbo, D. Sundman, H. Asplund, N. Jaldén, S. Dwivedi, Wireless urban propagation measurements at 2.44, 5.8, 14.8 & 58.68 GHz, in: 2017 XXXIInd General Assembly and Scientific Symposium of the International Union of Radio Science (URSI GASS), Aug 2017, pp. 1–4.

[MSE⁺16] Zachary T. Miers, Augustine Sekyere, John Ako Enohnyaket, Max Landaeus, Buon Kiong Lau, On Characteristic Modes of MIMO Terminals with Real Components, Technical Report TD-16-01008, Lille, France, May 2016.

[MT90] Silvano Martello, Paolo Toth, Knapsack Problems: Algorithms and Computer Implementations, John Wiley & Sons, Inc., New York, NY, USA, 1990.

[MV17] S. Mignardi, R. Verdone, On the performance improvement of a cellular network supported by an unmanned aerial base station, in: 2017 29th IEEE International Teletraffic Congress (ITC29), September 2017. Also available as TD(17)03068.

[MV18] S. Mignardi, R. Verdone, Joint path and radio resource management for UAVs supporting mobile radio networks, in: 2018 17th Annual Mediterranean Ad Hoc Networking Workshop (Med-Hoc-Net), June 2018. Also available as TD(18)07060.

[MVB⁺18] F. Mani, E.M. Vitucci, M. Barbiroli, F. Fuschini, V. Degli-Esposti, 26 GHz ray-tracing pathloss prediction in outdoor scenario in presence of vegetation, in: Proc. 12th European Conf. Ant. Prop. (EuCAP 2018), London, UK, Apr. 2018. Also available as TD(17)05050.

[MVBSR17] B. Montenegro-Villacieros, J. Bishop, S. Salous, X. Raimundo, Channel propagation experimental measurements and simulations at 52 GHz, in: 32nd General Assembly and Scientific Symposium of the International Union of Radio Science (GASS 2017), Montreal, Canada, August 2017. Also available as TD(16)02039.

[MVR09] R.K. Martin, J.S. Velotta, J.F. Raquet, Bandwidth efficient cooperative TDOA computation for multicarrier signals of opportunity, IEEE Transactions on Signal Processing 57 (6) (June 2009) 2311–2322.

[MVS19] B. Montenegro-Villacieros, S. Salous, Combined and independent statistics of clutter loss and building entry loss measurements at 26 GHz, Technical Report TD(19)10048, Oulu, Finland, May 2019.

[NA10] A. Nemra, N. Aouf, Robust INS/GPS sensor fusion for UAV localization using SDRE nonlinear filtering, IEEE Sensors Journal 10 (4) (Apr. 2010) 789–798.

[NBB19] J. Wagen, A. Nikodemski, F. Buntschu, G. Bovet, Some more results on MANET radio performances using the EMANE framework, Technical report, 2019. Also available as TD(19)09055.

[NC17] Andres Navarro, William Cruz, PDOA emitter location using game engines 3D ray based tools, in: COST IRACON (CA15104) 3rd Scientific Meeting, TD(17)03021, Lisbon, Portugal, Feb 1-3 2017, pp. 1–7. Also available as TD(17)03021.

[NCB18] U. Noreen, L. Clavier, A. Bounceur, LoRa-like CSS-based PHY layer, capture effect and serial interference cancellation, in: European Wireless 2018; 24th European Wireless Conference, May 2018. Also available as TD(18)07005.

[NCM17] Andrés Navarro, Narcís Cardona, Gerardo Martínez, Coexistence studies between LTE and narrowband mobile radio technologies in the 850 MHz band for Colombia, Technical Report TD(17)05048, Graz, Austria, September 2017.

[ND18] Thomas Kurner, Nils Dreyer, Michael Schweins, A V2X Extension for the Simulator for Mobile Networks, (SiMoNe), Technical Report TD(18)06017, Nicosia, Cyprus, 2018.

[NE19] Jessen Narrainen, Raffaele D. Errico, Channel characterization in industrial environment with high clutter, Technical Report TD(19)10034, Oulu, Finland, May 2019.

[Nel99] R. Nelson, An Introduction to Copulas, Springer-Verlag, New York, NY, 1999.

[NGAT17] M.G. Nilsson, C. Gustafson, T. Abbas, F. Tufvesson, A measurement-based multilink shadowing model for V2V network simulations of highway scenarios, IEEE Transactions on Vehicular Technology 66 (10) (Oct 2017) 8632–8643. Also available as TD(16)01040.

[NGAT18] M.G. Nilsson, C. Gustafson, T. Abbas, F. Tufvesson, A path loss and shadowing model for multilink vehicle-to-vehicle channels in urban intersections, Sensors 18 (12) (Dec 2018) 4433. Also available as TD(18)06043.

[NGN08] NGNM, Recommendation on SON and O&M Requirements, Whitepaper 1.23, Next Generation Mobile Networks Alliance, December 2008.

[NLVT17] C. Nelson, N. Lyamin, A. Vinel, F. Tufvesson, Geometry based channel models with cross- and auto-correlation for simulations of V2V wireless networks, Technical Report TD(17)04081, Lund, Sweden, May 2017.

[NLZ18] V. Nair, R. Litjens, H. Zhang, Assessment of the suitability of NB-IoT technology for ORM in smart grid, in: 2018 European Conference on Networks and Communications (EuCNC), June 2018. Also available as TD(18)07007.

[NM18] Karim M. Nasr, Klaus Moessner, Knapsack optimisation versus cell range expansion for mobility load balancing in dense small cells, in: 2018 European Conference on Networks and Communications, EuCNC 2018, Ljubljana, Slovenia, June 18-21, 2018, IEEE, 2018, pp. 1–9. Also available as TD(18)06021.

[NMP$^+$18] S.L.H. Nguyen, J. Medbo, M. Peter, A. Karttunen, K. Haneda, A. Bamba, R. D'Errico, N. Iqbal, C. Diakhate, J. Conrat, On the frequency dependency of radio channel's delay spread: analyses and findings from mmMAGIC multi-frequency channel sounding, in: 12th European Conference on Antennas and Propagation (EuCAP 2018), April 2018, pp. 1–5. Also available as TD(17)04069.

[NPJ15] Kishan N. Patel, Rutvij Jhaveri, A survey on emulation testbeds for mobile ad-hoc networks, Procedia Computer Science 45 (December 2015) 581.

[NS3] NS3, The NS3 network simulator, http://www.nsnam.org/. (Accessed 11 April 2019).

[NSB$^+$18] K.K. Nagalapur, E.G. Ström, F. Brännström, J. Carlsson, K. Karlsson, A simple method for robust vehicular communication with multiple non-ideal antennas, in: 2018 IEEE MTT-S International Conference on Microwaves for Intelligent Mobility (ICMIM), April 2018, pp. 1–5. Also available as TD(18)07065.

[NSM$^+$05] T. Nakano, T. Suda, M. Moore, R. Egashira, A. Enomoto, K. Arima, Molecular communication for nanomachines using intercellular calcium signaling, in: 5th IEEE Conference on Nanotechnology, 2005, vol. 2, July 2005, pp. 478–481.

[NVM18] Karim M. Nasr, Seiamak Vahid, Klaus Moessner, Knapsack optimisation for mobility load balancing in dense small cell deployments, in: Paulo Marques, Ayman Radwan, Shahid Mumtaz, Dominique Noguet, Jonathan Rodriguez, Michael Gundlach (Eds.), Cognitive Radio Oriented Wireless Networks, Springer International Publishing, Cham, 2018, pp. 337–346.

[NWB+18] A. Nikodemski, J. Wagen, F. Buntschu, C. Gisler, G. Bovet, Reproducing measured MANET radio performances using the EMANE framework, IEEE Communications Magazine 56 (10) (October 2018) 151–155. Also available as TD(18)08034.

[ODV17] Claude Oestges, Natalia Dementieva, Evgenii Vinogradov, Evaluation of Large-Scale Parameters in Urban Microcells at 3.8 GHz, Technical Report TD(17)04005, Lund, Sweden, May 2017.

[OFK18] Aleksandra Orzechowska, Joanna Fiedor, Pawel Kulakowski, FRET-based nanocommunication in organic structures of plant origin, Technical Report TD(18)08033, COST Action CA15104 (IRACON), Podgorica, Montenegro, October 2018.

[OMT+17] Giuseppe Oliveri, Mohamad Mostafa, Werner G. Teich, Jurgen Lindner, Hermann Schumacher, Advanced low power high speed nonlinear signal processing: an analog VLSI example, in: COST-IRACON, 2017, TD04070.

[OOKF68] Y. Okumura, E. Ohmori, T. Kawano, K. Fukuda, Field strength and its variability in VHF and UHF land mobile radio service, Review of the Electrical Communications Laboratories (Tokyo) 16 (Sep. 1968) 825–873.

[Ope] Open-source routing and network simulation. Available: http://www.brianlinkletter.com/open-source-network-simulators/.

[OR16] Francisco Javier Ordóñez, Daniel Roggen, Deep convolutional and LSTM recurrent neural networks for multimodal wearable activity recognition, in: Sensors, 2016.

[OV18] R. Onica, V. Bota, Carrier frequency-offset compensation by intercarrier symbol precoding at the transmitter in OFDM transmissions, in: 2018 International Conference on Communications (COMM), June 2018, pp. 165–170. Also available as TD(17)04041.

[P80] IEEE P802.11, LRLP topic interest group, http://www.ieee802.org/11/Reports/lrlp_update.htm. (Accessed January 2018).

[PAHG16] C. Patané Lötbäck, K. Arvidsson, M. Högberg, M. Gustafsson, Base station over-the-air testing in reverberation chamber, Technical Report TD-16-2006, Bluetest AB, Gothenburg, Sweden, October 2016.

[PB10] A. Pantelopoulos, N.G. Bourbakis, A survey on wearable sensor-based systems for health monitoring and prognosis, IEEE Transactions on Systems, Man, and Cybernetics, Part C (Applications and Reviews) 40 (1) (Jan 2010) 1–12.

[PCF17] S. Persia, C. Carciofi, M. Faccioli, NB-IoT and LoRA connectivity analysis for M2M/IoT smart grids applications, in: 2017 AEIT International Annual Conference, September 2017. Also available as TD(18)06024.

[PDT+19] D. Plets, W. Deprez, J. Trogh, L. Martens, W. Joseph, Joint received signal strength, angle-of-arrival, and time-of-flight positioning, in: 13th European Conference on Antennas and Propagation, Krakow, Poland, 31 March-5 April 2019, pp. 1–6. Also available as TD(19)09012.

[Ped18]   Troels Pedersen, Modeling of path arrival rate for in-room radio channels with directive antennas, IEEE Transactions on Antennas and Propagation 66 (9) (Sep. 2018) 4791–4805. Also available as TD(17)04047.

[Ped19]   Troels Pedersen, Stochastic multipath model for the in-room radio channel based on room electromagnetics, IEEE Transactions on Antennas and Propagation 67 (4) (April 2019) 2591–2603. Also available as TD(18)08037.

[Pet17]   M. Peter (Ed.), Measurement results and final mmMAGIC channel models, Technical report, May 2017, Deliverable 2.2 v.2, Technical report, ICT-671650 mmMAGIC project.

[PGB⁺11]   G. Piro, L.A. Grieco, G. Boggia, F. Capozzi, P. Camarda, Simulating LTE cellular systems: an open-source framework, IEEE Transactions on Vehicular Technology 60 (2) (Feb 2011) 498–513.

[PGMGPMI⁺18] J. Pascual-Garcia, J. Molina-Garcia-Pardo, M. Martinez-Ingles, J. Rodriguez, L. Juan-Llácer, Wireless channel simulation using geometrical models extrated from point clouds, in: 2018 IEEE International Symposium on Antennas and Propagation USNC/URSI National Radio Science Meeting, Nicosia, Cyprus, July 2018, pp. 83–84. Also available as TD(18)06036.

[PKM18]   Giannis Papas, Konstantinos Katzis, Luzango Mfupe, A Geo-Locational Spectrum Database for Cyprus - Support TVWS for rural Internet access in 5G, Technical report, Podgorica, Montenegro, October 2018.

[PMHI16]   J. Petajajarvi, K. Mikhaylov, M. Hamalainen, J. Iinatti, Evaluation of LoRa LPWAN technology for remote health and wellbeing monitoring, in: 2016 10th International Symposium on Medical Information and Communication Technology (ISMICT), March 2016, pp. 1–5.

[PMQ18]   M. Petitjean, S. Mezhoud, F. Quitin, Fast localization of ground-based mobile terminals with a transceiver-equipped UAV, in: 2018 IEEE 29th Annual International Symposium on Personal, Indoor and Mobile Radio Communications (PIMRC), Bologna, Italy, 9–12 September, 2018, pp. 323–327. Also available as TD(19)10036.

[POC19]   T. Pairon, C. Oestges, C. Craeye, Impact of beamforming on human shadowing in rich scattering environment: a numerical analysis, Technical report, Dublin, Ireland, Jan 2019, TD(19)09038.

[PP02]   Athanasios Papoulis, S. Unnikrishna Pillai, Probability, Random Variables, and Stochastic Processes, 4th edition, McGraw-Hill, New York, NY, USA, 2002.

[PPMK19]   Ladislav Polak, Martin Potocnak, Jiri Milos, Tomas Kratochvil, On the Coexistence of LTE and LoRa in the 2.4 GHz ISM Band, Technical Report TD(19)09034, Dublin, Ireland, January 2019.

[PPP15]   5G PPP, 5G and the factories of the future, White Paper, October 2015.

[PPT⁺18]   N. Podevijn, D. Plets, J. Trogh, L. Martens, P. Suanet, K. Hendrikse, W. Joseph, Tdoa-based outdoor positioning with tracking algorithm in a public lora network, Wireless Communications and Mobile Computing (2018). Also available as TD(18)08009.

[PR16]   M. Priesler, A. Reichman, Resource allocation in OFDMA wireless mesh networks, in: ICOF 2016; 19th International Conference on OFDM and Frequency Domain Techniques, VDE, August 2016, pp. 1–5.

[Pre]   PredicTAKE repository. Available: https://gitlab.forge.hefr.ch/predictake.

[PRH⁺15] R. Pryss, M. Reichert, J. Herrmann, B. Langguth, W. Schlee, Mobile crowd sensing in clinical and psychological trials – a case study, in: IEEE 28th International Symposium on Computer-Based Medical Systems, IEEE, 2015, pp. 23–24.

[PRLS15] R. Pryss, M. Reichert, B. Langguth, W. Schlee, Mobile crowd sensing services for tinnitus assessment, therapy, and research, in: IEEE International Conference on Mobile Services, IEEE, 2015, pp. 352–359.

[PS17] Fernando Velez, Pedro Silva, Communications for Public Protection and Disaster Relief: Overview and Vision Toward the Future, Technical Report TD(17)03003, Lisbon, Portugal, 2017.

[PSAGP⁺19] Sofia Perez-Simbor, Carlos Andreu, Concepcion Garcia-Pardo, Matteo Frasson, Narcis Cardona, UWB path loss models for ingestible devices, IEEE Transactions on Antennas and Propagation (January 2019), Early Access. Also available as TD(19)09071.

[PSF12] Troels Pedersen, Gerhard Steinböck, Bernard H. Fleury, Modeling of reverberant radio channel using propagation graphs, IEEE Transactions on Antennas and Propagation 60 (12) (Aug. 2012) 5978–5988.

[PSF14] Troels Pedersen, Gerhard Steinböck, Bernard H. Fleury, Modeling of outdoor-to-indoor radio channels via propagation graphs, in: Proc. General Assembly and Scientific Symposium (URSI GASS), 2014 XXXIth URSI, Aug. 2014.

[PSH⁺15] Juho Poutanen, Jussi Salmi, Katsuyuki Haneda, Veli-Matti Kolmonen, Fredrik Tufvesson, Pertti Vainikainen, Propagation characteristics of dense multipath components, IEEE Antennas and Wireless Propagation Letters 9 (Aug. 2015) 791–794.

[PSP18] Fernando Velez, Pedro Silva, Jon Peha, Overview of Disaster-Resilient Public Safety Service Provision, Technical Report TD(18)06012, Nicosia, Cyprus, 2018.

[PT12] S. Payami, F. Tufvesson, Channel measurements and analysis for very large array systems at 2.6 GHz, in: 2012 6th European Conference on Antennas and Propagation (EuCAP), Prague, Czech Republic, Mar. 2012, pp. 433–437.

[PT15] A. Papaiz, A.M. Tonello, Particle filtering with weight reshaping for opportunistic angle of arrival estimation in a vehicular scenario, in: 2015 IEEE 5th International Conference on Consumer Electronics - Berlin (ICCE-Berlin), Sep. 2015, pp. 145–149.

[PT16] A. Papaiz, A.M. Tonello, Azimuth and elevation dynamic tracking of UAVs via 3-axial ULA and particle filtering, International Journal of Aerospace Engineering 2016 (2016), 9 pp.

[PTB⁺17] D. Plets, J. Trogh, S. Benaissa, W. Joseph, L. Martens, Internet-of-animals: wireless communication and location tracking, in: IRACON 4th MC and Technical Meeting - Lund 2017, May 2017, pp. 1–24. Also available as TD(17)04035.

[PTCB15] V. Pham-Xuan, D. Trinh-Xuan, M. Condon, C. Brennan, Rapid convergent iterative solver for computing two-dimensional random rough surface scattering, in: 2015 International Conference on Advanced Technologies for Communications (ATC), Oct. 2015, pp. 63–67. Also available as TD(19)09060.

[PVP16]   R.R. Paulo, F.J. Velez, G. Piro, Impact of Transmitter Power in Packet Schedulers Performance in LTE HetNets, Technical report TD(16)01031, Lille, France, May-June 2016.

[PVP18a]   R.R. Paulo, F.J. Velez, G. Piro, Design of coordinated HeNB deployments, in: 2018 IEEE 87th Vehicular Technology Conference (VTC Spring), June 2018, pp. 1–6. Also available as TD(18)06018.

[PVP18b]   R.R. Paulo, F.J. Velez, G. Piro, Performance evaluation and packet scheduling in HeNB deployments, in: 2018 IEEE 88th Vehicular Technology Conference (VTC Fall), August 2018. Also available as TD(18)08025.

[PWK^{+}15]   M. Peter, R.J. Weiler, T. Kuhne, B. Goktepe, J. Serafimoska, W. Keusgen, Millimeter-wave small-cell backhaul measurements and considerations on street-level deployment, in: 2015 IEEE Globecom Workshops (GC Wkshps), Dec. 2015, pp. 1–6.

[QGZT16]   F. Quitin, V. Govindaraj, X. Zhong, W.P. Tay, Virtual multi-antenna array for estimating the angle-of-arrival of a RF transmitter, in: 2016 IEEE 84th Vehicular Technology Conference (VTC-Fall), IEEE, 2016, pp. 1–5. Also available as TD(16)01021.

[QHD17]   Francois Quitin, Francois Horlin, Philippe De Doncker, A synchronization-free method for estimating TDOA: technique and proof-of-concept, in: COST IRACON (CA15104) 3rd Scientific Meeting, TD(17)03013, Lisbon, Portugal, Feb 1-3 2017, pp. 1–7. Also available as TD(17)03013.

[Qui17]   F. Quitin, A synchronization-free method for estimating TDOA: technique and proof-of-concept, Technical Report TD(17)03013, Lisbon, Portugal, Feb 2017.

[RAN18]   3GPP RAN4, Verification of radiated multi-antenna reception performance of user equipment (UE) (release 15), Technical Report 3GPP TR 37.977 V15.0.0, 3GPP, September 2018.

[Rap89]   T.S. Rappaport, Indoor radio communications for factories of the future, IEEE Communications Magazine 27 (5) (may 1989) 15–24.

[RBM^{+}18]   D. Reyes, M. Beach, E. Mellios, J. Haine, M. Rumney, Novel over-the-air test method for 5G mmWave devices with beam forming capabilities, in: Proc. IEEE Global Communications Conference Workshops (GC Wkshps), Abu Dhabi, UAE, December 2018. Also available as TD(18)08001.

[RBR^{+}19]   D. Reyes, M. Beach, M. Rumney, J. Haine, E. Mellios, Novel over-the-air test method for 5G millimetre wave devices, based on elliptical cylinder reflectors, in: Proc. IEEE 90th Vehicular Technology Conference (VTC Fall), Hawaii, USA, September 2019. Also available as TD(19)10004.

[RBZ17]   Olivier Renaudin, Thomas Burgess, Thomas Zemen, Ray-tracing based fingerprinting for indoor localization, in: COST IRACON (CA15104) 5th Scientific Meeting, TD(17)05024, Graz, Austria, Sept. 12-14 2017, pp. 1–10. Also available as TD(17)05024.

[RCBO19]   Piotr Rajchowski, Krzysztof K. Cwalina, Olga Błaszkiewicz, Alicja Olejniczak, A review of localization in OFDM based systems, in: COST IRACON (CA15104) 10th Scientific Meeting, TD(19)10051, Oulu, Finland, May 27-29 2019, pp. 1–2. Also available as TD(19)10051.

[RCC17a]   B. Rouzbehani, L.M. Correia, L. Caeiro, A fair mechanism of virtual radio resource management in multi-RAT wireless Het-Nets, Montreal, Canada, October 2017. Also available as TD(17)04001.

[RCC17b] B. Rouzbehani, L.M. Correia, L. Caeiro, A modified proportional fair radio resource management scheme in virtual RANs, Oulu, Finland, June 2017. Also available as TD(17)03037.

[RCC18a] B. Rouzbehani, L.M. Correia, L. Caeiro, An energy efficient service-based RANslicing technique in virtual wireless networks, Technical Report TD(18)08019, Podgorica, Montenegro, October 2018.

[RCC18b] B. Rouzbehani, L.M. Correia, L. Caeiro, An optimised RRM approach with multi-tenant performance isolation in virtual RANs, Bologna, Italy, September 2018. Also available as TD(18)06013.

[RCC18c] B. Rouzbehani, L.M. Correia, L. Caeiro, Radio resource and service orchestration for virtualised multi-tenant mobile Het-Nets, Barcelona, Spain, April 2018. Also available as TD(17)05011.

[RCC18d] B. Rouzbehani, L.M. Correia, L. Caeiro, An SLA-based method for radio resource slicing and allocation in virtual RANs, Technical Report TD(18)07034, Cartagena, Spain, May 2018.

[RCK+17] Piotr Rajchowski, Krzysztof K. Cwalina, Pawel T. Kosz, Jacek Stefanski, Jaroslaw Sadowski, Research and analysis of high-speed data transmission radio link designed for maritime environment, in: COST-IRACON, 2017, TD05014.

[RCM+19] P. Rajchowski, K.K. Cwalina, J. Magiera, A. Olejniczak, P.T. Kosz, A. Czapiewska, R. Burczyk, K. Kowalewski, J. Sadowski, S.J. Ambroziak, AEGIS – mobile device for generating electromagnetic curtain for special applications and countering the threats of RCIED, Telecommunication Review and Telecommunication News 6 (june 2019) 366–369 (in Polish). Also available as TD(19)11032.

[RCS19] Piotr Rajchowski, Krzysztof K. Cwalina, Jaroslaw Sadowski, Person tracking in a ferry environment using ultra-wide band radio interface, in: COST IRACON (CA15104) 9th Scientific Meeting, TD(19)09018, Dublin, Ireland, Jan 16-19 2019, pp. 1–3. Also available as TD(19)09018.

[RDJK19] A. Rajaram, R. Dinis, D.N.K. Jayakody, N. Kumar, Receiver design to employ simultaneous wireless information and power transmission with joint CFO and channel estimation, IEEE Access 7 (2019) 9678–9687. Also available as TD(19)09057.

[RdW+18] I. Rashdan, F. de Ponte Müller, W. Wang, M. Schmidhammer, S. Sand, Vehicle-to-pedestrian channel characterization: wideband measurement campaign and first results, in: 12th European Conference on Antennas and Propagation (EuCAP 2018), April 2018, pp. 1–5. Also available as TD(18)06009.

[Rec17a] Recommendation ITU-R P.1238-9, Propagation data and prediction methods for the planning of indoor radiocommunication systems and radio local area networks in the frequency range 300 MHz to 100 GHz, https://www.itu.int/rec/R-REC-P.1238-9-201706-I/en, June 2017.

[Rec17b] Recommendation ITU-R P.1411-9, Propagation data and prediction methods for the planning of short-range outdoor radiocommunication systems and radio local area networks in the frequency range 300 MHz to 100 GHz, https://www.itu.int/rec/R-REC-P.1411-9-201706-I/en, June 2017.

[Rec17c] Recommendation ITU-R P.2108-0, Prediction of clutter loss, https://www.itu.int/rec/R-REC-P.2108/en, June 2017.

[Rec17d] Recommendation ITU-R P.2109-0, Prediction of building entry loss, https://www.itu.int/rec/R-REC-P.2109/en, June 2017.

[Rec19] Recommendation ITU-R P.2001, A general purpose wide-range terrestrial propagation model in the frequency range 30 MHz to 50 GHz, https://www.itu.int/rec/R-REC-P.2001/en, Aug. 2019.

[RGW⁺18] S. Ruffieux, C. Gisler, J. Wagen, F. Buntschu, G. Bovet, Take — tactical ad-hoc network emulation, in: 2018 International Conference on Military Communications and Information Systems (ICMCIS), May 2018, pp. 1–8. Also available as TD(18)07033.

[RHMA18] Orit Rozenblit, Yoram Haddad, Yisroel Mirsky, Rina Azoulay, Machine learning methods for SIR prediction in cellular networks, Physical Communication 31 (2018) 239–253.

[Ric05] A. Richter, Estimation of Radio Channel Parameters: Models and Algorithms, PhD thesis, Technische Universität Ilmenau, Fakultät für Elektrotechnik und Informationstechnik, Ilmenau, Germany, 2005.

[RJS16] M. Rumney, Y. Jing, M. Salmeron, Validation of SCME wave fields, Technical Report TD-16-02026, Keysight Technologies Ltd., Edinburgh, UK, October 2016.

[RK16] D.M. Rose, T. Kurner, Effects of hyper-dense small-cell network deployments on a realistic urban environment, in: 2016 IEEE 84th Vehicular Technology Conference (VTC-Fall), Sep. 2016, pp. 1–5.

[RKH17] M. Rumney, P. Kyösti, A. Hekkala, Channel models, requirements and test methods for 3GPP NR, Technical Report TD-17-04046, Keysight Technologies Ltd., Edinburgh, UK, May 2017.

[RKKe15] L. Rachowski, P. Kyösti, K. Kusume, T. Jämsä (Eds.), METIS channel models, Technical report, May 2015, Deliverable 1.4 v.1.3. Technical report, ICT-317669 METIS project.

[RKS17] U. Raza, P. Kulkarni, M. Sooriyabandara, Low power wide area networks: an overview, IEEE Communications Surveys and Tutorials 19 (2) (2017) 855–873.

[RlBMB11] J. Rodriguez, I. De la Bandera, P. Munoz, R. Barco, Load balancing in a realistic urban scenario for LTE networks, in: 2011 IEEE 73rd Vehicular Technology Conference (VTC Spring), IEEE, May 2011, pp. 1–5.

[RML⁺18] R. Rudd, J. Medbo, F. Lewicki, F. Chaves, I. Rodriguez Larrad, The development of the new ITU-R model for building entry loss, in: Proc. 12th European Conference on Antennas and Propagation (EuCAP 2018), London, UK, 2018.

[RMSS15] Theodore S. Rappaport, George R. MacCartney, Mathew K. Samimi, Shu Sun, Wideband millimeter-wave propagation measurements and channel models for future wireless communication system design, IEEE Transactions on Communications 63 (9) (2015) 3029–3056.

[RPaC19] J.S. Romero-Peã, N. Cardona, Applicability limits of simplified human blockage models at 5G mm-Wave frequencies, Technical Report TD(19)09089, Dublin, Ireland, jan 2019.

[RPL⁺13] F. Rusek, D. Persson, B.K. Lau, E.G. Larsson, T.L. Marzetta, O. Edfors, F. Tufvesson, Scaling up MIMO: opportunities and challenges with very large arrays, IEEE Signal Processing Magazine 30 (1) (Jan. 2013) 40–60.

[RPRR⁺18]   V.M. Rodrigo-Peñarrocha, L. Rubio, J. Reig, L. Juan-Llácer, Juan Pascual-García, J.M. Molina-García-Pardo, Millimeter wave channel measurements in an intra-wagon environment, Technical Report TD(18)07040, Cartagena, Spain, May 2018.

[RRK07]   Cássio B. Ribeiro, Andreas Richter, Visa Koivunen, Joint angular-and delay-domain MIMO propagation parameter estimation using approximate ML method, IEEE Transactions on Signal Processing 55 (10) (2007) 4775–4790.

[RRKS14]   Michael Rushanan, Aviel D. Rubin, Denis Foo Kune, Colleen M. Swanson, SoK: security and privacy in implantable medical devices and body area networks, in: 2014 IEEE Symposium on Security and Privacy, SP 2014, Berkeley, CA, USA, May 18-21, 2014, 2014, pp. 524–539.

[RS18]   S. Rozum, J. Sebesta, SIMO RSS measurement in bluetooth low power indoor positioning system, in: 2018 28th International Conference Radioelektronika (RADIOELEKTRONIKA), Prague, Czech Republic, Apr 2018, pp. 1–5. Also available as TD(18)08026.

[RSS17]   Piotr Rajchowski, Jacek Stefański, Jarosław Sadowski, New method for increasing precision of position estimation of a moving person in hybrid inertial navigation system, in: COST IRACON (CA15104) 5th Scientific Meeting, TD(17)05015, Graz, Austria, Sept. 12-14 2017, pp. 1–6. Also available as TD(17)05015.

[RT12]   Kristi M. Robusto, Stewart G. Trost, Comparison of three generations of ActiGraph™ activity monitors in children and adolescents, Journal of Sports Sciences 30 (13) (2012) 1429–1435, PMID: 22857599.

[Rud17]   R. Rudd, Urban and O2I loss measurements at 28 GHz, Technical Report TD(17)03076, Lisbon, Portugal, February 2017.

[Rum16a]   M. Rumney, OTA testing challenges at mmWave frequencies, Technical Report TD-16-01046, Keysight Technologies Ltd., Edinburgh, UK, May 2016.

[Rum16b]   Moray Rumney, Building testability into mmWave 5G, Technical Report TD(16)02028, Keysight Technologies Ltd., Edinburgh, UK, October 2016.

[Rum18a]   M. Rumney, Channel modelling and spatial performance requirements for 3GPP mmWave new radio, Technical Report TD-18-06047, Keysight Technologies Ltd., Edinburgh, UK, January 2018.

[Rum18b]   M. Rumney, Next steps in channel modelling and OTA test methods for 5G, Technical Report TD-18-07051, Keysight Technologies Ltd., Edinburgh, UK, May 2018.

[Rum18c]   M. Rumney, Recent developments in mmWave requirements and test methods for 5G, Technical Report TD-18-08035, Rumney Telecom Ltd., Edinburgh, UK, October 2018.

[Rum19]   M. Rumney, Channel model developments at 3GPP and implications on device testing, Technical Report TD-19-09090, Rumney Telecom Ltd., Edinburgh, UK, January 2019.

[RVG18]   A.R. Ramos, F.J. Velez, G. Gardašević, Experimental IoT testbed for testing the 6TiSCH and RPL coexistence, in: Third EAI International Conference on IoT in Urban Space Urb-IoT 2018, November 2018. Also available as TD(18)07052.

[RVG19]   A.R. Ramos, F.J. Velez, G. Gardašević, Update on the Evaluation of Source Routing Minimum Cost Forwarding Protocol over 6TiSCH Applied to the OpenMote-B Platform, Technical Report TD(19)09082, Dublin, Ireland, January 2019.

[RWM18]   A. Reichman, S. Wayer, M.P. Moreno, Resource allocation in wireless mesh networks, in: 2018 IEEE International Conference on the Science of Electrical Engineering in Israel (ICSEE), IEEE, December 2018, pp. 1–5.

[RWP18]   A. Reichman, S. Wayer, M. Priesler, Resource Allocation in Wireless Mesh Networks, Technical Report TD(18)07074, IRACON Cost Action CA15104, Cartagena, Spain, May 2018.

[S+16]   Paul Spaanderman, et al., D2.1 use cases and application requirements, HIGHTS Deliverable, March 2016.

[SA18]   M.N. Sial, J. Ahmed, Analysis of K-tier 5G heterogeneous cellular network with dual-connectivity and uplink–downlink decoupled access, Telecommunication Systems 67 (4) (April 2018) 669–685.

[SAJG17] G. Steinböck, M. Alm, T. Jämsä, M. Gustafsson, Simulation study of the polarization behavior for a WTTx scenario at 28 GHz, Technical Report TD(17)03058, Lisbon, Portugal, feb 2017.

[SBHN19] Lawrence Sayer, Mark Beach, Geoff Hilton, Andrew Nix, Ray tracing for antenna arrays, Technical report, Oulu, Finland, May 2019, TD(19)10038.

[Sch17a] S. Scholz, Combining dynamic clustering and scheduling for coordinated multi-point transmission in LTE, in: 2017 IEEE 28th Annual International Symposium on Personal, Indoor, and Mobile Radio Communications (PIMRC), Oct 2017, pp. 1–7.

[Sch17b] Sebastian Scholz, System level evaluation of dynamic base station clustering for coordinated multi-point in future cellular networks, Technical Report TD:(17)03038, Lisbon, Portugal, Feb. 2017.

[SCK+18] G. Sommerkorn, D. Czaniera, M. Käske, C. Schneider, R. Thomä, V2V/V2R channel measurements on a highway at 2.53 GHz: a delay and Doppler discussion, Technical Report TD(18)07038, Cartagena, Spain, May 2018.

[SCL+18] Anne Savard, Laurent Clavier, Christophe Loyez, Francois Danneville, Alain Cappy, Sara Hedayat, Ultra low power wake-up radio with neuro-based detector and stochastic resonance, in: COST-IRACON, 2018, TD06048.

[SCM16]   A. Stanciu, M.N. Cirstea, F.D. Moldoveanu, Analysis and evaluation of PUF-based SoC designs for security applications, IEEE Transactions on Industrial Electronics 63 (9) (Sep. 2016) 5699–5708.

[SCR18]   Sana Salous, Yusheng Cao, Xavier Raimundo, Fixed link long term measurements, Technical Report TD(18)07049, Cartagena, Spain, May 2018.

[SCR19]   S. Salous, Y. Cao, X. Raimundo, Impact of precipitation on millimetre wave fixed links, in: Proc. of the 13th IEEE European Conference on Antennas and Propagation (EuCAP 2019), Krakow, Poland, March 2019. Also available as TD(16)02004.

[SdPMSR18] M. Schmidhammer, F. de Ponte Müller, S. Sand, I. Rashdan, Detection and localization of non-cooperative road users based on propagation measurements at C-band, in: 12th European Conference on Antennas and Propagation (EuCAP 2018), London, UK, Apr 2018. Also available as TD(18)06011.

[SDS+18]   M. Soliman, Y. Dawoud, E. Staudinger, S. Sand, A. Schuetz, A. Deko-
           rsy, Influences of train wagon vibrations on the mmWave wagon-to-wagon
           channel, in: 12th European Conference on Antennas and Propagation (Eu-
           CAP 2018), 2018, pp. 1–5. Also available as TD(17)05017.

[SF11]   Dmitriy Shutin, Bernard H. Fleury, Sparse variational Bayesian SAGE al-
           gorithm with application to the estimation of multipath wireless channels,
           IEEE Transactions on Signal Processing 59 (8) (2011) 3609–3623.

[SG18]   K. Samouylov, Y. Gaydamaka, Resource Allocation and Sharing for Het-
           erogeneous Data Collection over Conventional 3GPP LTE and Emerging
           NB-IoT Technologies, Technical Report TD(18)07031, IRACON Cost Ac-
           tion CA15104, Cartagena, Spain, May 2018.

[SGM+16]   Gerhard Steinbock, Mingming Gan, Paul Meissner, Erik Leitinger, Klaus
           Witrisal, Thomas Zemen, Troels Pedersen, Hybrid model for reverberant
           indoor radio channels using rays and graphs, IEEE Transactions on Anten-
           nas and Propagation 64 (9) (Sep. 2016) 4036–4048.

[SGSC19]   K. Samouylov, Y. Gaidamaka, E. Sopin, L.M. Correia, Simulation of SLA-
           based radio resource slicing and allocation in virtual RANs, Technical Re-
           port TD(19)10040, Oulu, Finland, May 2019.

[SGV18]   D. Simić, G. Gardašević, D. Vasiljević, Preliminary Results of Port-
           ing RIOT OS to OpenMote-B Hardware Platform, Technical Report
           TD(18)08043, Podgorica, Montenegro, October 2018.

[Sha58]   C.E. Shannon, Channels with side information at the transmitter, IBM Jour-
           nal of Research and Development 2 (4) (October 1958) 289–293.

[SHF+19]   K. Saito, P. Hanpinitsak, W. Fan, J. Takada, G.F. Pedersen, Frequency char-
           acteristics of diffuse scattering in SHF band in indoor environments, in:
           2019 URSI Asia-Pacific Radio Science Conference (AP-RASC), March
           2019, pp. 1–1. Also available as TD(18)08040.

[SHiT+18] Kentaro Saito, Panawit Hanpinitsak, Jun-ichi Takada, Wei Fan, Gert F.
           Pedersen, Frequency dependency analysis of SHF band directional propa-
           gation channel in indoor environment, in: European Wireless 2016; 22nd
           European Wireless Conference, April 2018, pp. 1–6. Also available as
           TD(17)04045.

[SHP+18]   Ossi Saukko, Aki Hekkala, Ville Pitkäkangas, Anttoni Porri, Marjo
           Heikkilä, 3D map implementation and visualization for radio propaga-
           tion simulation, Technical report, Podgorica, Montenegro, Oct. 2018,
           TD(18)08013.

[SHP+20]   V. Semkin, J. Haarla, T. Pairon, C. Slezak, S. Rangan, V. Viikari, C. Oest-
           ges, Analyzing radar cross section signatures of diverse drone models at
           mmWave frequencies, IEEE Access 8 (2020) 48958–48969. Also available
           as TD(19)09051.

[SHRT17]   G. Sommerkorn, S. Häfner, M. Röding, R. Thomä, Synthesis of realistic
           bistatic range profiles, Technical Report TD(17)03049, Lisbon, Portugal,
           Feb 2017.

[SiTK17]   Kentaro Saito, Jun-ichi Takada, Minseok Kim, Dense multipath component
           characteristics in 11-GHz-band indoor environments, IEEE Transactions on
           Antennas and Propagation 65 (9) (Sep. 2017) 4780–4789. Also available as
           TD(16)02014.

[SJG16] Gerhard Steinböck, Tommi Jämsä, Mattias Gustafsson, Study of dominant path probability models for 5G 3GPP channel model, Technical report, May 2016. Also available as TD(16)01064.

[SJKB19] T. Skoric, S. Jovanovic, K. Katzis, D. Bajic, Four Pillars of IoE in Health, Technical Report TD(19)10003, COST Action CA15104 (IRACON), Oulu, Finland, May 2019.

[SJM18] N. Sharma, D.N.K. Jayakody, M. Magarini, On-demand ultra-dense cloud drone network: opportunities, challenges, and benefits, IEEE Communications Magazine 56 (8) (2018) 85–91. Also available as TD(18)06007.

[Ska17] Vitaly Skachek, Coding for packet losses in a broadcast channel with side information, in: COST-IRACON, 2017, TD05038.

[SKP19] S. Rozum, J. Kufa, L. Polak, Bluetooth low power portable indoor positioning system using SIMO approach, in: 42nd Int'l Conference on Telecommunications and Signal Processing (TSP19), Budapest, Hungary, July 2019.

[SLZ15] R. Shafin, L. Liu, J. Zhang, DoA estimation and RMSE characterization for 3D massive-MIMO/FD-MIMO OFDM system, in: Proc. of IEEE Global Communications Conference (GLOBECOM), Dec. 2015, pp. 1–6.

[SM19] C. Buratti, E.M. Vitucci, F. Fuschini, R. Verdone, S. Mignardi, M.J. Arpaio, Performance Evaluation of UAV-Aided Mobile Networks by means of Ray Launching generated REMs, 2019. Available as TD(19)11029.

[SMM+17] T. Skoric, O. Mohamoud, B. Milovanovic, N. Japundzic-Zigon, D. Bajic, Binarized cross-approximate entropy in crowdsensing environment, Computers in Biology and Medicine 80 (2) (2017) 137–147. Also available as TD(17)04014.

[SMVB20] S. Salous, B. Montenegro-Villacieros, J. Bishop, Building entry loss and clutter loss at 26 GHz, in: Proc. of the 14th IEEE European Conference on Antennas and Propagation (EuCAP 2020), Copenhagen, Denmark, March 2020. Also available as TD(19)10048.

[SPCN18] Sana Salous, Hugo Parrott, Costas Constantinou, Yuri Nechayev, On body and on body to off body communication in the 60 GHz band for health monitoring scenarios, in: Proc. EuCAP'18 - 12th European Conference on Antennas and Propagation, London, UK, April 2018.

[SPF+13] Gerhard Steinböck, Troels Pedersen, Bernard H. Fleury, Wei Wang, Ronald Raulefs, Distance dependent model for the delay power spectrum of in-room radio channels, IEEE Transactions on Antennas and Propagation 61 (8) (Aug. 2013) 4327–4340.

[SPF+15] Gerhard Steinböck, Troels Pedersen, Bernard Henri Fleury, Wei Wang, Ronald Raulefs, Experimental validation of the reverberation effect in room electromagnetics, IEEE Transactions on Antennas and Propagation 63 (5) (Apr. 2015) 2041–2053.

[SR19] Gowshigan Selvarasa, Antti Roivainen, Comparison of propagation parameters between deterministic models and semi deterministic model in urban street canyon scenario, Technical report, Oulu, Finland, May 2019, TD(19)10043.

[SRH+14] Shu Sun, Theodore S. Rappaport, Robert W. Heath, Andrew Nix, Sundeep Rangan, MIMO for millimeter-wave wireless communications: beamforming, spatial multiplexing, or both?, IEEE Communications Magazine 52 (12) (2014) 110–121.

[SSA+19] S. Schieler, C. Schneider, C. Andrich, M. Dobereiner, J. Luo, A. Schwind, P. Wendland, R.S. Thoma, G. Del Galdo, OFDM waveform for distributed radar sensing in automotive scenarios, in: 2019 16th European Radar Conference (EuRAD), Oct 2019, pp. 225–228. Also available as TD(19)09061.

[SSC19] E. Sopin, K. Samouylov, L.M. Correia, The analysis of the computation offloading scheme with two-parameter offloading criterion in fog computing, Technical Report TD(19)11034, Gdańsk, Poland, September 2019.

[SST16] K. Shi, E. Sillekens, B.C. Thomsen, 246 GHz digitally stitched coherent receiver, in: Optical Fiber Communication Conference, Anaheim, US, March 2016.

[SSTV18] Bruno C. Silva, Sofia C. Sousa, Emanuel Teixeira, Fernando J. Velez, Insights on Spectrum Sharing in Heterogeneous Networks with Small Cells, Technical Report TD(18)07020, Cartagena, Spain, May, June 2018.

[ST18] B. Salamat, A.M. Tonello, Novel trajectory generation and adaptive evolutionary feedback controller for quadrotors, in: Proc. of IEEE Aerospace, 2018.

[STKK19] M. Schweins, J. Talvitie, M. Koivisto, T. Kürner, Position awareness for drones to facilitate beamforming, Technical Report TD(19)10027, Oulu, Finland, May 2019.

[STV+19] S. Shikhantsov, A. Thielens, G. Vermeeren, E. Tanghe, P. Demeester, L. Martens, G. Torfs, W. Joseph, Hybrid ray-tracing/FDTD method for human exposure evaluation of a massive MIMO technology in an industrial indoor environment, IEEE Access 7 (2019) 21020–21031. Also available as TD(19)09025.

[SUL19] Saúl A. Torrico, Cuneyt Utku, Roger H. Lang, The study of the probability density function of the total intensity in a trunk dominated forest, Technical report, Oulu, Finland, May 2019, TD(19)10009.

[SV87] Adel A.M. Saleh, Reinaldo A. Valenzuela, A statistical model for indoor multipath propagation, IEEE Journal on Selected Areas in Communications 5 (2) (1987) 128–137.

[SVC19] T. Pairon, V. Viikari, V. Semkin, J. Haarla, C. Oestges, Analyzing the radar cross section signatures of the diverse drone models, Technical Report TD(19)10045, Oulu, Finland, may 2019.

[SVHR17] S.C. Sousa, F.J. Velez, K. Huq, J. Rodriguez, Scenarios and architectures for RRM and optimization of heterogenous networks, Technical Report TD:(17)05006, Graz, Austria, Sept. 2017.

[SVP17] Sofia Sousa, Fernando J. Velez, Jon M. Peha, Impact of propagation model on capacity in small-cell networks, in: Performance Evaluation of Computer and Telecommunication Systems (SPECTS), 2017 International Symposium on, IEEE, 2017, pp. 1–8.

[SW10] Yuan Shen, M.Z. Win, Fundamental limits of wideband localization; part I: a general framework, IEEE Transactions on Information Theory 56 (10) (October 2010) 4956–4980.

[Syk16] Jan Sykora, Distributed consensus estimator of hierarchical network transfer function in WPNC networks, in: COST-IRACON, 2016, TD02020.

[Syk17] Jan Sykora, Achievable rates of hierarchical bit-wise mapped higher-order network coding maps in H-MAC channel with relative fading, in: COST-IRACON, 2017, TD05029.

[Syk18a]   Jan Sykora, H-MAC channel phase estimator with hierarchical data deci-
           sion aided feed-back gradient solver, in: COST-IRACON, 2018, TD08010.

[Syk18b]   Jan Sykora, Hierarchical decoding with bit-wise soft-aided H-SODEM and
           iterative double-loop processing, in: COST-IRACON, 2018, TD07025.

[SZHA15]   Prabodini Semasinghe, Kun Zhu, Ekram Hossain, Alagan Anpalagan,
           Game theory and learning techniques for self-organization in small cell
           networks, in: Alagan Anpalagan, Mehdi Bennis, Rath Vannithamby (Eds.),
           Design and Deployment of Small Cell Networks, Cambridge University
           Press, 2015, pp. 242–283.

[SZZ19]   D. Loeschenbrand, G. Kail, M. Schiefer, S. Zelenbaba, M. Hofer, T. Zemen,
           Spatial properties of industrial wireless ultra-reliable low-latency commu-
           nication MIMO links, in: Asilomar Conference on Signals, Systems, and
           Computers (ASILOMAR), Pacific Grove (CA), Nov 2019. Also available
           as TD(19)10007.

[TAC17]   Kenan Turbic, Slawomir J. Ambroziak, Luis M. Correia, Characteristics
           of the polarised off-body channel in indoor environments, EURASIP Jour-
           nal on Wireless Communications and Networking 2017 (1) (October 2017)
           174. Also available as TD(16)02002.

[TAC18a]   Kenan Turbic, Slawomir J. Ambroziak, Luis M. Correia, A body-
           shadowing model for off-body and body-to-body communications, in: Proc.
           URSI'18 - Baltic URSI Symposium, Poznan, Poland, May 2018. Also
           available as TD(18)06031.

[TAC18b]   Kenan Turbic, Slawomir J. Ambroziak, Luis M. Correia, Fading character-
           istics for dynamic body-to-body channels in indoor and outdoor environ-
           ments, in: Proc. EuCAP'18 - 12th European Conference on Antennas and
           Propagation, London, UK, April 2018.

[TACB18]   Kenan Turbic, Slawomir J. Ambroziak, Luis M. Correia, Marko Beko,
           Wideband Channel Measurements for Polarised Indoor Off-Body Com-
           munications, Technical Document TD(18)07063, COST Action CA15104
           (IRACON), Cartagena, Spain, May 2018.

[TACB19]   Kenan Turbic, Slawomir J. Ambroziak, Luis M. Correia, Marko Beko,
           Wideband off-body channel characteristics with dynamic user, in: Proc.
           EuCAP'19 - 13th European Conference on Antennas and Propagation,
           Krakow, Poland, March 2019. Also available as TD(19)09013.

[TADG+19] R.S. Thomä, C. Andrich, G. Del Galdo, M. Döbereiner, M.A. Hein, M.
           Käske, G. Schäfer, S. Schieler, C. Schneider, A. Schwind, P. Wendland, Co-
           operative passive coherent location: a promising 5G service to support road
           safety, IEEE Communications Magazine 57 (9) (September 2019) 86–92.
           Also available as TD(19)09003.

[TBDG17]   S. Tomic, M. Beko, R. Dinis, J.P. Gomes, Target tracking with sensor
           navigation using coupled RSS and AoA measurements, Sensors 17 (11)
           (November 2017) 2690–2711. Also available as TD(17)03009.

[TBL98]   S.A. Torrico, H.L. Bertoni, R.H. Lang, Modeling tree effects on path loss in
           a residential environment, IEEE Transactions on Antennas and Propagation
           46 (6) (June 1998) 872–880.

[TC20a]   Kenan Turbic, Luis M. Correia, Effects on polarization characteristics of
           off-body channels with dynamic users, in: Proc. WCNC'20 - IEEE Wireless
           Communications and Networking Conference, Seoul, South Korea, April
           2020. Also available as TD(19)10035.

[TC20b]   Kenan Turbic, Luis M. Correia, Influence of user dynamics on small-scale fading characteristics in off-body channels, in: Proc. ICC'20 - IEEE International Conference on Communications, Dublin, Ireland, June 2020. Also available as TD(19)11015.

[TCB17]   Kenan Turbic, Luis M. Correia, Marko Beko, Influence of User's Motion on Signal Depolarisation in Off-Body Channel, Technical Document TD(17)03044, COST Action CA15104 (IRACON), Lisbon, Portugal, February 2017.

[TCB18]   Kenan Turbic, Luis M. Correia, Marko Beko, A mobility model for wearable antennas on dynamic users, IEEE Access 6 (1) (December 2018) 63635–63648. Also available in TD(17)04043 and TD(17)05054.

[TCB19]   Kenan Turbic, Luis M. Correia, Marko Beko, A channel model for polarised off-body communications with dynamic users, IEEE Transactions on Antennas and Propagation 67 (11) (November 2019) 7001–7013. Also available in TD(17)05054 and TD(18)08011.

[TDM11]   Lorenzo Taponecco, A.A. D'Amico, Umberto Mengali, Joint TOA and AOA estimation for UWB localization applications, IEEE Transactions on Wireless Communications 10 (7) (2011) 2207–2217.

[Tec19]   Keysight Technologies, DUT test zone size evaluation for UMi and InO at FR2, Technical Report 3GPP R4-1906830, Keysight Technologies, Reno, USA, May 2019. Also available as TD(19)10057.

[Tei19]   Werner G. Teich, From iterative threshold decoding to a low-power high-speed analog VLSI decoder implementation, in: COST-IRACON, 2019, TD09068.

[TEVY16] Li Tian, Vittorio Degli Esposti, Enrico M. Vitucci, Xuefeng Yin, Semi deterministic radio channel modeling based on graph theory and ray tracing, IEEE Transactions on Antennas and Propagation 64 (6) (Jun. 2016) 2475–2486. Also available as TD(16)01068.

[TGL+14] Emmeric Tanghe, Davy P. Gaillot, Martine Lienard, Luc Martens, Wout Joseph, Experimental analysis of dense multipath components in an industrial environment, IEEE Transactions on Antennas and Propagation 62 (7) (July 2014) 3797–3805.

[TGR16]   J. Tasic, M. Gusev, S. Ristov, A medical cloud, in: 2016 39th International Convention on Information and Communication Technology, Electronics and Microelectronics (MIPRO), May 2016, pp. 400–405.

[Thu18]   P. Thubert, An Architecture for IPv6 over the TSCH mode of IEEE 802.15.4, Technical Report draft-ietf-6tisch-architecture-19, December 2018.

[TIM17]   TIMON - enhanced real time services for optimized multimodal mobility relying on cooperative networks and open data, H2020 Project. Grant Agreement No. 636220, 2017. https://www.timon-project.eu/.

[TIPV+16] V. Toldov, R. Igual-Pérez, R. Vyas, A. Boé, L. Clavier, N. Mitton, Experimental evaluation of interference impact on the energy consumption in wireless sensor networks, in: 2016 IEEE 17th International Symposium on a World of Wireless, Mobile and Multimedia Networks (WoWMoM), June 2016. Also available as TD(16)01052.

[TL18]   Saul A. Torrico, Roger H. Lang, Clutter loss in a vegetated residential environment – part-i, Technical report, Podgorica, Montenegro, Oct 2018, TD(18)08016.

[TL19]   S.A. Torrico, R.H. Lang, Comparison of physics-based propagation prediction model and measurements in a vegetated residential area at 3.5 GHz and 5.8 GHz, Radio Science 54 (11) (Nov. 2019) 1046–1058. Also available as TD(18)06001.

[TMR⁺14] R. Di Taranto, S. Muppirisetty, R. Raulefs, D. Slock, T. Svensson, H. Wymeersch, Location-aware communications for 5G networks: how location information can improve scalability, latency, and robustness of 5G, IEEE Signal Processing Magazine 31 (6) (Nov 2014) 102–112.

[TPMJ15] Jens Trogh, David Plets, Luc Martens, Wout Joseph, Advanced real-time indoor tracking based on the Viterbi algorithm and semantic data, International Journal of Distributed Sensor Networks 11 (10) (2015).

[TPMJ17] J. Trogh, D. Plets, L. Martens, W. Joseph, Bluetooth low energy based localization and tracking for livestock monitoring, in: 8th European Conference on Precision Livestock Farming (EC-PLF), September 2017.

[TPS⁺19] J. Trogh, D. Plets, E. Surewaard, M. Spiessens, M. Versichele, L. Martens, W. Joseph, Outdoor 3G location tracking of mobile devices in cellular networks, EURASIP Journal on Wireless Communications and Networking 115 (December 2019). Also available as TD(18)08006.

[TSC12]   IEEE Standard for Local and metropolitan area networks–Part 15.4: Low-Rate Wireless Personal Area Networks (LR-WPANs) Amendment 1: MAC sublayer, IEEE Std 802.15.4e-2012 (Amendment to IEEE Std 802.15.4-2011), April 2012, pp. 1–225.

[TSRC18] J.L. Towers, S. Salous, X. Raimundo, A. Cheema, Building entry loss for traditional and thermally efficient houses between 0.4 and 73 GHz, in: 12th European Conference on Antennas and Propagation (EuCAP 2018), April 2018, pp. 1–4.

[TSVG17] N. Temene, C. Sergiou, V. Vassiliou, C. Georgiou, Utilizing Mobile Nodes for Network Recovery in Wireless Sensor Networks, Technical Report TD(17)04040, Lund, Sweden, May 2017.

[TUY14]   Mohsen Nader Tehrani, Murat Uysal, Halim Yanikomeroglu, Device-to-device communication in 5G cellular networks: challenges, solutions, and future directions, IEEE Communications Magazine 52 (5) (2014) 86–92.

[TV18]   Emanuel Teixeira, Fernando J. Velez, Cost/revenue trade-off of small cell networks in the millimetre wavebands, in: IEEE 87th Vehicular Technology Conference: VTC2018, IEEE, 2018.

[TZJD17] V. Timcenko, N. Zogovic, M. Jevtic, B. Dordevic, An IoT business environment for multi objective cloud computing sustainability assessment framework, in: ICIST 2017 Proceedings, vol. 1, 2017.

[UG19]   M. Ulmschneider, C. Gentner, RANSAC for exchanging maps in multipath assisted positioning, in: 2019 IEEE International Conference on Industrial Cyber Physical Systems (ICPS), May 2019, pp. 275–281. Also available as TD(19)10012.

[UGJD17] M. Ulmschneider, C. Gentner, T. Jost, A. Dammann, Association of transmitters in multipath-assisted positioning, Singapore, Singapore, December 2017. Also available as TD(17)04003.

[UJWK18] P. Unterhuber, T. Jost, W. Wang, T. Kürner, Measurement based spatial characteristics of MPCs in train-to-train propagation, in: 12th European Conference on Antennas and Propagation (EuCAP 2018), April 2018, pp. 1–5. Also available in TD(17)05005.

[ULG18] M. Ulmschneider, D.C. Luz, C. Gentner, Exchanging transmitter maps in multipath assisted positioning, in: 2018 IEEE/ION Position, Location and Navigation Symp. (PLANS), Monterey, CA, USA, April 2018. Also available as TD(18)07059.

[UPA13] Bige Unluturk, Ecehan Pehlivanoglu, Ozgur Akan, Molecular channel model with multiple bit carrying molecules, July 2013, pp. 79–83.

[URW19] P. Unterhuber, I. Rashdan, M. Walter, Path loss models and fading statistics for C-band train-to-train communication, Technical Report TD(19)11031, Gdańsk, Poland, September 2019.

[US17] Kristian Ulshöfer, Sebastian Scholz, Dynamic Resource Partitioning between Massive Broadband and Machine Type Communication in 5G Networks, Technical Report TD(17)070342, Lisbon, Portugal, February 2017.

[USS⁺17] P. Unterhuber, S. Sand, M. Soliman, B. Siebler, A. Lehner, T. Strang, M. d'Atri, F. Tavano, D. Gera, Wideband propagation in train-to-train scenarios - measurement campaign and first results, in: 2017 11th European Conference on Antennas and Propagation (EUCAP), March 2017, pp. 3356–3360. Also available as TD(16)02005.

[VBJO17] Evgenii Vinogradov, Aliou Bamba, Wout Joseph, Claude Oestges, Physical-statistical modeling of dynamic indoor power delay profiles, IEEE Transactions on Wireless Communications 16 (10) (Oct 2017) 6493–6502. Also available as TD(16)01017.

[VC03] Fernando J. Velez, Luis M. Correia, Optimisation of mobile broadband multi-service systems based in economics aspects, Wireless Networks 9 (5) (Sep 2003) 525–533.

[VCMV12] Fernando J. Velez, Orlando Cabral, Francisco Merca, Vasos Vassiliou, Service characterization for cost/benefit optimization of enhanced UMTS, Telecommunication Systems 50 (1) (Apr 2012) 31–45.

[Ves91] Manuel Nieto Vesperinas, Scattering and Diffraction in Physical Optics, 1991.

[VFB⁺18] E.M. Vitucci, F. Fuschini, M. Barbiroli, M. Zoli, V. Degli-Esposti, A study on the performance of over-roof-top propagation models in dense urban environment, in: 12th European Conference on Antennas and Propagation (EuCAP 2018), London, UK, Apr. 2018, pp. 1–4. Also available as TD(16)01079.

[VFPK18] G. Villemaud, F. Hutu, P. Belloche, F. Kninech, Wireless transmission in ventilation (HVAC) ducts for the internet of things and smarter buildings: proof of concept and specific antenna design, in: Proc. 12th European Conference on Antennas and Propagation (EuCAP 2018), London, UK, april 2018. Also available as TD(18)06030.

[VG16] D. Vasiljević, G. Gardašević, Performance evaluation of OpenWSN operating system on open mote platform for industrial IoT applications, in: 2016 International Symposium on Industrial Electronics (INDEL), November 2016.

[VH19] U.T. Virk, K. Haneda, Modeling human blockage at 5G millimeter wave frequencies, IEEE Transactions on Antennas and Propagation (October 2019). Also available as TD(18)07050.

[VH20] U.T. Virk, K. Haneda, Modeling human blockage at 5G millimeter-wave frequencies, in: IEEE Transactions on Antennas and Propagation, vol. 68, March 2020, pp. 2256–2266. Also available as TD(18)07050.

[Vie]   Vienna LTE-A simulators. Available: https://www.nt.tuwien.ac.at/research/
            mobile-communications/vccs/vienna-lte-a-simulators/.

[VLJ97]   Reinaldo A. Valenzuela, Orlando Landron, D.L. Jacobs, Estimating local
            mean signal strength of indoor multipath propagation, IEEE Transactions
            on Vehicular Technology 46 (1) (February 1997) 203–212.

[VM18]   R. Verdone, S. Mignardi, Joint aerial-terrestrial resource management in
            UAV-aided mobile radio networks, IEEE Network 32 (5) (September 2018)
            70–75. Also available as TD(18)07062.

[VNC⁺15]   F.J. Velez, A. Nadziejko, A. Lyhne Christensen, S. Oliveira, T. Rodrigues,
            V. Costa, M. Duarte, F. Silvia, J. Gomes, Experimental Characterization
            of WSNs Applied to Swarms of Aquatic Surface Drones, in: Proc. of
            10th Conference on Telecommunications - Conftele, September 2015. Also
            available as TD(16)01032.

[VNHW18]   U.T. Virk, S.L.H. Nguyen, K. Haneda, J. Wagen, On-site permittivity es-
            timation at 60 GHz through reflecting surface identification in the point
            cloud, IEEE Transactions on Antennas and Propagation 66 (7) (July 2018)
            3599–3609. Also available as TD(17)04025.

[VPP97]   Michaela C. Vanderveen, Constantinos B. Papadias, Arogyaswami Paulraj,
            Joint angle and delay estimation (JADE) for multipath signals arriving at an
            antenna array, IEEE Communications Letters 1 (1) (1997) 12–14.

[VRL⁺17]   Joao Vieira, Fredrik Rusek, Erik Leitinger, Xuhong Li, Fredrik Tufves-
            son, Deep convolutional neural networks for massive MIMO fingerprint-
            based positioning, in: COST IRACON (CA15104) 4th Scientific Meeting,
            TD(17)04033, Lund, Sweden, May 29-31 2017, pp. 1–7. Also available as
            TD(17)04033.

[vT02]   Harry L. van Trees, Optimum Array Processing. Part IV of Detection, Esti-
            mation and Modulation Theory, Wiley, 2002.

[VZL⁺17]   Sathya N. Venkatasubramanian, Chunqing Zhang, Leo Laughlin, Katsuyuki
            Haneda, Mark A. Beach, Geometry-based modelling of self-interference
            channels for outdoor scenarios, Technical Report TD(17)04024, Lund,
            Sweden, May 2017.

[WA19]   Szymon Wiszniewski, Slawomir J. Ambroziak, Characterization of Slow
            and Fast Fading in Off-Body Communication at 2.45 GHz with Space Di-
            versity Scheme in Indoor Environment, Technical Document TD(19)09016,
            COST Action CA15104 (IRACON), Dublin, Ireland, January 2019.

[Wag17]   J. Wagen, Point clouds: some results of simple scattering and diffrac-
            tion models at V/UHF in CH, Technical report, Graz, Austria, Sep 2017,
            TD(17)05044.

[Wag18]   J. Wagen, On simple scattering and diffraction models for channel model
            predictions, Technical report, Nicosia, Cyprus, Jan 2018, TD(18)06045.

[Wag19a]   J. Wagen, On scattering and specular reflection formulations for mod-
            elling using point cloud maps, Technical report, Dublin, Ireland, Jan 2019,
            TD(19)09079.

[Wag19b]   Jean-Freédéric Wagen, On simple scattering models and the GGX direc-
            tional model for point cloud predictions, Technical report, Gdansk, Poland,
            Sep. 2019, TD(19)11013.

[WAHdPR⁺18] K. Witrisal, C. Antón-Haro, J.A. del Peral-Rosado, G.S. Granados, R.
            Raulefs, E. Leitinger, S. Grebien, T. Wilding, D. Dardari, E.S. Lohan, H.

Wymeersch, J.J. Floch, et al., Whitepaper on new localization methods for 5G wireless systems and the internet-of-things, Technical report, COST Action IC15104 "IRACON", April 2018.

[WB88] J. Walfisch, H. Bertoni, A theoretical model of UHF propagation in urban environments, IEEE Transactions on Antennas and Propagation 36 (Dec. 1988) 1788–1796.

[WGL$^+$18] Thomas Wilding, Stefan Grebien, Erik Leitinger, Ulrich Mühlmann, Klaus Witrisal, Single-anchor, multipath-assisted indoor positioning with aliased antenna arrays, in: 2018 52nd Asilomar Conference on Signals, Systems, and Computers, IEEE, Oct 2018, pp. 525–531. Also available as TD(19)09063.

[WGLM$^+$17] H. Wang, M. Garcia-Lozano, E. Mutafungwa, X. Yin, S. Ruiz, Performance Comparison of Up-link and Down-link Techniques under DUDe Strategy for Heterogeneous Networks, Technical Report TD(17)03014, IRACON Cost Action CA15104, Lisbon, Portugal, February 2017.

[WGMW18] Thomas Wilding, Stefan Grebien, Ulrich Mühlmann, Klaus Witrisal, Accuracy bounds for array-based positioning in dense multipath channels, Sensors 18 (12) (Dec 2018) 4249. Also available as TD(19)09062.

[who] World Health Organization.

[WK18a] J. Wagen, K. Krużelecki, On simple scattering and diffraction models using point cloud maps for channel model or coverage predictions, Technical report, Cartagena, Spain, Jun 2018, TD(18)07032.

[WK18b] J. Wagen, K. Krużelecki, An update on: on simple scattering and diffraction models using point cloud maps for channel model or coverage predictions, Technical report, Podgorica, Montenegro, Oct 2018, TD(18)08032.

[WLA$^+$17] Y.P. Wang, X. Lin, A. Adhikary, A. Grövlen, Y. Sui, Y. Blankenship, J. Bergman, H.S. Razaghi, A primer on 3GPP narrowband Internet of Things, IEEE Communications Magazine 55 (3) (March 2017) 117–123.

[WLHM16] Klaus Witrisal, Erik Leitinger, Stefan Hinteregger, Paul Meissner, Bandwidth scaling and diversity gain for ranging and positioning in dense multipath channels, IEEE Wireless Communications Letters 5 (4) (August 2016) 396–399. Also available as TD(16)01057.

[WML$^+$16] Klaus Witrisal, Paul Meissner, Erik Leitinger, Yuan Shen, Carl Gustafson, Fredrik Tufvesson, Katsuyuki Haneda, Davide Dardari, Andreas Molisch, Andrea Conti, Moe Z. Win, High-accuracy localization for assisted living: 5G systems will turn multipath channels from foe to friend, IEEE Signal Processing Magazine 33 (2) (March 2016) 59–70.

[WMW19] Thomas Wilding, Ulrich Mühlmann, Klaus Witrisal, Human Body Influence on UWB Channels: Evaluation of Channel Measurements, Technical Document TD(19)11020, COST Action CA15104 (IRACON), Gdansk, Poland, September 2019.

[WSGD$^+$17] Henk Wymeersch, Gonzalo Seco-Granados, Giuseppe Destino, Davide Dardari, Fredrik Tufvesson, 5G mm-Wave positioning for vehicular networks, IEEE Wireless Communications 24 (6) (2017) 80–86.

[WSM$^+$19] M. Walter, D. Shutin, D.W. Matolak, N. Schneckenburger, T. Wiedemann, A. Dammann, Analysis of non-stationary 3D air-to-air channels using the theory of algebraic curves, IEEE Transactions on Wireless Communications 18 (8) (Aug 2019) 3767–3780. Also available as TD(19)11036.

[WVH16] J. Wagen, U.T. Virk, K. Haneda, Measurements based specular reflection formulation for point cloud modelling, in: 2016 10th European Conf. Ant. Prop. (EuCAP 2016), Davos, Switzerland, Apr. 2016, pp. 1–5. Also available as TD(16)01060.

[WVO17] Charles Wiame, Luc Vandendorpe, Claude Oestges, Stochastic geometry based coverage estimation using realistic urban shadowing models, Technical report, May 2017. Also available as TD(17)03019.

[Wyn75] A.D. Wyner, The wire-tap channel, The Bell System Technical Journal 54 (8) (May 1975) 1355–1387.

[YAH+19] M. Yang, B. Ai, R. He, L. Chen, X. Li, J. Li, B. Zhang, C. Huang, Z. Zhong, A cluster-based three-dimensional channel model for vehicle-to-vehicle communications, IEEE Transactions on Vehicular Technology 68 (6) (June 2019) 5208–5220. Also available as TD(19)10016.

[YBF+14] Avi Yaeli, Peter Bak, Guy Feigenblat, Sima Nadler, Haggai Roitman, Gilad Saadoun, Harold J. Ship, Doron Cohen, Omri Fuchs, Shila Ofek-Koifman, et al., Understanding customer behavior using indoor location analysis and visualization, IBM Journal of Research and Development 58 (5/6) (2014) 1–19.

[YL17] Irfan M. Yousaf, Buon Kiong Lau, Measurement Study of Cable Influence on Channels Inside Vehicles, Technical Report TD(17)04071, Lund, Sweden, May 2017.

[YLB16] Irfan Mehmood Yousaf, Buon Kiong Lau, Bjorn Bergqvist, Impact of Electromagnetic Interference on Vehicular Antenna Performance, Technical Report TD(16)01007, Lille, France, May 2016.

[YTC+19] M. Yusuf, E. Tanghe, F. Challita, P. Laly, D.P. Gaillot, M. Liénard, B. Lannoo, R. Berkvens, M. Weyn, L. Martens, W. Joseph, Experimental characterization of V2I radio channel in a suburban environment, in: 2019 13th European Conference on Antennas and Propagation (EuCAP), March 2019, pp. 1–5. Also available as TD(19)09002.

[ZACCL17] M. Zahid-Aslam, R. Charbonnier, Y. Corre, Y. Lostanlen, Blockage modelling for evaluation of a 60 GHz dense small-cell network performance, Technical Report TD(17)03035, Lisbon, Portugal, Feb 2017.

[Zan92] Jens Zander, Performance of optimum transmitter power control in cellular radio systems, IEEE Transactions on Vehicular Technology 41 (1) (February 1992) 57–62.

[ZCYJ15] R. Zhang, X. Cheng, L. Yang, B. Jiao, Interference graph-based resource allocation (InGRA) for D2D communications underlaying cellular networks, IEEE Transactions on Vehicular Technology 64 (8) (August 2015) 3844–3850.

[ZEC+19] C. Zheng, M. Egan, L. Clavier, G.W. Peters, J.M. Gorce, Copula-based interference models for IoT wireless networks, in: ICC 2019 - 53rd IEEE International Conference on Communications, May 2019, pp. 1–6.

[ZGL+17] Kun Zhao, Carl Gustafson, Qingbi Liao, Shuai Zhang, Thomas Bolin, Zhinong Ying, Sailing He, Channel Characteristics and User Body Effects in an Outdoor Urban Scenario at 15 and 28 GHz, Technical Document TD(17)04068, COST Action CA15104 (IRACON), Lund, Sweden, May 2017.

[Zha17] H. Zhang, 5G for Factories of the Future: an Introduction to H2020 Clear5G Project, Technical Report TD(17)05009, Graz, Austria, September 2017.

[ZHL17]   Thomas Zemen, Markus Hofer, David Loeschenbrand, Low-complexity equalization for orthogonal time and frequency signaling (OTFS), in: COST-IRACON, 2017, TD05004.

[ZHL+18]   E. Zöchmann, M. Hofer, M. Lerch, J. Blumenstein, S. Sangodoyin, H. Groll, S. Pratschner, S. Caban, D. Löschenbr, L. Bernadó, T. Zemen, A. Prokes, M. Rupp, C.F. Mecklenbräuker, A.F. Molisch, Statistical evaluation of delay and Doppler spread in 60 GHz vehicle-to-vehicle channels during overtaking, in: 2018 IEEE-APS Topical Conference on Antennas and Propagation in Wireless Communications (APWC), Sep. 2018, pp. 1–4. Also available as TD(19)09066.

[ZLB+17]   Chunqing Zhang, Leo Laughlin, Mark Beach, Kevin Morris, John Haine, On the effect of phase noise and local oscillator sharing on self-interference cancellation, in: COST-IRACON, 2017, TD04074.

[ZMG19]   E. Zöchmann, T. Mathiesen, G. Ghiaasi, A bandwidth scalable millimetre wave over-the-air test system with low complexity, Technical Report TD-19-11052, TU Wien, Vienna, Austria, September 2019.

[ZMK18]   Radovan Zentner, Nikola Mataga, Ana Katalinic, Arbitrary rotation of the antenna in global coordinates for geometry based radio channel models, Technical report, Nicosia, Cyprus, Jan. 2018, TD(18)06041.

[ZML+18]   E. Zöchmann, C.F. Mecklenbräuker, M. Lerch, S. Pratschner, M. Hofer, D. Löschenbrand, J. Blumenstein, S. Sangodoyin, G. Artner, S. Caban, T. Zemen, A. Prokeš, M. Rupp, A.F. Molisch, Measured delay and Doppler profiles of overtaking vehicles at 60 GHz, in: 12th European Conference on Antennas and Propagation (EuCAP 2018), April 2018, pp. 1–5. Also available as TD(19)09069.

[ZMM17]   Radovan Zentner, Nikola Mataga, Ana Katalinic Mucalo, Transfer matrix of ray tracing simulated MIMO radio channel, Technical report, Lisbon, Portugal, Feb 2017, TD(17)03026.

[Zog17]   Nikola Zogovic, Energy-efficiency vs. performance optimization in low-power wireless transmission, in: COST-IRACON, 2017, TD05025.

[ZTS+15]   Tao Zhou, Cheng Tao, Sana Salous, Zhenhui Tan, Liu Liu, Li Tian, Graph-based stochastic model for high-speed railway cutting scenarios, IET Microwaves, Antennas and Propagation 9 (15) (2015) 1691–1697.

[ZYL+18]   L. Zhou, Z. Yang, F. Luan, A.F. Molisch, F. Tufvesson, S. Zhou, Dynamic channel model with overhead line poles for high-speed railway communications, IEEE Antennas and Wireless Propagation Letters 17 (5) (May 2018) 903–906. Also available as TD(17)04042.

List of acronyms

2-D	Two-Dimensional
2G	Second Generation
3-D	Three-Dimensional
3G	Third Generation
3GPP	3rd Generation Partnership Project
4G	Fourth Generation
5G NR	5G New Radio
5G	Fifth Generation
6G	Sixth Generation
6LoWPAN	IPv6 over Low-Power Wireless Personal Area Network
6TiSCH	IPv6 over the TSCH
AAS	Advanced Antenna System
ABF	Analog Beamforming
ABS	Almost Blank Subframes
ACLR	Adjacent Channel Leakage power Ratio
ADC	Analog-to-Digital Converter
AEB	Angulation Error Bound
AI	Artificial Intelligence
AIC	Akaike Information Criterion
AOA	Angle Of Arrival
AOD	Angle Of Departure
AP	Access Point
APS	Angular Power Spectrum
AR	Augmented Reality
ARM	Advanced RISC Machine
ASA	Azimuth angular Spread of Arrival
ASD	Azimuth angular Spread of Departure
ASN	Absolute Slot Number
ATG	Air-to-Ground
AWGN	Additive White Gaussian Noise
BAN	Body Area Network
BBU	Base-Band Unit
BE	Best Effort
BFSK	Binary Frequency Shift Keying
BG	Best effort with minimum Guaranteed
BLE	Bluetooth Low Energy
BLER	Block Error Rate
BP	BackPressure
BR	Blocking Ratio
BRET	Bioluminescence Resonance Energy Transfer
BS	Base Station
BWDM	Multi-Band Body-Worn Distributed
CC	Congestion Control
CCDF	Complementary Cumulative Distribution Functions
CDF	Cumulative Distribution Function
CDL	Clustered Delay Line
CF	Compute-and-Forward

363

CHARQ	Cooperative Hybrid-ARQ
CIoT	Cognitive Internet of Things
CIR	Channel Impulse Response
CNIR	Carrier to Noise-plus-Interference Ratio
CoMP	Coordinated Multi-Point
COST	Co-operation in Science and Technology
COTS	Commercial Off-The-Shelf
CoV	Coefficient of Variation
CPU	Central Processing Unit
CQI	Channel Quality Indicator
CR	Cognitive Radio
C-RAN	Cloud Radio Access Network
CRE	Cell Range Expansion
CRLB	Cramér-Rao lower bound
CR-MANET	Cognitive Radio Mobile Ad-Hoc Network
CSE	Channel State Estimation
CSI	Channel State Information
CSMA	Carrier Sense Multiple Access
CSMA/CA	Carrier Sense Multiple Access/Collision Avoidance
CTF	Channel Transfer Function
D2D	Device-to-Device
DA	Distributed Array
DAC	Digital-to-Analog Converter
DAG	Directed Acyclic Graph
DFA	Detrended Fluctuations Analysis
DFD	Distributed Fault Detection
DI	Data information
DiS	Diffuse Scattering
DL	Down-Link
DM	Diffuse Multipath
DMC	Dense Multipath Component
DMSO	Dimethyl sulfoxide
DODAG	Destination Oriented DAG
DP	Direct-Path
DS	Delay Spread
DSA	Dynamic Spectrum Access
DSD	Doppler Spectral Density
DSFT	Discrete Symplectic Fourier Transform
DTN	Delay/Disruption-Tolerant Network
DTT	Digital Terrestrial Television
DUDe	DL and UL Decoupling
DuT	Device under Test
DVB-T2	Digital Video Broadcasting Terrestrial Access 2
E2E	End-to-End
EB	Enhanced Beacon
E-COOP	Energy efficient High-speed Cost Effective Cooperative Backhaul for LTE/LTE-A Small-cells
ED	Energy Detection
EH	Energy Harvesting
EHR	Electronic Health Records
eICIC	enhanced Inter Cell Interference Coordination
EIRP	Effective Isotropic Radiated Power
EM	Electro-Magnetic

EMANE	Extendable Mobile Ad hoc Network Emulator
EMF	Electro-Magnetic Field
eNB	evolved Node B
EPC	Evolved Packet Core
ER	Effective Roughness
ESA	Elevation angular Spread of Arrival
ESD	Elevation angular Spread of Departure
E-UTRA	Evolved UMTS Terrestrial Radio Access
EVM	Error Vector Magnitude
FC	Fusion Center
FH	Frequency-Hopping
FMCW	Frequency Modulated Continuous Wave
FoF	Factories of the Future
FPGA	Field-Programmable Gate Array
FR1	Frequency Range 1 (in 5G context: up to 7.125 GHz)
FR2	Frequency Range 2 (in 5G context: mmWaves 24.25–52.6 GHz)
FRET	Förster Resonance Energy Transfer
FS	Fixed Service
FSPL	Free Space Propagation Loss
FSS	Fixed Satellite Service
GB	Guaranteed Bitrate
GLSD	Geolocation Spectrum Database
GLSF	Generalized Local Scattering Function
GNSS	Global Navigation Satellite System
GO	Geometrical Optics
GoF	Goodness of Fit
GPS	Global Positioning System
GPU	Graphic Processing Unit
GSCM	Geometry-based Stochastic Channel Model
GSM	Global System for Mobile Communications
GUI	Graphical User Interface
GW	Gateway
GWNSyM	Generic Wireless Network System Modeler
HAP	High Altitude Platform
HAPS	High Altitude Platform Stations
HDF	Hierarchical Decode and Forward
HeNB	Home eNodeB
HetNet	Heterogeneous Networks
HO	Hand-over
HPC	High Performance Computing server
HR	High-Resolution
HSR	High-Speed Railway
HST	High-Speed Train
HVAC	Heating, Ventilation and Air-Conditioning
HW	Hardware
IBFD	In-band full-duplex
ICI	Inter Carrier Interference
ICNIRP	International Commission on Non-Ionizing Radiation Protection
IIoT	Industrial IoT
IMT-2020	International Mobile Telecommunication 2020
IMU	Inertial Measurement Unit
INR	Interference to Noise Ratio
IoE	Internet of Everything

IoT	Internet of Things
IP	Internet Protocol
IPv6	Internet Protocol version 6
IRACON	Inclusive Radio Communications
ISD	Inter-Site Distance
ISI	Inter-Symbol Interference
ISM	Industrial, Scientific, Medical
ITS	Intelligent Transportation System
ITU	International Telecommunication Union
JPR	Jammed-to-Protect Power Density
JSRA	Joint Scheduling and Routing Algorithm
JT	Joint Transmission
KEST	Kalman Enhanced Super Resolution Tracking
KL	Kullback-Leibler
kNN	k-Nearest Neighbors
KO	Knapsack Optimization
KPI	Key Performance Indicator
L3	Layer 3
LAP	Low Altitude Platform
LB	Load Balancing
LCR	Level-Crossing Rate
LED	Light Emitting Diode
LISP	Locator Identifier Separation Protocol
LO	Local Oscillator
LODT	Local Outlier Detection Test
LoRa	Long Range
LOS	Line of Sight
LPWAN	Low Power Wide Area Network
LSA	Licensed Spectrum Access
LSF	Local Scattering Function
LSP	Large-Scale Parameter
LSTM	Long Short-Term Memory
LTE	Long Term Evolution
LTE-TDD	LTE Time Division Duplex
LTE-V	LTE-Vehicle
M2M	Machine To Machine
MAC	Medium Access Control
MANET	Mobile Ad-Hoc Network
MANO	MANagement and Orchestration
MAP	Medium Altitude Platform
MBB	Massive Broadband
MBFA	Merging Brute Force Algorithm
MCF	Minimum Cost Forwarding
MCS	Mobile Crowdsensing
MEC	Mobile Edge Computing
MF	Matched Filter
MIMO OTA	OTA testing of MIMO equipment
MIMO	Multiple Input Multiple Output
MISO	Multiple-Input Single-Output
MKP	Multiple Knapsack Problem
ML	Maximum Likelihood
MLB	Mobility Load Balancing
MLP	Multi-Layer Perceptron

MMSE	Minimum Mean Squared Error
mMTC	massive Machine Type Communication
mmWave	millimeter-Wave
MNO	Mobile Network Operator
MO	Mobile Operator
MoCap	Motion Capture
MoM	Method of Moments
MOS	Mean Opinion Score
MP	Multi-Path
MPAC	Multi-Probe Anechoic Chamber
MPC	Multipath Component
MPI	Message Passing Interface
MPL	Mean Path Loss
MR	Mixed Reality
MRN	Moving Relay Node
MR-OFDM	Multi-Rate and multi-regional Orthogonal Frequency Division Multiplexing
MS	mobile station
MTC	Machine Type Communication
MUSIC	MUltiple SIgnal Classification
NB	Narrow-Band
NB-IoT	Narrow Band IoT
NB-LTE	Narrow Band LTE
NFP	Network Flying Platform
NFV	Network Function Virtualization
NLOS	Non Line of Sight
NOMA	Non-Orthogonal Multiple Access
NPRACH	NB-IoT Physical Random Access Channel
NPUSCH	NB-IoT Physical Uplink Shared Channel
NR	New Radio
NTN	Non-Terrestrial Networks
OAM	Operation Administration & Management
OBE	Out-of-Band Emissions
OFDM	Orthogonal Frequency Division Multiplexing
OFDMA	Orthogonal Frequency Division Multiple Access
OLOS	Obstructed Line of Sight
OOR	Open Overlay Router
ORM	Object Relational Mapping
OS	Operating System
OSM	Open Street Map
OTA	Over-the-Air
OTF	On-The-Fly
OTFS	Orthogonal Time and Frequency Signaling
PAAS	Platform as a Service
PAPR	Peak to Average Power Ration
PAS	Power Angular Spectrum
PB	Packet Block
PCA	Power Control Algorithm
PCell	Macro base station
PCHIP	Piecewise Cubic Hermite Interpolating Polynomial
PDF	Probability Density Function
PDP	Power Delay Profile
PDU	Packet Data Unit
PEM	Personal Exposure Meter

PG	Propagation graph
PHY	Physical Layer
PIFA	Planar-Inverted F Antenna
PL	Path Loss
PLNC	Physical Layer Network Coding
PLR	Packet Loss Rate
PMIPv6	Proxy Mobile IPv6
PMSE	Programme Making and Special Events
PN	Phase Noise
PO	Physical Optics
P-P	Point to Point
PPDR	Public Protection and Disaster Relief
PPP	Poisson Point Process
PS	Power Splitting
PWD	Personal Wireless Communication Device
QAM	Quadrature Amplitude Modulation
QCI	QoS Class Identifier
QLOS	Quasi-LoS
QoE	Quality of Experience
QoI	Quality of Information
QoS	Quality of Service
RA	Random Access
RAN	Radio Access Network
RAT	Radio Access Technology
RB	Resource Block
RC	Reverberation Chamber
RCIEDs	Radio Controlled Improvised Explosive Devices
RE	Resource Element
REB	Ranging Error Bound
REM	Radio Environmental Map
RF	Radio Frequency
RHM	Remote Health Monitoring
RMS	Root Mean Squared
RMSE	Root Mean Square Error
RPL	Routing Protocol for Low-Power and Lossy Networks
RRH	Remote Radio Head
RRM	Radio Resource Management
RRU	Radio Resource Unit
RSPG	Radio Spectrum Policy Group
RSRP	Reference Signal Received Power
RSS	Received Signal Strength
RSSI	Receiver Signal Strength Indicator
RT	Ray Tracing
RTE	Radiative Transfer Equation
RTP	Real Time Pricing
RTS	Radiated Two-Stage
RTT	Round Trip Time
Rx	Receiver
SAR	Specific Absorption Rate
SCell	Small Cell
SC-FDMA	Single Carrier Frequency Division Multiple Access
SCM	Spatial Channel Model
SCME	Spatial Channel Model Extended

SC-PTM	Single Cell Point-to-Multipoint
SDC	State-Dependent Channels
SDL	Supplemental Downlink
SDN	Software Defined Network
SDoF	Spatial Degree of Freedom
SDR	Software Defined Radio
SF	Spreading Factor
SHF	Super High Frequency
SI	Self-Interference
SIC	Successive Interference Cancellation
SIMO	Single-Input Multiple-Output
SINR	Signal-to-Interference-plus-Noise Ratio
SIR	Signal-to-Interference Ratio
SISO	Single-Input Single-Output
SLA	Service Level Agreement
SmartUMA	Smart University of Málaga
SMC	Specular Multipath Component
SNR	Signal-to-Noise Ratio
SoC	System-on-Chip
SON	Self Organizing Networks
SPB	Small Packet Blocks
SR	Source Routing
SRMCF	Source Routing for Minimum Cost Forwarding
STSM	Short Term Scientific Mission
SVC	Support Vector Classifiers
SVM	Support Vector Machine
SWIPT	Simultaneous Wireless Information and Power Transfer
T2I	Train-to-Infrastructure
T2T	Train-to-Train
TDD	Time Division Duplex
TDE	Time Delay Estimation
TDL	Tapped Delay Line
TDMA	Time Division Multiple Access
TDNN	Time Delay Neural Network
TDOA	Time Difference Of Arrival
TF	Time-Frequency
THz	Terahertz
TICK	Telegraf, InfluxDB, Chronograf, Kapacitor
TOA	Time Of Arrival
TRP	Total Radiated Power
TRUDI	Testing Environment for Vehicular Applications Running with Devices in the Loop
TSCH	Time-Slotted Channel Hopping
TVWS	TV White Spaces
TWDP	Two-Wave with Diffuse Power
Tx	Transmitter
UABS	Unmanned Aerial Base Station
UAV	Unmanned Aerial Vehicle
UDN	Ultra-Dense Network
UE	User Equipment
UHF	Ultra High Frequency
UL	Up-Link
ULA	Uniform Linear Array
UMa	Urban Macro

UMTS	Universal Mobile Telecommunications System
UTRAN	UMTS Terrestrial Radio Access Network
URA	Uniform Rectangular Array
URLLC	Ultra Reliable Low Latency Communication
USRP	Universal Software Radio Platform
UTD	Uniform Theory of Diffraction
UTRAN	UMTS Terrestrial Radio Access Network
UWB	Ultra WideBand
V2I	Vehicle-to-Infrastructure
V2P	Vehicle-to-Pedestrian
V2V	Vehicle-to-Vehicle
V2X	Vehicle-to-Everything
VANET	Vehicular Ad-hoc NETwork
VEFIE	Volume Electric Field Integral Equation
VIM	Virtualized Infrastructure Manager
VM	Virtual Machine
VNA	Vector Network Analyzer
VNF	Virtual Network function
VNO	Virtual Network Operator
VR	Visibility Region
ViR	Virtual Reality
VRRM	Virtual-RRM
VSAT	Very Small Aperture Terminal
WAN	Wide Area Network
WB	Wide Band
WBAN	Wireless Body Area Network
WBB	Wireless Broad Band
WCDMA	Wideband Code Division Multiple Access
WiMAX	Worldwide Interoperability for Microwave Access
WINNER	Wireless World Initiative New Radio
WLAN	Wireless Local Area Network
WMDs	Wearable Medical Devices
WMN	Wireless Mesh Networks
WPNC	Wireless Physical layer Network Coding
WSN	Wireless Sensor Network
WSS	Wide-Sense Stationary
WSSUS	Wide-Sense Stationary Uncorrelated Scattering
WTC	Wire Tap Channel
XR	Extended Reality
ZF	Zero Forcing

Index

Printed in the United States
by Baker & Taylor Publisher Services